Mathematics in Industry

The European Consortium for Mathematics in Industry

Volume 38

The *ECMI* subseries of the *Mathematics in Industry* series is a project of *The European Consortium for Mathematics in Industry. Mathematics in Industry* focuses on the research and educational aspects of mathematics used in industry and other business enterprises. Books for *Mathematics in Industry* are in the following categories: research monographs, problem-oriented multi-author collections, textbooks with a problem-oriented approach, conference proceedings. Relevance to the actual practical use of mathematics in industry is the distinguishing feature of the books in the *Mathematics in Industry* series.

More information about this subseries at https://link.springer.com/bookseries/4651

Michael Günther • Wil Schilders
Editors

Novel Mathematics Inspired by Industrial Challenges

Editors
Michael Günther
Applied Math. & Numerical Analysis
Bergische Universität Wuppertal
Wuppertal, Nordrhein-Westfalen
Germany

Wil Schilders
Mathematics
TU Eindhoven
Eindhoven, Noord-Brabant
The Netherlands

ISSN 1612-3956 ISSN 2198-3283 (electronic)
Mathematics in Industry
The European Consortium for Mathematics in Industry
ISBN 978-3-030-96175-6 ISBN 978-3-030-96173-2 (eBook)
https://doi.org/10.1007/978-3-030-96173-2

Mathematics Subject Classification (2020): 00A69, 68U99, 68T09

This Springer imprint is published by the registered company Springer Nature Switzerland AG
The registered company address is: Gewerbestrasse 11, 6330 Cham, Switzerland

To our loved ones

Preface

Mathematics is essential for innovations in industry and science. In many countries, books with success stories of mathematics have been published [1], reports by independent accountants have shown the high economic value of mathematics[2][3][4], and the Study Groups of Mathematics with Industry are spread all over the world [5]. Conferences like the biennial ECMI conference, the biennial SIAM CSE conference and the ICIAM conference organized every 4 years contain a large variety of applications of mathematics to challenges from industry and other sciences.

Despite all of these success stories, it is still claimed that these successes build on existing mathematics, that no new mathematics is generated and that mathematics for industry is not challenging at all. Some even feel that new mathematics is only created by brilliant theoretical mathematicians. A prominent and well-known example is formed by the number-theoretic results obtained by the famous mathematician Hardy in the 1940s, that are now the basis of all encryption algorithms used for financial transactions. Hardy himself would never have dreamed about this, in fact he would have considered it a nightmare that his methods are used for something practical.

With this book, we wish to demonstrate that mathematics for industry is challenging and extremely rewarding, leading to new mathematical methods and sometimes even to entirely new fields within mathematics. A nice example from our own experience is the solution of indefinite linear systems which originated from working on electronic circuit simulation. Solving the large linear systems associated with electronic circuits often led to problems with pivoting. Then the idea came up to use the fact that there are two variables in the problem: currents and voltages. This inspired us to re-order the matrix in terms of 2x2 blocks, coupling voltages to cur-

---

[1] https://www.eu-maths-in.eu/wp-content/uploads/2017/01/2011_FLMI-EU_IndustrialMaths-SuccessStories.pdf

[2] https://www.platformwiskunde.nl/wp-content/uploads/2016/10/Deloitte-rapport-20140115-Mathematical-sciences.pdf

[3] https://www.eu-maths-in.eu/wp-content/uploads/2016/02/2015-France-SocioEconomic_Impact_of_Mathematics.pdf

[4] http://www.eu-maths-in.eu/wp-content/uploads/2019/04/MathematicsImpactStudy_Spain.pdf

[5] https://ecmiindmath.org/study-groups/

rents. The success was immediate, no pivoting was required anymore. The method was generalised, so that it is applicable to all kinds of indefinite systems, and not only those coming from electronic circuits. It led to very nice new research results, and new methods for indefinite linear systems [6][7]. Another example is the development of methods within the field of model order reduction. This field has benefited much from demands of and developments in the electronics industry. Methods like PRIMA [8] and SPRIM [9] originated here, as did many other developments, but all methods are generally applicable. In recent years we observe a much wider variety of applications of model order reduction.

This book presents methods that fall into the category sketched in the foregoing paragraph. The starting point is always an industrial challenge. The chapters describe how the authors addressed the challenge and developed new methods that were initially specific for the application, but later formulated for general application. The book is split into two parts, one on engineering applications and one on stochastics and finance.

All chapters contained in this book clearly show that industrial challenges do lead to the development of new mathematical methods, or even completely new fields of mathematics, needed to address these challenges. The starting point maybe an application of existing mathematical methods, but when it is found that more is needed, or different methods, then the interaction between application and mathematics starts. Mathematicians can then on the one hand develop new mathematical techniques, on the other hand solve the challenges. This is extremely rewarding, it often leads to nice journal papers on the theoretical results, which subsequently are the starting point of a lot of further research inside the mathematics area. It also leads to papers in applied journals.

Concluding, we may say that "mathematics for industry" or, even broader, "applied mathematics", is much more than just applying existing mathematical methods to industrial problems. In many cases, the application of existing methods does not lead to the desired solution, and hence adaptations of existing methods or even entirely new methods need to be developed in order to effectively address the industrial challenge. In some cases, this has led to entirely new fields within mathematics. The interplay between mathematics and industry is, hence, beneficial for both. Industry benefits by having their problems solved and mathematics benefits because new methods are developed that are versatile in nature.

---

[6] H.S. Dollar, N.I.M. Gould, W.H.A. Schilders, A.J. Wathen: On iterative methods and implicit-factorization preconditioners for regularized saddle-point systems, SIAM J. Matr. Anal. Appl. (27) 170–189 (2006)

[7] W.H.A. Schilders: Solution of indefinite linear systems using an LQ decomposition for the linear constraints, Linear Algebra and its Applications 431:30-4 381–395 (2009)

[8] A. Odabasioglu, M. Celik and L.T. Pileggi: PRIMA: passive reduced-order interconnect macromodeling algorithm, IEEE Trans. Comp. Aid. Dsg. Int. Circ. Syst. 17:8 645–654 (1998)

[9] R.W. Freund: Structure-Preserving Model Order Reduction of RCL Circuit Equations, in: W.H.A. Schilders, H.A. van der Vorst and J. Rommes (eds): Model Order Reduction: Theory, Research Aspects and Applications, Springer Verlag, Heidelberg, 51–75 (2008)

The book contains two parts on applications in *Computational Science and Engineering* and *Data Analysis and Finance*. It should be remarked that all authors have been asked to start and end their chapter with a brief description of why their chapter fits into this volume: explaining which industrial challenges have been instrumental for their inspiration, and which methods have been developed as a result.

Wuppertal and Eindhoven,
Spring 2021

Michael Günther
Wil Schilders

Acknowledgements

We are grateful to Dr. Jörg Mittelsten Scheid for his generous grant to the Bergische Universität Wuppertal, which had made possible the Mittelsten Scheid Visiting Professorship of Prof.dr. W.H.A. Schilders in the winter semester 2020/2021 at the Faculty of Mathematics and Natural Sciences. In the course of this semester, the work on this book could be advanced a decisive piece, which would not have been possible so easily without Prof. Schilders' stay in Wuppertal.

We are grateful to Dr. Harshit Bansal for his valuable help in a number of LaTex and layout problems.

Contents

List of Contributors

Martin Arnold
Martin Luther University Halle-Wittenberg, Institute of Mathematics, 06099 Halle (Saale), Germany, e-mail: martin.arnold@mathematik.uni-halle.de

Andreas Bartel
Bergische Universität Wuppertal, School of Mathematics and Natural Sciences, IMACM, Gaußstraße 20, D-42119 Wuppertal, e-mail: bartel@math.uni-wuppertal.de

A. Bermúdez
Dpto. de Matemática Aplicada & Instituto de Matemáticas (IMAT) & Instituto Tecnológico de Matemática Industrial (ITMATI), Universidade de Santiago de Compostela, ES-15782 Santiago de Compostela, Spain, e-mail: alfredo.bermudez@usc.es

Kai Bittner
University of Applied Sciences of Upper Austria, Softwarepark 11, Hagenberg im Mühlkreis, 4232, Austria, e-mail: Kai.Bittner@fh-hagenberg.at

Jean-Daniel Boissonnat
Université Côte d'Azur Inria, e-mail: jean-daniel.boissonnat@inria.fr

Hans Georg Brachtendorf
University of Applied Sciences of Upper Austria, Softwarepark 11, Hagenberg im Mühlkreis, 4232, Austria, e-mail: brachtd@fh-hagenberg.at

Frédéric Chazal
Inria Saclay, e-mail: frederic.chazal@inria.fr

Nicola Demo
Mathematics Area, mathLab, SISSA, International School of Advanced Studies, via Bonomea 265, I-34136 Trieste, Italy, e-mail: ndemo@sissa.it

Manuel Febrero-Bande

Department of Statistics, Mathematical Analysis and Optimization, Universidade de Santiago de Compostela, e-mail: manuel.febrero@usc.es

Roland W. Freund
Department of Mathematics, University of California at Davis, One Shields Avenue, Davis, California 95616, USA, e-mail: freund@math.ucdavis.edu

D. Gómez
Dpto. de Matemática Aplicada & Instituto de Matemáticas (IMAT) & Instituto Tecnológico de Matemática Industrial (ITMATI), Universidade de Santiago de Compostela, ES-15782 Santiago de Compostela, Spain, e-mail: mdolores.gomez@usc.es

Wenceslao González-Manteiga
Department of Statistics, Mathematical Analysis and Optimization, Universidade de Santiago de Compostela, e-mail: wenceslao.gonzalez@usc.es

Michael Günther
Bergische Universität Wuppertal, School of Mathematics and Natural Sciences, IMACM, Gaußstraße 20, D-42119 Wuppertal, e-mail: guenther@math.uni-wuppertal.de

J. Kienitz
Fachbereich Mathematik und Naturwissenschaften, Bergische Universität Wuppertal / The African Institute for Financial Markets and Risk Management (AIFMRM), University of Cape Town / Quaternion Risk Management GmbH, e-mail: jkienitz@uni-wuppertal.de

Oleg N. Kirillov
Northumbria University, NE1 8ST Newcastle upon Tyne, UK, e-mail: oleg.kirillov@northumbria.ac.uk

Jan Kleinert
Hochschule Bonn-Rhein-Sieg, Grantham-Allee 20, D-53757 Sankt Augustin, German Aerospace Center, Simulation and Software Technology, Linder Höhe, D-51147 Cologne, e-mail: jan.kleinert@h-brs.de, jan.kleinert@dlr.de

T.A. McWalter
The African Institute for Financial Markets and Risk Management (AIFMRM), University of Cape Town / Faculty of Science, Department of Statistics, University of Johannesburg, e-mail:

Bertrand Michel
Ecole Centrale de Nantes, e-mail: Bertrand.Michel@ec-nantes.fr

Andrea Mola
Mathematics Area, mathLab, SISSA, International School of Advanced Studies, via Bonomea 265, I-34136 Trieste, Italy, e-mail: andrea.mola@sissa.it

Manuel Oviedo de la Fuente
Technological Institute for Industrial Mathematics and Department of Statistics,

Mathematical Analysis and Optimization, Universidade de Santiago de Compostela, e-mail: manuel.oviedo@usc.es

M. Piñeiro
Dpto. de Matemática Aplicada, Universidade de Santiago de Compostela, ES-15782 Santiago de Compostela, Spain, e-mail: marta.pineiro@usc.es

E. Platen
Finance Discipline Group and School of Mathematical and Physical Sciences, University of Technology Sydney / The African Institute for Financial Markets and Risk Management (AIFMRM), University of Cape Town, e-mail:

Roland Pulch
Institute of Mathematics and Computer Science, University of Greifswald, Walther-Rathenau-Str. 47, Greifswald, 17489, Germany, e-mail: roland.pulch@uni-greifswald.de

Gianluigi Rozza
Mathematics Area, mathLab, SISSA, International School of Advanced Studies, via Bonomea 265, I-34136 Trieste, Italy, e-mail: gianluigi.rozza@sissa.it

R. Rudd
The African Institute for Financial Markets and Risk Management (AIFMRM), University of Cape Town, e-mail:

P. Salgado
Dpto. de Matemática Aplicada & Instituto de Matemáticas (IMAT) & Instituto Tecnológico de Matemática Industrial (ITMATI), Universidade de Santiago de Compostela, ES-15782 Santiago de Compostela, Spain, e-mail: mpilar.salgado@usc.es

Bernd Simeon
Felix-Klein-Zentrum, TU Kaiserslautern, D-67663 Kaiserslautern, Germany, e-mail: simeon@mathematik.uni-kl.de

Marco Tezzele
Mathematics Area, mathLab, SISSA, International School of Advanced Studies, via Bonomea 265, I-34136 Trieste, Italy, e-mail: marco.tezzele@sissa.it

Ferdinand Verhulst
Mathematisch Instituut, PO Box 80.010, 3508TA Utrecht, Netherlands, e-mail: F.Verhulst@uu.nl

Part I
Computational Science and Engineering

Computational Science and Engineering has become a very flourishing area of research. Amongst the conferences organised by SIAM, the CSE series attracts the highest number of participants nowadays. Also the attendence of ECMI and ICIAM conferences is growing rapidly in the past decade. Engineers are about the top users of mathematical software, and hence of mathematical methods. In turn, they provide many challenges and also many opportunities for mathematicians to adapt their methods or construct even entirely new ones. Clearly, this is very advantageous for the development of mathematical methods and theories.

Let us, for example, consider ordinary differential equations (ODE). In the last few decades, the theory of numerical methods for general (non-stiff and stiff) ordinary differential equations has reached a certain maturity, and excellent general-purpose codes, mainly based on Runge–Kutta methods or linear multistep methods, have become available. The motivation for developing structure preserving algorithms for special classes of problems came independently from such different areas of research as astronomy, molecular dynamics, mechanics, theoretical physics, and numerical analysis as well as from other areas of both applied and pure mathematics. It turned out that the preservation of geometric properties of the flow not only produces an improved qualitative behaviour, but also allows for a more accurate long-time integration than with general-purpose methods. Such methods belong to the more general class of mimetic methods. An important shift of view-point came about by ceasing to concentrate on the numerical approximation of a single solution trajectory and instead to consider a numerical method as a discrete dynamical system which approximates the flow of the differential equation – and so the geometry of phase space comes back again through the window. This view allows a clear understanding of the preservation of invariants and of methods on manifolds, of symmetry and reversibility of methods, and of the symplecticity of methods and various generalizations. It has brought an enormously rich additional theory into the area of mathematical methods for ODEs.

The chapters in this first part on "Engineering applications" describe similar developments, inspired by different branches of industry. By sheer coincidence, most of the chapters in this first part of the book originate from the electronics industry. On the one hand, this is owing to the connections of the two editors, who have been very active in this area. On the other hand, the electronics industry has been, and still is, extremely important for the progress of mankind in recent decades. Hence, it is not surprising that many researchers have felt attracted to this branch of industry.

The first chapter is by Bartel and Günther. They observed the need, in the electronics industry, for so-called multirate methods. It turns out that electronic circuits often have different parts of the circuit being latent, and other parts being active. In principle, using the same time steps for both parts would be a waste of computational effort. The latent part could do with much larger time steps, until this part of the circuit becomes active again. This observation, as well as the desire to be able to carry out such time stepping procedure, has led to multirate methods. A lot of research has been carried out in this area, and the methods developed are clearly not only applicable to electronic circuits. The chapter contains a very detailed and concise description of the developments, as well as applications other than those from the electronics industry.

The chapter by Roland Freund on model order reduction does not need much further introduction, we discussed MOR already in the previous section. Freund has been at the basis of the enormous developments in the area of MOR, with the seminal paper on PVL [10], written together with Peter Feldmann. It established a connection of methods for model order reduction, then still in their infancy, and numerical linear algebra. As argued already in the previous section, from then on MOR benefited much from the electronics industry, as the continuous miniaturisation of semiconductor devices led to many challenges. Even now, challenges from the electronics industry are an inspiration for new developments in the MOR area.

Martin Arnold concentrates on stability problems frequently encountered when solving systems of differential-algebraic equations. Well-known methods such as multi-rate methods or waveform relaxation may suffer from so-called exponential instabilities. Arnold developed an entirely new framework for the analysis of such instability as well as for the convergence behaviour, which led to new methods that do not suffer from the aforementioned problems. As a result, a totally new theory of methods for coupled problems within system dynamics has been developed.

The next chapter in this part is by a team of Spanish researchers around Alfredo Bermudez, pioneer in the area of mathematical methods for industrial problems, that concentrates on electromagnetic models. The problem at hand is to calculate the steady-state solution of an electromagnetic problem. This can often take many days, due to the fact that the transient are dying out only very slowly. Engineers had the desire to be able to reach the steady state much faster, and to this end new numerical methods were developed that can handle this task. The chapter sketches the historical development, as well as the final result of the search for such novel methods. In many areas nowadays this kind of challenge is encountered, and the ideas underlying the methods in this chapter can immediately be used to achieve the desired speed-up of the simulations. A related method can be found in [10].

In the chapter by Bittner, Brachtendorf and Pulch, the emphasis is on so-called MPDEs, in full multirate partial differential equations. These MPDE are of hyperbolic type, which have a physically meaningful solution only along one characteristic curve. The problem is analysed in-depth, and it is shown that the theory is applicable also to other fields where such MPDE may occur. The chapter contains several examples that illustrate the challenge very nicely.

The lust but one final chapter in this part on engineering applications is by a group of Italian researchers around Gianluigi Rozza, well-known for the many contributions in the area of model order reduction and reduced basis methods. The chapter discusses an integrated data-driven computational pipeline, containing several model order reduction techniques, for addressing industrial challenges. It contains many different ingredients, amongst which dimension reduction of the parameter space. The framework developed has been put in a general context, and hence the methodologies are usable in many different application fields.

---

[10] https://pure.tue.nl/ws/files/97050457/publicsummary_giovannideluca.pdf

The final chapter by Oleg N. Kirillov and Ferdinand Verhulst is an example how the the interplay between applications and progress in mathematics may take place. Starting point is a curious phenomenon observed in the natural sciences, destabilization by dissipation — usually it is the other way round. The solution of this problem, which involved the eigenvalue calculus of matrices, finally was one source for the mathematical theory of structural stability of matrices. An important consequence of these results is that in a large number of problem-fields one can now predict and characterise precisely this type of instability.

Multirate Schemes

—

An Answer of Numerical Analysis to a Demand from Applications

Andreas Bartel and Michael Günther

Abstract In science and engineering, simulation tasks often involve numerical time integration of differential equations. Usually, these systems contain different time constants of the involved components and/or right-hand side. This multirate behavior may be caused by coupling subsystems in multiphysics problems acting on different time scales. Such a behavior does already occur if one deals with just single-physics problems: for example, the activity level of components in electrical networks may strongly vary depending on the according functional purpose, physics or time; another example is given in lattice QCD, where the equations of motion may depend on weak and strong forces, which demand to sample these forces with different frequencies to gain the same rate of approximation.

To be efficient or to speed up simulation of highly complex coupled systems is necessary for many design and optimization work flows. To this end, numerical integration schemes have to be adapted to exploit this multirate behavior. One idea proposed by Rice in 1960 are multirate schemes, which use different step sizes adapted to the various activity levels. In the last 50 years, the methodology of numerical time integration schemes has been advanced in a constant interplay between the demands defined by the need of exploiting multirate behavior in different fields of applications and the development of tailored multirate schemes to answer these demands.

1 Introduction

In many technical applications, ranging from electric circuits to multibody systems and in particular for multiphysics simulation, the governing set of differential equations is characterized by a multirate behavior in time domain: that is to say, some parts of the right-hand side follow a fast dynamics, whereas the other parts are char-

Andreas Bartel · Michael Günther
Bergische Universität Wuppertal, School of Mathematics and Natural Sciences, IMACM, Gaußstraße 20, D-42119 Wuppertal, e-mail: `{bartel, guenther}@math.uni-wuppertal.de`

M. Günther, W. Schilders (eds.), *Novel Mathematics Inspired by Industrial Challenges*, Mathematics in Industry 38, https://doi.org/10.1007/978-3-030-96173-2_1

acterized by a comparatively slow dynamics. Often, the slower parts represents even a much larger part/partition of the overall system in question. A traditional, standard time integration scheme has to resolve the fastest dynamics (according to the given tolerance). Thus the time stepping is adapted to these fast components, and yields rather small time steps for the slower components, which results an an oversampling revealing only superfluous detail. Now, to be efficient, a numerical (time) integration scheme needs to exploit this multirate potential.

For the moment, let us consider the case of an initial-value problem of ordinary differential equations $\dot{w} = h(a,b)$

$$\dot{w} = h(t,w), \qquad w(t_0) = w_0 \tag{1}$$

with $h : \mathbb{R} \times \mathbb{R}^n \to \mathbb{R}^n$ assumed to be Lipschitz continuous in w. We denote the unique solution of (1) at time point t by $w(t; w_0)$. Furthermore, we assume that in h we have some multirate potential, i.e., some coordinates or summands of h are slower than some remaining ones.

On the one hand, the interpretation of fast and slow coordinates in h leads to a partitioning of the unknowns $w^\top = (y_S^\top, y_F^\top)$ in slow $y_S(t) \in \mathbb{R}^m$ and fast components $y_S(t) \in \mathbb{R}^{n-m}$. In this way, the ODE (1) is transferred to a partitioned system:

$$\begin{aligned} \dot{y}_S &= f_S(t, y_S, y_F), \;\; y_S(t_0) = y_{S,0}, \\ \dot{y}_F &= f_F(t, y_S, y_F), \; y_F(t_0) = y_{F,0}, \end{aligned} \tag{2}$$

with corresponding slow right-hand side $f_S : \mathbb{R} \times \mathbb{R}^m \times \mathbb{R}^{n-m} \to \mathbb{R}^m$ and fast right-hand side $f_F : \mathbb{R} \times \mathbb{R}^m \times \mathbb{R}^{n-m} \to \mathbb{R}^{n-m}$. This is termed *component-wise partitioning*. On the other hand, one can split the right-hand side:

$$\dot{w} = h_s(t, w) + h_f(t, w), \qquad w(t_0) = y_0 \tag{3}$$

into slow and fast terms. This induces an additive splitting of the unknown $w = w_s + w_f$ into slow and fast varying parts w_s and w_f. This is referred to as *right-hand side partitioning*. Note that both formulations are equivalent in the sense that each component-wise partitioned system can be rewritten as a right-hand side partitioned system and vice versa.

Multiorder as multirate. There are many ways to exploit multirate behavior in such systems. One idea is to use multi-order methods. Here a single method with a single step size is employed for the whole system. To adapt to the activity level, the order of the method is modified accordingly. This class comprises, for example, the schemes MURX [17] and MUR8 [16] by Engstler and Lubich. The first method (MURX) is based on Richardson extrapolation of the explicit Euler scheme. Thereby the computation of the extrapolation tableau is stopped if a component is accurate enough. The latter MUR8 uses low-order methods embedded in a high-order method, where the update of the slow components is deactivated after a first few function evaluations.

Multiple step sizes as multirate. To our knowledge, the first method based on exploiting multirate behavior by adapting step sizes to the activity level of components was derived by Rice [33] in 1960 for missile simulations. This method is based on Runge-Kutta schemes and employs a so-called compound step. Later, Gear and Wells [21] proposed an alternative approach based on linear multistep methods as well as extrapolation and interpolation.

The work at hand focuses on the multiple step size approaches. In Section 2, we discuss the ideas of Rice and Gear/Wells. The class of extra- and interpolation coupling is strongly linked to waveform relaxation/dynamic iteration schemes [3, 12, 29], which we treat in Section 3. In fact, dynamic iteration enables the general application to multiphysics systems. Subsequently, we treat some applications. Section 4 covers electric circuits simulation, where the compound-step approach has been successfully applied to develop multirate strategies for single-physics problems. Then molecular dynamics is discussed in Section 5, in which the multirate potential has to be exploited, without violating the preservation of geometric properties. For this case, operator splitting has turned out to be the right framework for developing multirate schemes, as properties of the base scheme are easily preserved. Finally, we draw a conclusion and give an outlook on open problems.

2 Strategies for multirate and convergence

Before discussing in detail the methods by Rice [33] as well as by Gear and Wells [21], we first focus on the common idea behind these approaches: the combination of basic numerical integration schemes with extra- and interpolation techniques to solve (2) with a small step size h for the fast variable z and large step size $H = M \cdot h$ for the slow y. Here, the multirate factor is a fixed number $M \in \mathbb{N}$.

2.1 Combining extra- and interpolation for multirate properly

Depending on the sequence of the computation of the unknowns y_S and y_F, one distinguishes the following three versions of extra-/and interpolation techniques for a macro step from $\bar{t}$ to $\bar{t}+H$ [11]:

1) *fully-decoupled approach:* fast and slow variables are integrated in parallel using in both cases extrapolated waveforms; computations are based on information from the initial data of the current macro step which is given at $\bar{t}$;
2) *slowest-first approach [21]:* in the first step, the slow variables are integrated, using an extrapolated waveform of y_F based on the information available at $\bar{t}$; this is used to evaluate the coupling variable y_F in the slow system on the current macro step $[\bar{t}, \bar{t}+H]$. In the second step, M micro steps are performed to integrate the fast variables y_F from $\bar{t}$ to $\bar{t}+H$, using an interpolated waveform

of the coupling variable y_S. Thereby the interpolation is based on information from the current macro step $[\bar{t}, \bar{t}+H]$.

3) *fastest-first approach [21]:* now, in the first step, M micro steps are performed to integrate the fast variables on the current macro step $[\bar{t}, \bar{t}+H]$. To this end, the coupling variable y_S are extrapolated based on the information available at $\bar{t}$. In the second step, one macro step is performed to integrate the slow variables y_S from $\bar{t}$ to $\bar{t}+H$. Thereby, the interpolated waveform of y_F is computed based on the information from the current macro step $[\bar{t}, \bar{t}+H]$ and this is used to evaluate the coupling variable y_F.

We have the following main result for the decoupled case:

Theorem 2.1 *We consider* (2) *with both right-hand sides f_S and f_F Lipschitz continuous in both variables y_S and y_F. Furthermore, we consider an arbitrary macro step from $\bar{t} \to \bar{t}+Mh$ and respective initial values $y_S(\bar{t}) = y_{S,\bar{t}}, y_F(\bar{t}) = y_{F,\bar{t}}$. Let $M \in \mathbb{N}$ be a fixed multirate factor be given and two integration schemes of order p be applied: a first scheme for one macro step of size $H = M \cdot h$ for the slow y_S and a second scheme is applied for M steps of size h for fast y_F. If we use an integration scheme of order p and combine this with an extrapolation procedure of order $p-1$, the overall scheme has order p.*

Proof For the coupled system (2) with initial data $y_S(\bar{t}) = y_{S,\bar{t}}, y_F(\bar{t}) = y_{F,\bar{t}}$, we refer to the unique solution by $(y_S(t; y_{S,\bar{t}}, y_{F,\bar{t}})^\top,\ y_F(t; y_{S,\bar{t}}, y_{F,\bar{t}})^\top)$. Now, we replace the coupled system (2) by a modified system which decouples both parts:

$$\begin{aligned} \dot{y}_S &= f_S(t, y_S, \widetilde{y}_F) =: \widetilde{f}_S(t, y_S), \quad y_S(\bar{t}) = y_{S,\bar{t}}, \\ \dot{y}_F &= f_F(t, \widetilde{y}_S, y_F) =: \widetilde{f}_F(t, y_F), \; y_F(\bar{t}) = y_{F,\bar{t}}, \end{aligned} \tag{4}$$

where $\widetilde{y}_S$ and $\widetilde{y}_F$ are extrapolations of order $p-1$, i.e.,

$$y_S(t) - \widetilde{y}_S(t) = \mathcal{O}(H^p) \text{ and } y_F(t) - \widetilde{y}_F(t) = \mathcal{O}(H^p) \qquad \text{for any } t \in [\bar{t}, \bar{t}+Mh]. \tag{5}$$

We denote the unique solution of (4) by $(\widehat{y}_S(t; y_{S,\bar{t}}, y_{F,\bar{t}})^\top,\ \widehat{y}_F(t; y_{S,\bar{t}}, y_{F,\bar{t}})^\top)$.

Next, we solve the decoupled system (4) with two numerical integration schemes of order p, with M step sizes h applied to y_F and one step size $H = M \cdot h$ applied to y_S. The numerical solution obtained at $t^* = \bar{t}+Mh$ is denoted by $(y_{S,H}(t^*), y_{F,H}(t^*))^\top$.

For the difference between the numerical multirate approximation and the exact solution at t^*, the triangle inequality yields

$$\begin{pmatrix} \|y_{S,H}(t^*) - y_S(t^*)\| \\ \|y_{F,H}(t^*) - y_F(t^*)\| \end{pmatrix} \leq \begin{pmatrix} \|y_{S,H}(t^*) - \widehat{y}_S(t^*)\| \\ \|y_{F,H}(t^*) - \widehat{y}_F(t^*)\| \end{pmatrix} + \begin{pmatrix} \|\widehat{y}_S(t^*) - y_S(t^*)\| \\ \|\widehat{y}_F(t^*) - y_F(t^*)\| \end{pmatrix}. \tag{6}$$

The first term on the right-hand side represents the error of the applied integration schemes. Employing for both coordinates an integration scheme of order p, with one macro step of size $H = M \cdot h$ for the slow y_S and M steps of size h for fast y_F, we have

$$\begin{pmatrix} \|y_{S,H}(t^*) - \widehat{y}_S(t^*)\| \\ \|y_{F,H}(t^*) - \widehat{y}_F(t^*)\| \end{pmatrix} \leq \begin{pmatrix} c_S \\ \frac{c_F}{M^{p+1}} \end{pmatrix} H^{p+1} \tag{7}$$

with respective leading error coefficients c_S and c_F.

For the second term on the right-hand side (6), we get from the Lipschitz continuity of f_S, f_F with corresponding constants ($L_{i,j}$ for f_i w.r.t. y_j)

$$\begin{pmatrix} \|\widehat{y}_S(t^*) - y_S(t^*)\| \\ \|\widehat{y}_F(t^*) - y_F(t^*)\| \end{pmatrix} \leq \int_{\bar{t}}^{t^*} \begin{pmatrix} \|f_S\left(\tau, \widehat{y}_S(\tau), \widetilde{y}_F(\tau)\right) - f_S\left(\tau, y_S(\tau), y_F(\tau)\right)\| \\ \|f_F\left(\tau, \widetilde{y}_S(\tau), \widehat{y}_F(\tau)\right) - f_F\left(\tau, y_S(\tau), y_F(\tau)\right)\| \end{pmatrix} d\tau$$
$$\leq \int_{\bar{t}}^{t^*} \begin{pmatrix} L_{S,S}\|\widehat{y}_S(\tau) - y_S(\tau)\| + L_{S,F}\|\widetilde{y}_F(\tau) - y_F(\tau)\| \\ L_{F,S}\|\widetilde{y}_S(\tau) - y_S(\tau)\| + L_{F,F}\|\widehat{y}_F(\tau) - y_F(\tau)\| \end{pmatrix} d\tau$$

actually a decoupled estimate. Using that $\widetilde{y}_S$ and $\widetilde{y}_F$ are approximations of order $p-1$ (5) and respective Lipschitz constants L_S, L_F of the corresponding extrapolation operators, we find

$$\begin{pmatrix} \|\widehat{y}_S(t^*) - y_S(t^*)\| \\ \\ \|\widehat{y}_F(t^*) - y_F(t^*)\| \end{pmatrix} \leq \begin{pmatrix} \frac{L_{S,F} \cdot L_F}{H}^{p+1} + L_{S,S} \int_{\bar{t}}^{t^*} \|\widehat{y}_S(\tau) - y_S(\tau)\| d\tau \\ \\ L_{F,S} \cdot L_S \cdot H^{p+1} + L_{F,F} \int_{\bar{t}}^{t^*} \|\widehat{y}_F(\tau) - y_F(\tau)\| d\tau \end{pmatrix}$$

Applying Gronwall's lemma, it follows:

$$\begin{pmatrix} \|\widehat{y}_S(t^*) - y_S(t^*)\| \\ \|\widehat{y}_F(t^*) - y_F(t^*)\| \end{pmatrix} \leq \begin{pmatrix} L_{S,F} L_F \, e^{L_{S,S}(t^* - \bar{t})} \, H^{p+1} \\ L_{F,S} L_S \, e^{L_{F,F}(t^* - \bar{t})} \, H^{p+1} \end{pmatrix}.$$

Finally combining this with the integration error (7) into the split error (6), we obtain that the multirate scheme has consistency order p for the compound step from $\bar{t}$ to $\bar{t} + Mh$ (on the macro step level). □

Remark 2.1 a) One can show that the order p of the underlying numerical integration scheme is preserved for the slowest- and fastest-first approach, too, provided that the order of the used extra-/interpolation schemes is of order $p-1$ at least [11]. These methods are referred to as *extrapolation/interpolation-based multirate scheme*.

b) Notice for a working multirate scheme, we still have to define the extrapolation/interpolation routines. Furthermore, arbitrary high orders of the extra-/interpolation are not possible in the one-step-method context.

Corollary 2.1 *Under the assumptions of Thm. 2.1, we have an overall multirate scheme of convergence order p if*

a) we use one-step integration schemes.
b) we use multistep schemes, where both schemes are 0*-stable.*

Summing up, the art of defining multirate (multistep) schemes lies in implicitly defining the extrapolation and interpolation procedures used within the scheme to be of high enough order.

Corollary 2.2 *We consider the following initial values problem (IVP)*

$$\begin{aligned} \dot{y}_S &= f_S(t, y_S, y_F, z_S, z_F), \; y_S(t_0) = y_{S,0}, & \dot{y}_F &= f_F(t, y_S, y_F, z_S, z_F), \; y_F(t_0) = y_{F,0}, \\ 0 &= g_S(t, y_S, y_F, z_S, z_F) & 0 &= g_F(t, y_S, y_F, z_S, z_F) \end{aligned}$$

of coupled semi-explicit DAEs with slow subsystem (y_S, z_S) *and fast subsystem* (y_F, z_F). *Futhermore, we assume that the overall system is index-1 and both subsystems are index-1 with Lipschitz continuous* f_λ, g_λ $(\lambda \in \{S, F\})$ *with uniform Lipschitz constants on any macro step* $[\bar{t}, \bar{t}+Mh]$ *in* $[t_0, T]$. *Applying a multirate method, which is for coupled ODEs of order p (Thm. 2.1) and the algebraic variables* z_S, z_F *are always consistently computed (i.e., implicit), then the method has still order p.*

Proof The proof is a direct consequence of the recursion estimate Lemma 3.1 in [2] for dynamic iteration schemes of coupled index-1 DAE systems, if only one iteration is considered and the initial iteration error is given by the extrapolation/interpolation error. □

2.2 Linear multistep methods

How to define the extrapolated and interpolated approximations? Linear multistep schemes are based on polynomial interpolation using information of possibly several previous steps. Here, these methods are advantageous over one-step schemes. This can be seen as follows: let us assume, we use a linear K-step scheme for the slow part, and a k-step scheme for the fast variables. Accordingly, the extrapolation can be based on K macro step approximations for the slow variable and k micro step values for the fast variables. This yields a global error of order $\min\{K, k\}$ (Thm. 2.1) if the respective schemes are at least of order K and k.

Based on preliminary work by Gear [20], Orailoglu [31] and Wells [49], the first comprehensive study and still fundamental work on multirate linear multistep methods was published in 1984 by Gear and Wells [21]. It covers efficiency considerations, convergence and error analysis, absolute stability and numerical test results for an Adams-type based algorithm. In addition, the concepts of slowest-first and fastest-first methods were first introduced and discussed. At a first glance, the fastest first strategy seems to be advantageous. The extrapolation error in the computation of the fast part is acceptable, as one is extrapolating over an interval of size H, which is tailored to the activity level of the slow part. In the slowest first strategy, we are extrapolating the fast variables over many micro step h to compute the slow variables, which may only be tolerable if the coupling of the fast into the slow part, measured by $\|\partial f_S / \partial y_F\|$, is small. As a rule of thumb, this quantity will be usually small, as otherwise a high level of activity would be transferred from the fast into the slow part, and the slow part would not be slow anymore. Certainly, this is only a rule of thumb, since there exist — although academic — counterexamples [21].

However, considering an adaptive step sizes selection based on error control, the slowest first strategy can become advantageous. For instance, if a secondly performed macro step $[t, t+H]$ (slow variables) fails, the macro step size H has to be decreased and the computation has to be repeated. If the approximations of the micro steps (fast variable) are not stored (for instance, due to memory reasons) or if the computation of the fast variables is based on extrapolated values on the failed macro step size H_{old}, these values have to be recomputed. In contrast, the slowest first strategy does not introduce any problems: if an integration of the fast part fails, one only has to repeat it with a smaller step size. The necessary information to interpolate the slow part does not change. See [20, 21] for further details.

Remark 2.2 Some systems do not allow for a static partitioning into fast and slow subsystems (e.g. the inverter chain benchmark [23]). In these settings, slow components can become fast (wake up) and vice versa. Thus, the partitions in fast and slow need to be adapted over time. This causes a rather large step size modification, which is more problematic for multistep methods, see e.g. [26], than for one-step schemes.

Remark 2.3 Other linear-multistep approaches have been discussed, for instance, by Verhoeven et al. [48] and [32] (both BDF-based), or by Sandu and Constantinescu (Adams-based) [34]. Stabilty issues have been addressed first by Skelboe [40] and by Skelboe and Andersen [41].

2.3 Runge-Kutta schemes

In one-step schemes, only information of the last step is available. This limits the approximation order to 1, as only constant or linear interpolation is available in the multirate scheme with extra- or interpolation (cf. Thm. 2.1). Of course, the use of approximations at previous steps (in the extra- and interpolation) would turn the one-step scheme into a multistep scheme.

About twenty years before Gear's paper on multirate multistep schemes, Rice proposed a solution to overcome this problem for one-step schemes [33]: split Runge-Kutta schemes. Firstly, these schemes perform one large Runge-Kutta compound step with macro step size H for the joint system, but use only the result for the slow variables as approximates. Secondly, to get the approximations for the fast part in all micro steps of the macro step, the fast part is then integrated using interpolated information of the slow variables based on the Runge-Kutta increments of the macro step. In 2008, this approach has been applied by Verhoeven to BDF schemes [48], see also Section 4.

We give some details for the split Runge-Kutta schemes [33] on system (2). The numerical approximation $y_{S,H}(t^\star)$ is given by one explicit Runge-Kutta step from $\bar{t}$ to $t^\star = \bar{t} + H$:

$$y_{S,H}(t^\star) = y_{S,\bar{t}} + \sum_{i=1}^{s} b_i k_i,$$

$$k_i = H f_S\Big(\bar{t} + \sum_{j=1}^{i-1} a_{i,j} H,\ y_{S,\bar{t}} + \sum_{j=1}^{i-1} a_{i,j} k_j,\ y_{F,\bar{t}} + \sum_{j=1}^{i-1} a_{i,j} l_j\Big), \qquad (i = 1,\ldots,s),$$

$$l_i = H f_F\Big(\bar{t} + \sum_{j=1}^{i-1} a_{i,j} H,\ y_{S,\bar{t}} + \sum_{j=1}^{i-1} a_{i,j} k_j,\ y_{F,\bar{t}} + \sum_{j=1}^{i-1} a_{i,j} l_j\Big), \qquad (i = 1,\ldots,s),$$

using internal stages l_i for the active part, which will not be used later on. This is referred to as compound step and it employs the coefficients $b_i, a_{i,j}$ with s stages.

Secondly, for the M micro-steps, another Runge-Kutta scheme with coefficients $\widetilde{b}_i, \widetilde{a}_{i,j}$ and $\widetilde{s}$ stages is used to compute $y_{F,H}(\bar{t}_i + lh)$ for y_F at micro grid points $\bar{t} + \lambda h$ for $\lambda = 1,\ldots,M$:

$$y_{F,H}\big(\bar{t} + (\lambda+1)h\big) = y_{F,H}(\bar{t} + \lambda h) + \sum_{i=1}^{\widetilde{s}} \widetilde{b}_i k_i^\lambda, \qquad (\lambda = 1,\ldots,M-1)$$

$$k_i^\lambda = h f_F\Big(\bar{t} + \lambda h + \sum_{j=1}^{i-1} \widetilde{a}_{i,j} h,\ \widehat{y}_S\Big(\bar{t} + \lambda h + \sum_{j=1}^{i-1} \widetilde{a}_{ij} h\Big),\ y_{F,H}(\bar{t} + \lambda h) + \sum_{j=1}^{i-1} \widetilde{a}_{i,j} k_j^\lambda\Big),$$

$(i = 1,\ldots,\widetilde{s})$. Hereby, the values $\widehat{y}_S(\bar{t} + \lambda h + \sum_{j=1}^{i-1} \widetilde{a}_{ij})$ of the slow components at grid points $\bar{t} + \lambda h + \sum_{j=1}^{i-1} \widetilde{a}_{ij} h$ are approximated by a dense output of the Runge-Kutta approximation of the compound step, i.e.,

$$\widehat{y}_S(\bar{t} + \theta H) := y_{S,\bar{t}} + \sum_{i=1}^{s} b_i(\theta) k_i,$$

such that it holds

$$\max_{0 \le \theta \le 1} \|\widehat{y}(\bar{t} + \theta H) - y(\bar{t} + \theta H)\| \le c_S H^{p-1}.$$

for all $0 \le \theta \le 1$ and some constant $c_S > 0$.

Remark 2.4 Note that the increments l_i according to y_F are only used within the computation of k_i. Consequently, for an explicit Runge-Kutta scheme, l_s does not have to be computed.

Remark 2.5 Other multirate one-step schemes have been developed and analyzed, for example, by Savcenco et al. [35] and by Günther and Rentrop [23], both based on Rosenbrock-Wanner methods (ROW), by Sandu and Günther [25] based on GARK methods, and by Striebel et al [42] based on ROW methods for index-1 DAE systems, and others.

2.4 Overview on multirate strategies

To conclude this section, we state two main multirate strategies which exist for component-wise splitting:

- *Extra-/Interpolation based multirate schemes*: here, one computes the split variables one after the other. We have *slowest-first* and *fastest-first*, where the coupling variables are extrapolated or interpolated based on previously computed approximations. One may use any numerical integration scheme as basis scheme of such multirate schemes; the order of the scheme p is preserved provided that the extra-/interpolation is at least of order $p-1$.
- *Compound-step based multirate methods*: here, slow and fast variables are jointly computed using one macro step (compound step); then, the fast approximation is disregarded and replaced by M micro steps using the fast dynamics and dense output of the slow variables (*compound-fast approach*).

As the dense output for the coupling term $y_S(t)$ is available on the whole macro step, one may discard the idea of using M micro steps of the same step size h and use the different step sizes according to the step size prediction of the numerical integration scheme. The latter approach, combined with the compound step, is called *mixed multirate* [4].

The *time stepping approach* introduced by Savcenco, Hundsdorfer and Verwer [35] is a generalization of the compound-step approach. First, an approximation for all components after one macro step is computed. For those components not accurate enough the computation is redone with smaller steps. The refinement is recursively continued until the error estimator is below a given tolerance for all components.

One criticism of this compound step approach is the use of the macro step size H also for the fast component y_F inside the computation of the new approximate for the slow variables y_S. One may overcome this problem by combining the macro step for the slow part with the first micro step of the active part (*generalized compound-step approach*). As micro and macro step are interwoven in this case, additional coupling conditions have to be fulfilled for the coefficients of one-step methods to preserve the order of the method for the slow components. This approach has been introduced by Kværnø and Rentrop in [28] for Runge-Kutta schemes. Corresponding further methods are based on the W-method [9] or on generalized additive Runge-Kutta schemes [25].

If right-hand side splitting is concerned, operator splitting might be the method of choice. For the split system (3), the idea reads as follows: suppose that the slow dynamics h_s is characterized by an expensive evaluation, whereas the fast dynamics h_f can be cheaply evaluated. In this case, one may develop an operator splitting approach where the slow dynamics is solved only once on a macro step, and the fast systems M-times during one macro step. See Section 5 for some more details, in particular, in the context of geometric integration.

3 Dynamic iteration and multiphysics

Due to downscaling in electric devices and due to higher accuracy requests to numerical simulation, more and more coupled problems need to be studied and thus simulated for industrial applications. This gives naturally rise to multiphysics problems.

In modular time integration of such multiphysics problems, different subsystems are modeled and simulated by different simulation packages, see e.g. [3, 12, 50]. This allows to exploit the multirate potential, which is caused by different time scales of the subsystems, efficiently within a waveform relaxation approach. We describe this in the following.

To start with, we consider again the component-wise partitioned ODE (2), which comprises a slow subsystem in y_S and fast subsystem in y_F. Furthermore, we assume that the solution or an approximation $y_{S,H}, y_{F,H}$ is available on $[t_0, \bar{t}]$. The coupled initial value problem (2) can be solved iteratively on a time window $[\bar{t}, \bar{t}+H]$ employing old iterates $(i-1)$ and current iterates (i) and respective splitting functions F_S, F_F to encode usage of old and current iterates inside the coupled system. This reads:

$$\begin{aligned} \dot{y}_S^{(i+1)} &= F_S(t, y_S^{(i+1)}, y_S^{(i)}, y_F^{(i+1)}, y_F^{(i)}), & y_S^{(i+1)}(\bar{t}) &= y_S^{(0)}(\bar{t}), \\ \dot{y}_F^{(i+1)} &= F_F(t, y_S^{(i+1)}, y_S^{(i)}, y_F^{(i+1)}, y_F^{(i)}), & y_F^{(i+1)}(\bar{t}) &= y_F^{(0)}(\bar{t}), \end{aligned} \tag{8}$$

where the initial waveforms $y^{(0)}, z^{(0)}$ are given by extrapolating from the previous time window:

$$\begin{pmatrix} y_{S,H} \\ y_{F,H} \end{pmatrix}\Big|_{[\bar{t}-H,\bar{t}]} \to \begin{pmatrix} y_S^{(0)} \\ y_F^{(0)} \end{pmatrix}\Big|_{[\bar{t},\bar{t}+H]} := \Phi\left(\begin{pmatrix} y_{S,H} \\ y_{F,H} \end{pmatrix}\Big|_{[\bar{t}-H,\bar{t}]}\right).$$

Here, Φ denotes an extrapolation operator and let L_Φ be the respective Lipschitz constant. Furthermore the splitting functions shall fulfill Lipschitz conditions with respect to all components and are consistent, and the consistency with IVP (2):

$$F_S(t, y_S, y_S, y_F, y_F) = f_S(t, y_S, y_F) \quad \text{and} \quad F_F(t, y_S, y_S, y_F, y_F) = f_F(t, y_S, y_F).$$

Depending on the choice of the splitting functions, one may define different instants of waveform relaxation schemes: $Y = (t, y_S^{(i+1)}, y_S^{(i)}, y_F^{(i+1)}, y_F^{(i)})$

- Picard iteration:

$$F_S(Y) = f_S(t, y_S^{(i)}, y_F^{(i)}), \quad F_F(Y) = f_F(t, y_S^{(i)}, y_F^{(i)}).$$

- Jacobi-type iteration:

$$F_S(Y) = f_S(t, y_S^{(i+1)}, y_F^{(i)}), \quad F_F(Y) = f_F(t, y_S^{(i)}, y_F^{(i+1)}). \tag{9}$$

- Gauß-Seidel type iteration (slowest-first):

$$F_S(Y) = f_S(t, y_S^{(i+1)}, y_F^{(i)}), \quad F_F(Y) = f_F(t, y_S^{(i+1)}, y_F^{(i+1)}). \tag{10}$$

- Gauß-Seidel type iteration (fastest-first):

$$F_S(Y) = f_S(t, y_S^{(i+1)}, y_F^{(i+1)}), \quad F_F(Y) = f_F(t, y_S^{(i)}, y_F^{(i+1)}). \tag{11}$$

The iterates converge monotonically to the exact solution provided that the macro step size H is small enough.

If (9) is used, then both new iterates $y_S^{(i+1)}$ and $y_F^{(i+1)}$ in (8) can be computed in parallel, with extrapolation. If the exact integration of (10) or (11) is replaced by a numerical integration, the computation of the new iterates $y_S^{(i+1)}$, $y_F^{(i+1)}$ is equivalent to applying an extra-/interpolation based multirate scheme (to slow y_S and fast y_F component): (10) represents slowest-first and (11) fastest-first setting And vice versa, extra-/interpolation based multirate is equivalent to stopping the iteration of the waveform-relaxation after the first step. See also Section 4 for further discussion.

IVPs of coupled DAE systems (cf. Cor. 2.2)

$$\begin{aligned} \dot{y}_S &= f_S(t, y_S, y_F, z_S, z_F), \; y_S(t_0) = y_{S,0}, & \dot{y}_F &= f_F(t, y_S, y_F, z_S, z_F), \; y_F(t_0) = y_{F,0}, \\ 0 &= g_S(t, y_S, y_F, z_S, z_F) & 0 &= g_F(t, y_S, y_F, z_S, z_F) \end{aligned} \tag{12}$$

arise, for example, in circuit simulation or in electro-thermal coupling (see Section 4). A corresponding dynamic iteration scheme needs splitting functions F_L, G_L, F_A and G_A, which fulfill Lipschitz conditions with respect to all arguments and are consistent, i.e.,

$$\begin{aligned} F_\star(t, y_S, y_S, y_F, y_F, z_S, z_S, z_F, z_F) &= f_\star(t, y_S, y_F, z_S, z_F), \\ G_\star(t, y_S, y_S, y_F, y_F, z_S, z_S, z_F, z_F) &= g_\star(t, y_S, y_F, z_S, z_F) \quad \text{with } \star \in \{S, F\}. \end{aligned}$$

Then, the dynamic iteration scheme reads: for $\star \in \{S, F\}$

$$\begin{aligned} y_\star^{(i+1)} &= F_\star(t, y_S^{(i+1)}, y_S^{(i)}, y_F^{(i+1)}, y_F^{(i)}, z_S^{(i+1)}, z_S^{(i)}, z_F^{(i+1)}, z_F^{(i)}), \\ 0 &= G_\star(t, y_S^{(i+1)}, y_S^{(i)}, y_F^{(i+1)}, y_F^{(i)}, z_S^{(i+1)}, z_S^{(i)}, z_F^{(i+1)}, z_F^{(i)}). \end{aligned} \tag{13}$$

Again for the actual time window $[\bar{t}, \bar{t} + H]$, initial waveforms $y_L^{(0)}, y_A^{(0)}, z_L^{(0)}, z_A^{(0)}$ are given by extrapolation of the waveform approximates of the last window $[\bar{t} - H, \bar{t}]$. In contrast to coupled ODE systems, monotone convergence can no longer be guaranteed by choosing the window step size H small enough. In addition, one has to fulfill two additional contractivity conditions, see e.g. [3]). For the Gauss-Seidel type approaches, this reads:

- *Convergence within one window:*

$$\alpha < 1, \tag{14}$$

whereas the contractivity constant α is given by the sum of the Schur complent-like quantity

$$\max_{\substack{\tau, \text{ with} \\ \bar{t} \leq \bar{t}+\tau H \leq \bar{t}+H}} \left\| \begin{pmatrix} \frac{\partial G_S}{z_S^{(i+1)}} & \frac{\partial G_S}{z_F^{(i+1)}} \\ \frac{\partial G_F}{z_S^{(i+1)}} & \frac{\partial G_F}{z_F^{(i+1)}} \end{pmatrix}^{-1} \cdot \begin{pmatrix} \frac{\partial G_S}{z_S^{(i)}} & \frac{\partial G_S}{z_F^{(i)}} \\ \frac{\partial G_F}{z_S^{(i)}} & \frac{\partial G_F}{z_F^{(i)}} \end{pmatrix} \right\|$$

and a term of order $\mathscr{O}(H)$, which tends to zero for the limit $H \to 0$.

- *Stable error propagation from window to window:*

$$L_\Phi \alpha^k < 1 \tag{15}$$

with k iterations in the current window and Lipschitz constant L_Φ of the extrapolation operator. Hence, depending on the L_Φ, more than one iteration may be necessary, though convergence within one window is given.

If we replace the exact solution of (13) by a numerical integration and stop after the first iteration, again an extra-/interpolation based multirate scheme is defined, now for coupled DAE systems (12). Note that stability can only be guaranteed if (15) holds; here, additional iterations of the multirate scheme can be necessary to obtain a convergent scheme.

Remark 3.1 Note that in the case of DAE-ODE coupling, i.e., when the second subsystem to be solved in a Gauss-Seidel type approach is defined by an ODE, then $\alpha = 0$ (14), since both splitting functions G_S and G_F do not depend on old iterates of the algebraic variables. In this case, one can show that the dynamic iteration is convergent of first order in the macro step size H (cf. [5]). In Section 4.3, a electro-thermal problems, which yields a DAE-ODE coupling is discussed. Depending on the fine structure of the coupling, up to second order convergence is possible, see [5].

Multirate schemes on basis of dynamic iteration schemes have been successfully applied to multiphysics problems. See, for example, field-circuit coupling [36, 37], electro-thermal coupling refined network modelling [6].

4 Applications in circuit simulation

Large integrated electrical networks are usually modeled via differential algebraic equations (e.g., see [22]): for a charge-flux oriented formulation, we have compactly

$$\mathscr{F}(t, x, \frac{d}{dt} w(x)) = 0, \tag{16}$$

in terms of node potential and currents (through voltage defining elements) x, charges and fluxes $w = q(x)$ as well as time t. In (16), the matrix $\frac{d}{dw}\mathscr{F} \cdot \frac{d}{dx} w$ is generally not regular (i.e., in this case, (16) represents a DAE). In fact, these networks are built up by numerous coupled subcircuits of different functionality. As the subcircuits constitute different functional units, the overall system often shows multirate behavior To exploit this multirate potential, such a subcircuit partitioning has to be taken into account.

4.1 Partitioned network modeling

The subcircuits are modeled independently and composed to one macro system by connecting respective terminals, i.e., each pair of connected (terminal) nodes merge to a one node of the coupled system. This can be modeled by inserting virtual voltage sources for each pair of connected (terminal) nodes, where virtual means zero applied voltage. This approach preserves the macro circuit's block structure and produces additional variables: the branch currents u through the coupling voltage sources. These currents are implicitly determined by the property, that the node potentials of each pair of connected boundary nodes have to coincide. Now, we assume given a macro circuit composed of $\lambda = 1,\ldots,r$ subcircuits each of type (16) and coupled via the algebraic constraints of virtual voltages sources. This reads:

$$\mathscr{F}_\lambda(t,x,\frac{d}{dt}w_\lambda(x_\lambda),u) = 0, \qquad (\lambda = 0,\ldots,r) \tag{17a}$$

$$\mathscr{G}(x_1,\ldots,x_r) = 0, \tag{17b}$$

with x_λ internal node potentials and voltages, internal fluxes and charges $w_\lambda = q_\lambda(x_\lambda)$ of subsystem λ. Several index-1 conditions can be formulated for system (17) and its subsystems: [10]:

(C1) The overall system (17) has index 1 (with respect to $x_1,\ldots,x_r,u$).

(C2) All systems (17a) define index-1 systems with respect to x_{λ} (and u given as input).

(C3) For any $\lambda \in \{1,\ldots,r\}$, the λ-th system (17a) with coupling condition (17b) has index-1 with respect to x_λ, and u given all other x_i (i.e.,$i \neq \lambda$) as input).

In analogy to the procedure described in [18], topological conditions to guarantee the index conditions (C1)-(C3) can be derived. Given the above index-1 conditions, the overall model (17) can be transformed into algebraically coupled semi-explicit systems

$$\left.\begin{aligned} \dot{y}_\lambda(t) &= f_\lambda(t,z_\lambda,u) \\ 0 &= h_\lambda(t,y_\lambda,z_\lambda,u) \end{aligned}\right\} \quad \lambda = 1,\ldots,r$$
$$0 = g(z_1,\ldots,z_r) \tag{18}$$

with algebraic variables z_λ as node potentials and inner currents of system λ as well as differential variables y_λ defining the charges and fluxes (of system λ). In particular, setting (18) is the basis for developing more efficient schemes.

Remark 4.1 Notice we have two sets of coupled DAEs: one abstractly coupled DAEs (12), one with dedicated coupling equation (18), which represents networks, which are coupled via virtual voltage sources. In fact, for the latter case, the coupling condition (18-3) is linear in all arguments.

4.2 Multirate schemes

To get a mixed multirate scheme for coupled index-1 DAEs of type (18) (semi-explicit), we first regard the case $r = 2$ with F (fast) and S (slow) scale. It is natural to assume the coupling variable u (defined by $0 = g$) behaves slowly like y_S and z_S. This amounts to the following structure:

$$\begin{aligned} \dot{y}_S &= f_S(t, z_S, u), & \dot{y}_F &= f_F(t, z_F, u), \\ 0 &= h_S(t, y_S, z_S, u), & 0 &= h_F(t, y_F, z_F, u), \\ 0 &= g(z_S, z_F). \end{aligned} \tag{19}$$

With the index assumption (C3), the slow subsystem $[\dot{y}_S = f_S, 0 = h_S, 0 = g]$ is index-1 for algebraic variables z_S and u; in addition, the fast subsystem $[\dot{y}_A = f_A, 0 = h_A]$ is of index-1 with respect to z_F if (C2) holds.

Now, we can apply multirate schemes to solve system (19). We present two options:

- *Compound-step approach:* Verhoeven et al. have developed a series of multirate schemes [45] for coupled network equations: slowest-first [46] and compound-step approach [48], equipped with step size prediction for both macro and micro step size and error control [48]. His approach does not demand a given sub-circuit partitioning, but allows for automatic partitioning of the overall network equations [47].
- *Mixed multirate:* Striebel et al. [42] have developed a mixed multirate scheme on the basis of Rosenbrock-Wanner methods, which uses a generalized compound step to (jointly) compute macro step for the slow part and the first micro step of the fast part. The remaining micro steps of the active part are done by mixed multirate. The scheme allows for more than two time scales in a hierarchical setting by nesting compound steps and later micro-steps in a way that at each time merely a two-level multirate scheme is engaged [43].

Remark 4.2 In circuit simulation, the slow variable is usually referred to as latent and the fast variable is referred to as active.

4.3 Thermal-electric coupling — silicon on insulator

The transistor technology, silicon on insulator (SOI), introduces a thin insulator layer within the device. This layer can made of an oxide, which has the purpose to electrically decouple the substrate and the channel area, see Fig. 1 for a sketch of such a device. Furthermore, it confines the channel to a rather small area and thus lowers parasitic (electrical) capacitance. In the end, the power consumption can be reduced and frequencies enhanced. However, since the electric insulator is often also a thermal insulator, the thermal coupling gains importance and needs to be investigated, see, e.g., [44]. Thus the SOI technology also drives the development of thermal-electric simulation.

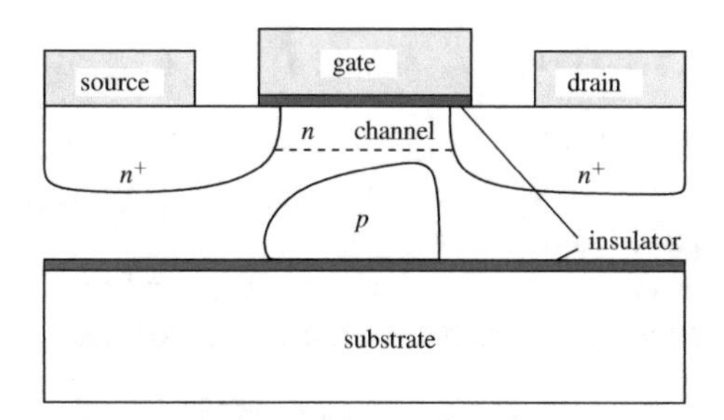

Fig. 1: Sketch of an SOI device.

Such a thermal-electric system is a multiphyics problem. On larger spatial units, one often obtains a multirate system: On the one hand, the electric subsystem switches on a very fast scale (e.g. on GHz scale). On the other hand, the temperature is dissipated and conducted slowly; e.g. the heat diffusivity in silicon is at 300K about $87\,10^{-6}\mathrm{m}^2/\mathrm{s}$. This yields a multirate setting and demands for an according simulation strategy.

In [7, 8], an according modeling and a simulation strategy was addressed. To this end, an accompanying (thermal) network (AN) for an electric network was proposed. The AN allows the connection of simple, thermal elements. These elements are spatially one dimensional and thus enable heat conduction along dedicated 1d structures.

Coupled thermal-electric model. The standard electric network model is a DAE of type (16), see e.g. [24] for details. To enable a thermal-electric coupling, we introduce a set of parameters p (which are going to be temperature dependent). This slightly generalizes the network equations (16) to the following initial value problem (for given p):

$$\overline{\mathscr{F}}(t,x,\frac{d}{dt}w(x);p)=0, \qquad \text{with consistent initial value } x(t_0)=x_0. \tag{20}$$

Thermally, the AN model comprises a set of heat conduction equations for distributed temperatures: $T:[t_0,t_e]\times[0,1]\to\mathbb{R}^k$; all 1d elements shall have a normalized spatial dimension $[0,1]$. Furthermore, we have lumped temperatures $\theta:[t_0,t_e]\to\mathbb{R}^m$, which denote the temperatures at the coupling nodes of the AN.

This can be cast into the following type of equations (cf. [8])

$$\begin{aligned} \partial_t T &= \partial_{rr} T - (T - T_{\text{env}}) + P_1, \\ \partial_t \theta &= g(\partial_r T(0,t), \partial_r T(1,t)) - (\theta - T_{\text{env}}) + P_2 \end{aligned} \tag{21}$$

with source terms P_1, P_2 (for the power dissipation) and material coefficients normalized to one, for simplicity of notation. In addition, we have boundary conditions (or coupling conditions) for the AN:

$$T(0,t) = M_0 \theta, \qquad T(1,t) = M_1 \theta$$

with suitable matrices M_0, M_1 identifying coupling of the 1d-elements. Finally, we need to supply suitable initial conditions: $\theta(t_0) = \theta_0 \geq T_{\text{env}}$, $T(t_0,x) = T_0(x) \geq T_{\text{env}}$, where T_{env} is a given ambient temperature.

Now, the coupling conditions are as follows. On the one hand, the source terms P_i indicate the dissipated powers, which are a function of the network variables x:

$$P_i = P_i(x) \geq 0 \qquad (\text{for } i = 1,2).$$

On the other hand, the electric network does depend on the temperatures T, θ, via the network parameters:

$$p = p(T, \theta).$$

Later on, the accompanying thermal network was generalized to patches, where two spatial dimensional are involved, see [1, 14].

Dedicated multirate strategy for simulation. First, we consider a method of lines approach, where we apply a suitable spatial discretization to the thermal part. Due to the structure of the AN equation (21), we can obtain an ODE model for the thermal part, and thus, a coupled DAE-ODE system.

For the time integration, we apply a dynamic iteration, cf. Section 3, with windowing. Due to the DAE-ODE coupling, the contraction factor $\alpha = 0$ (14) and thus we have linear convergence of the dynamic iteration scheme [6].

Practically, we first solve the electric subsystem. To exploit the multirate setting with fast electric signals and slow thermal adjustments, we pursue an energy coupling, which is based on averaging, cf. [6, 15]. That is, instead of coupling via the fast changing instantaneous power dissipation, one can compute the total dissipated power E^H of electric subsystem during a certain time interval $[\bar{t}, \bar{t} + H]$. This is an energy, and may be computed by simply adding according energy variables E to the network variables and appending the network equations (20) with corresponding differential equations

$$\dot{E} = P(x).$$

Then, the total amount of the dissipated power is added to the thermal network as averaged quantity: $P_i = E^H / H$. This way, the total amount of the dissipated (electric) energy is added to the thermal system. The addition of the actual power as source term to the heat conduction might happen with a little time shift on the scale of the fast system. Of course, this is for a slow heat system of minor importance. The reverse coupling needs to adjust the temperature dependent network parameters, i.e., an updated device temperature needs to be assigned. To this end, the heat equation is solved. Since we expect on the $H-$scale only minor temperature changes, we can even skip any iterations of the dynamic scheme and compute the preceding communication step (e.g. $[\bar{t}+h, \bar{t}+2H]$), i.e., solve again first the electric subsystem with the new temperatures. This is referred to as multirate co-simulation, since both subsystems may be solved on their time scale and no overhead of iterations occurs.

This multirate co-simulation strategy was successfully applied to solve a ring oscillator circuit, see [7]. In fact, it is reasonable to exchange data after one step of the slow subsystem, i.e., the communication step size can be chosen as the inherent step size of the solve subsystem. As a further enhancement, one can apply a dedicated midpoint rule to solve the slow (heat) subsystem and gain formally a second order method, see [6].

5 Molecular dynamics

In molecular dynamics, one has to solve initial-value problems of the type

$$\dot{w} = h(w), \qquad w(t_0) = w_0 \tag{22a}$$

with

$$w := \begin{pmatrix} q \\ p \end{pmatrix}, \quad h(w) := J^{-1}\nabla H(q,p), \quad J := \begin{pmatrix} 0 & -1 \\ 1 & 0 \end{pmatrix}, \quad w_0 := \begin{pmatrix} q_0 \\ p_0 \end{pmatrix} \tag{22b}$$

with positions $q : [t_0, T] \to \mathbb{R}^n$, momenta $p : [t_0, T] \to \mathbb{R}^n$ and Hamiltonian $H : \mathbb{R}^n \times \mathbb{R}^n \to \mathbb{R}$. The separable Hamiltonian $H(q,p) = T(p) + V(q)$ is composed of a kinetic energy T and a potential energy V.

The standard numerical integration scheme applied to Hamiltonian systems is the Störmer-Verlet (or leap-frog) scheme, which we refer to as $\psi_{h_0}^H$ for step size h_0. In fact, we can describe $\psi_{h_0}^H$ for (22a-22b) via operator splitting based on explicit Euler steps φ_h for the components q and p:

$$\psi_{h_0}^H(q_0, p_0) := \varphi_{h_0/2}^V \circ \varphi_{h_0}^T \circ \varphi_{h_0/2}^V(q_0, p_0). \tag{23}$$

Here, φ_h^T advances the kinetic energy $H = T$ with an explicit Euler step of size h and φ_h^V analogously the potential energy $H = V$. In fact, this numerical approximation (23) solves exactly the shadow Hamiltonian $\widehat{H}$, which differs from the original Hamiltonian H by the following expression:

$$\widehat{H} - H = -\frac{h_0^2}{24}([V,[V,T]] + 2[T,[V,T]]) + \mathcal{O}(h_0^4).$$

If the forces derived from the Hamiltonian are hierarchical and the larger forces turn out to be cheap to compute, the integration of molecular dynamics can be accelerated by means of multirate schemes. For example in lattice quantum chromodynamics [27] holds: the gauge force is cheap to evaluated and largest in size the fermionic force is expensive but smaller in size. Thus, let us assume that the potential energy consists of two parts V_1 and V_2, which differ as follows:

- V_1: the contribution
$$f_1(q) := J^{-1}\nabla V_1(q),$$
to the force is strong and with a fast dynamics, but its evaluation is cheap;
- V_2: the contribution
$$f_2(q) := J^{-1}\nabla V_2(q),$$
to the force is weak with a slow dynamics, but its evaluation is expensive.

Consequently, one may split the ODE (22) with respect to the right-hand side:

$$h_s(w,t) := f_2(q), \qquad h_f(w,t) := f_1(q) + J^{-1}\nabla T(p),$$

where we have assumed that the dynamics related to the kinetic energy is fast, but cheap to be evaluated. Thus, we have the structure of (3).

Naturally, one idea to exploit this multirate behavior is an according evaluation: within one macro step H evaluate the slow, expensive part h_s (only) once; the fast, cheap part h_f several times. However, the multirate approach has to preserve the symplectic and time-reversible structure of the Hamiltonian flow. Hence, an operator splitting approach appears to be favorable, as the composition of symplectic and time reversible schemes is again symplectic and time reversible.

For the right-hand side split ODE (3), a multirate method $\widetilde{\psi}^H_{h_0}(q_0, p_0)$ [38] with macro step size h_0 and m_1 mirco steps of step size $h_1 = h_0/m_1$ is obtained from

$$\widetilde{\psi}^H_{h_0}(q_0, p_0) := \varphi^{V_2}_{h_0/2} \circ \left(\psi^{T+V_1}_{h_1}\right)^{m_1} \circ \varphi^{V_2}_{h_0/2}(q_0, p_0),$$

where we used the Störmer-Verlet scheme ψ^H_h of the single-rate case with $H := T + V_1$ and step size h_1. This multirate scheme conserves now the following shadow Hamiltonian $\widehat{H}$, which differs from the original Hamiltonian H by

$$\widehat{H} - H = -\frac{h_0^2}{24}\Big(([V_2,[V_2,T]] - 2[V_1,[V_2,T]] - 2[T,[V_2,T]] + \\ \frac{1}{m_1^2}([V_1,[V_1,T]] - 2[T,[V_1,T]])\Big) + \mathscr{O}(h_0^4).$$

Remark 5.1 Let us assume that the dynamics level of V_1 over V_2 is proportional to m_1, then all commutators involving one or two instances of V_1 are properly scaled with $1/m_1^2$ besides the commutator $[V_1,[V_2,T]]$. However, if the Störmer-Verlet scheme for m_1 inner micro-steps is replaced by a force-gradient scheme [30], the commutator

$[V_1,[V_2,T]]$ is eliminated, and the shadow Hamiltonian for this multirate version [39] is given by

$$H+\left(\frac{h_0}{m_1}\right)^2\left(\frac{1}{96}[V_1,[T,V_1]]+\frac{1}{48}[T,[T,V_1]]\right)+\mathcal{O}(h_0^4).$$

This multirate approach can be applied to Hamiltonians with a further splitting of the potential energy in a hierarchical manner as follows: let us assume that the Hamiltonian is given by

$$H(q,p)=T(p)+\sum_{l=1}^{N}V_l(q).$$

If the energies are ordered such that the computational cost are increasing, while at the same time the strength of the associated forces is decreasing, a multirate integration based on the Störmer-Verlet scheme can be defined [38]: using macro step size h_0 and micro step size $h_1=h_0/M$ proceeds as follows:

$$\varphi_{h_0}(q_0,p_0)=\varphi^{H_2}_{h_0/2}\left(\varphi^{H_1}_{h_1}\right)^{m_1}\varphi^{H_2}_{h_0/2},$$

with $H_1(q,p)=T(p)+\sum_{l=1}^{N-1}V_l(q)$ and $H_2(q,p)=V_N(q)$. This scheme can be nested, by introducing a next finer step size $h_2=h_1/m_2$ and further splitting H_1 into

$$H_1(q,p)=H_{11}(q,p)+H_{12}(q,p)$$

with $H_{11}(q,p)=T(p)+\sum_{l=1}^{N-2}V_l(q)$ and $H_{12}(q,)=V_{N-1}(q)$ in order to replace $\varphi^{H_1}_{h_1}$ above by

$$\varphi^{H_1}_{h_1}=\varphi^{H_{12}}_{h_1/2}\left(\varphi^{H_{11}}_{h_2}\right)^{m_2}\varphi^{H_{12}}_{h_1/2}.$$

This procedure can be applied recursively to obtain N different step size ratios at the end, corresponding to the activity levels of the N potential energies $V_l(q)$.

6 Conclusion and outlook

In many applications, the governing differential equations are characterized by strongly varying time scales. Numerical time integration of these type of models demands integration schemes able to exploit this behavior, not only for efficiency reason, but quite often to enable numerical simulation at all.

In this paper, we have discussed two basic classes of multirate schemes based on extra-/interpolation and compound-step approaches: the pioneering work by Rice for missile applications (compound step approach for one-step methods) and the seminal work by Gear and Wells driven by electrical circuit application (extra-/interpolation for multi-step methods). For the latter, we characterized the overall convergence properties for coupled ODE systems.

Another trend in applications — refined modeling yielding coupled systems consisting of multiphysical subsystems — initialized further development in dynamic iteration approaches. They deliver another class of (full) multirate schemes by stopping the iteration after the first sweep. We discussed the basics of this approach and verified the efficiency of such multirate schemes by inspecting the coupled thermal-electric problem, where we gave a tailored numerical algorithm.

If the multirate behavior is not given by varying solution components, but by different characteristics of the right-hand side, an operator splitting approach is the method of choice to derive multirate schemes, especially if preservation of properties such as in geometric integration is mandatory. Here, molecular dynamics serves as an example of applications in this field.

Summing up, the need for the simulation of complex, time consuming applications inspired the numerical analysis to develop a new class of methods, multirate methods. And this class is still growing. In turn, the enhancement of the algorithm's efficiency enabled an increase in complexity of applications.

The increasing interest in parallel-in-time schemes in the last years offers a new way to speed up simulation time by the help of multirate schemes. Here the combination of parallel-in-time schemes with dynamic iteration and multirate schemes seems to be promising, see e.g. [19]. This will be an interesting topic for research on multirate schemes in the coming years.

References

[1] Alì, G., Bartel, A., Culpo, M., De Falco, C.: Analysis of a PDE thermal element model for electrothermal circuit simulation. In: J. Roos and L.R.J. Costa (eds.), Scientific Computing in Electrical Engineering (SCEE 2008). Springer, 2010, 273–280.

[2] Alì, G., Bartel, B., Günther, M., Romano, V., Schöps, S.: Simulation of coupled PDAEs: Dynamic iteration and multirate simulation. In: M. Günther (ed.): Coupled Multiscale Simulation and Optimization in Nanoelectronics, Springer Berlin, 2015, 103–156.

[3] Arnold, M., Günther, M.: Preconditioned Dynamic Iteration for Coupled Differential-Algebraic Systems. BIT **41**:1 (2001), 1–25.

[4] Bartel, A.: Multirate ROW Methods of Mixed Type for Circuit Simulation. In: van Rienen, U., Günther, M., Hecht, D. (eds.): Scientific Computing in Electrical Engineering. Lecture Notes in Computational Science and Engineering **18**, Springer-Verlag, Berlin, 2001, 241–249.

[5] Bartel, A., Brunk, M, Schöps, S.: On the convergence rate of dynamic iteration for coupled problems with multiple subsystems, J. Comp. Appl. Math. **262** (2014), 14–24.

[6] Bartel, A.: Partial Differential-Algebraic Models in Chip-Design — Thermal and Semiconductor Problems. PhD-Thesis TU Munich, 2004. Fortschritt-Berichte, VDI Verlag, Düsseldorf.
[7] Bartel, A., Feldmann, U.: Modeling and Simulation for Thermal-Electric Coupling in an SOI-Circuit. In: A.M. Anile, G. Alì, G. Mascali (eds.): Scientifc Computing in Electrical Engineering (SCEE-2004 Proceedings), Springer, Berlin, 2006, 27–32.
[8] Bartel, A., Günther, M.: From SOI to abstract electric-thermal-1D multiscale modeling for first order thermal effects. Math. Comp. Modell. Dyn. Syst **9**:1 (2003), 25–44.
[9] Bartel, A., Günther, M.: A multirate W-method for electrical networks in state–space formulation. J. of Comput. Appl. Math. **147** (2002), 411–425.
[10] Bartel, A., Günther, M.: PDAEs in Refined ElectricalNetwork Modeling. SIAM Review **60**:1 (2018), 59–91.
[11] Bartel, A. and Günther, M.: Inter/extrapolation-based multirate schemes – a dynamic-iteration perspective. Submitted for publication (ArXiV math.NA 2001.02310).
[12] Burrage, K.: Parallel and sequential methods for ordinary differential equations. Oxford Science Publications, 1995.
[13] Ciuprina, G., Ioan D. (eds.): Scientific Computing in Electrical Engineering. Mathematics in Industry. Springer, Berlin, 2007.
[14] Culpo, M.: Numerical Algorithms for System Level Electro-Thermal Simulation. PhD-Thesis at BU Wuppertal, 2009. http://elpub.bib.uni-wuppertal.de
[15] Deml, Ch., Türkes, P.: Fast Simulation Technique for Power Electronic Circuits with Widely Different Time Constants. IEEE Trans. on Industry Applications **35**:3 (1999), 657–662.
[16] Engstler Ch., Lubich, Ch.: MUR8: a multirate extension of the eighth-order Dormand-Prince method, Appl. Num. Math. **25**:2–3 (1997), 185–192.
[17] Engstler Ch., Lubich Ch.: Multirate extrapolation methods for differential equations with different time scales, Computing **85**:2 (1997), 173–185.
[18] Estévez Schwarz, D., Tischendorf, C.: Structural Analysis for Electric Circuits and Consequences for MNA, Int. J. of Circuit Theory and Applications **28** (2000), 131–162.
[19] Gander, M.J., Kulchytska-Ruchka, I., Niyonzima, I., Schöps, S.: A New Parareal Algorithm for Problems with Discontinuous Sources. SIAM J. on Scientific Computing **41**:2 (2019), B375–B395.
[20] Gear, C.W.: Automatic multirate methods for ordinary differential equations. In: Proceedings IFIP, North-Holland, Amsterdam 1980, 712-722.
[21] Gear, C.W., Wells, D.: Multirate linear multistep methods. BIT **24** (1984), 484–502.
[22] Günther, M., Feldmann, U.: CAD-based electric-circuit modeling in industry I: mathematical structure and index of network equations Surveys on Mathematics for Industry 8:2 (1999), 97-130.
[23] Günther M., Rentrop P.: Multirate ROW methods and latency of electric circuits, Appl. Num. Math. **13** (1992), 83–102.

[24] Günther, M., Rentrop, P.: Numerical Simulation of Electrical Circuits, GAMM-Mitteilungen **23**:1 (2000).
[25] Günther M., Sandu, A.: Multirate generalized additive Runge Kutta methods Numerische Mathematik **133** (2016), 497–524.
[26] Hairer, E., Wanner, G.: Solving Ordinary Differential Equations II, Springer, Berlin, 2002.
[27] Knechtli, F., Günther, M., Peardon. M.: Lattice Quantum Chromodynamics Practical Essentials, Springer Briefs in Physics, 2016.
[28] Kværnø, A., Rentrop, P.: Low order multirate Runge-Kutta methods in electric circuit simulation, Preprint No. 99/1, IWRMM, Universität Karlsruhe (TH), 1999.
[29] Lelarasmee, E. Ruehli, A.E, Sangiovanni-Vincentelli, A.L.: The Waveform Relaxation Method for Time-Domain Analysis of Large Scale Integrated Circuits, IEEE Transactions on Computer-Aided Design of Integrated Circuits and Systems **1**:3 (1982), 131–145.
[30] Omelyan, I.P., Mryglod, I.M., Folk, R.: Symplectic analytically integrable decomposition algorithms: classification, derivation, and application to molecular dynamics, quantum and celestial mechanics, Comput. Phys. Commun. **151** (2003), 272–314.
[31] Orailoglu, A.: A multirate ordinary differential equation integrator. Master Thesis UIUCDCS-R-79-959, University of Illinois at Urbana-Champaign, Dept. of Computer Science, 1979.
[32] Rodríguez-Gómez, G., González-Casanova, P., Martínez-Carballido, J.: Computing general companion matrices and stability regions of multirate methods, Int. J.for Numerical Methods in Engineering **61** (2004), 255–273.
[33] Rice, J.R.: Split Runge-Kutta method for simultaneous equations, J. Res. Nat. Bur. Standards 64B (1960), 151–170.
[34] Sandu, A., Constantinescu, E.M.: Multirate Explicit Adams Methods for Time Integration of Conservation Laws. J. of Scientific Computing **38**:2 (2009), 229–249.
[35] Savcenco, V., Hundsdorfer. W., Verwer, J.G.: A multirate time stepping strategy for stiff ordinary differential equations, BIT **47** (2007), 137–155.
[36] Schöps, S.: Multiscale Modeling and Multirate Time-Integration of Field/Circuit Coupled Problems of Elektrotechnik. (PhD Thesis, Univ. Wuppertal). VDI Verlag, Düsseldorf, 2011.
[37] Schöps, S., Bartel, A., De Gersem, H.: Multirate Time Integration of Field/Circuit Coupled Problems by Schur Complements. In: Michielsen, B., Poirier, J.-R. (eds.): Scientific Computing in Electrical Engineering SCEE 2010, Springer Berlin Heidelberg, 2012, 243–251.
[38] Sexton, J.C., Weingarten, D.H.: Hamiltonian evolution for the hybrid Monte Carlo algorithm. Nucl. Phys. B **380**:3 (1992), 665–677.
[39] Shcherbakov, D., Ehrhardt, M., Günther M., Peardon, M.: Force-gradient nested multirate methods for Hamiltonian systems. Comput. Phys. Commun. **187** (2015), 91–97.

[40] Skelboe, S.: Stability properties of backward differentiation multirate formulas, Appl. Num. Math. **5** (1989), 151–160.
[41] Skelboe S., Andersen, U.: Stability properties of backward Euler multirate formulas, SIAM J. Sci. Stat. Comput. **10** (1989), 1000–1009.
[42] Striebel, M., Bartel A., Günther, M.: A multirate ROW-scheme for index-1 network equations, Appl. Num. Math. **59**:3–4 (2009), 800–814.
[43] Striebel, M., Günther, M.: Hierarchical mixed multirating in circuit simulation. In: [13], 221–228.
[44] Su, L.T., Chung, J.E., Antoniadis, D.A., Goodson K.E., Flik, M.I.: Measurement and modeling of self-heating in SOI nMOSFET's, IEEE Trans. on Electron Devices **41**:1, 69–75, 1994.
[45] Verhoeven, A., El Guennouni, A., ter Maten, E.J.W., Mattheij, R.M.M.: A general compound multirate method for circuit simulation problems. In: Scientific Computing in Electrical Engineering, Springer, 2006, 143–149.
[46] Verhoeven, A., ter Maten, E.J.W., Mattheij, R.M.M. Tasic, B.: Stability analysis of the BDF slowest-first multirate methods, Int. J. of Comput. Math. **84**:6 (2007), 895–923.
[47] Verhoeven, A., Tasic, B., Beelen, T.G.J., ter Maten, E.J.W. Mattheij, R.M.M.: Automatic partitioning for multirate methods. In: [13], 229–236.
[48] Verhoeven, A., Tasic, B., Beelen, T.G.J., ter Maten, E.J.W., Mattheij, R.M.M.: BDF compound-fast multirate transient analysis with adaptive stepsize control, J. Numer. Anal. Ind. Appl. Math **3** (2008), 275–297.
[49] Wells, D.R.: Multirate linear multistep methods for the solution of systems of ordinary differential equationsequation integrator, PhD Thesis UIUCDCS-R-82-1093, Unviersity of Illinois at Urbana-Champaign, Dept. of Computer Science, 1982.
[50] White J.K., Sangiovanni-Vincentelli A.: Waveform Relaxation. In: Relaxation Techniques for the Simulation of VLSI Circuits. Kluwer International Series in Engineering and Computer Science **20**, Springer, Boston, MA, 1987.

Electronic Circuit Simulation and the Development of New Krylov-Subspace Methods

Roland W. Freund

Abstract Ever since the 1960s, the semiconductor industry has heavily relied on simulation in order to analyze and verify the design of integrated circuits before actual chips are manufactured. Over the decades, the algorithms and tools of circuit simulation have evolved in order to keep up with the ever-increasing complexity of integrated circuits, and at certain points of this evolution, new simulation techniques were required. Such a point was reached in the early 1990s, when a new approach was needed to efficiently and accurately simulate the effects of the ever-increasing amount of on-chip wiring on the proper functioning of the chip. The industry's proposed solution for this task, the AWE approach, worked well for small- to moderate-size networks of on-chip wiring, but suffered from numerical issues for larger networks. It turned out that for the special case of networks with single inputs and single outputs, these problems can be remedied by exploiting the connection between AWE and the classical Lanczos algorithm for single starting vectors. However, the general case of on-chip wiring involves networks with multiple inputs and outputs, and so a Lanczos-type algorithm was needed that could handle such multiple starting vectors. Since no such extension existed, a new band Lanczos algorithm for multiple starting vectors was developed. It turned out that this new band approach can also be employed to devise extensions of other Krylov-subspace methods. In this chapter, we describe the band Lanczos algorithm and the band Arnoldi process and how their developments were driven by the need to efficiently and accurately simulate the effects of on-chip wiring of integrated circuits.

Roland W. Freund
Department of Mathematics, University of California at Davis,
One Shields Avenue, Davis, California 95616, USA
e-mail: freund@math.ucdavis.edu

M. Günther, W. Schilders (eds.), *Novel Mathematics Inspired by Industrial Challenges*,
Mathematics in Industry 38, https://doi.org/10.1007/978-3-030-96173-2_2

1 Introduction

Since the invention of integrated circuits in the late 1950s, the semiconductor industry has succeeded in manufacturing chips with ever-decreasing feature size and ever-increasing complexity. As a result, the number of transistors on state-of-the-art chips evolved from tens of transistors on a single chip in the late 1950s to tens of billions of transistors on a single chip in 2019. Already early in this evolution, it became apparent that computer simulation is indispensable in order to analyze and verify circuit designs before actual chips are manufactured. The methods, tools, and software used for such simulations are referred to as *electronic circuit simulation* or simply *circuit simulation*; see, e.g., [33]. The basic framework of circuit simulation was created in the 1960s and early 1970s, culminating in Nagel's SPICE circuit simulator [1]; accounts of these developments can be found in [31, 35]. Most of the circuit simulators in use today are variants or derivatives of SPICE.

1.1 The Central Numerical Task in Circuit Simulation

Circuit simulation uses the *lumped-element approach* to model integrated circuits as networks of idealized electrical circuit elements, such as resistors, capacitors, inductors, diodes, and transistors. The branches of such a network model correspond to the circuit elements, and the nodes of the network correspond to the interconnections of the circuit elements. The electrical performance of the network model is characterized by three types of equations. *Kirchhoff's current law* (KCL) states that for each node of the network, the currents flowing in and out of that node sum up to zero. *Kirchhoff's voltage law* (KVL) states that for each closed loop of the network, the voltage drops along that loop sum up to zero. *Branch constitutive relations* (BCRs) are equations that characterize the electrical performance of the idealized electrical circuit elements. For example, the BCR of a linear resistor is Ohm's law. The BCRs are linear equations for simple devices, such as linear resistors, capacitors, and inductors, and they are nonlinear equations for more complex devices, such as diodes and transistors. In general, the BCRs involve first time-derivatives of the unknowns and are thus first-order *ordinary differential equations* (ODEs). On the other hand, the KCLs and KVLs are linear algebraic equations that only depend on the topology of the network. The KCLs, KVLs, and BCRs can be summarized as a system of first-order, in general nonlinear, *differential-algebraic equations* (DAEs) of the form

$$\frac{\mathrm{d}}{\mathrm{d}t} q(x,t) + f(x,t) = 0, \tag{1}$$

together with suitable initial conditions. Here, f and q are vector-valued functions, each with N scalar component functions[1], and the unknown $x = x(t)$ is a vector-valued function of length N the entries of which are the circuit variables at time t. We stress that (1) is a system of DAEs rather than ODEs due to the fact that all KCLs and KVLs and the BCRs of some elements (e.g., resistors) are algebraic equations. In particular, the Jacobian $E = \mathrm{D}_x q(x,t)$ of $q(x,t)$ with respect to x is a singular matrix in general.

The numerical solution of systems (1) is the central task in circuit simulation. This is a very challenging task for a number of reasons. The electrical performance of circuits typically involves vastly different time scales, resulting in equations (1) that exhibit *stiffness* in general. Only a small fraction of the huge arsenal of methods for solving nonstiff ODEs are suitable for stiff DAEs. In particular, implicit methods need to be used. These are computationally expensive since the solution of a system of N algebraic equations for N unknowns is required at each time step. Moreover, since these systems are nonlinear in general, some variant of Newton's method needs to be employed, which in turn involves the solution of a system of N linear algebraic equations for N unknowns at each Newton step. Finally, the number N of circuit variables is so large that special algorithms for large-scale matrix computations need to be used in order to make the numerical solution of systems (1) feasible.

1.2 Large-Scale Matrix Computations and Krylov-Subspace Methods

The archetype of a matrix computation is the numerical solution of linear systems of equations

$$M z = b. \tag{2}$$

Here, M is a given $N \times N$ matrix, b is a given vector of length N, and z is the unknown solution vector of (2). For small to moderately large N, the standard approach for computing z is Gaussian elimination, which is based on factoring M into a product of a lower-triangular matrix L and an upper-triangular matrix U. For problems (2) with large N that actually arise in meaningful applications, the matrices M usually exhibit special structures, such as sparsity. An $N \times N$ matrix M is called *sparse* if only a small fraction of its N^2 entries are nonzero. The problem (2) is said to be *large-scale* if its solution z can be computed only by employing algorithms that exploit the special structure of M.

As we discussed in Sect. 1.1, the numerical solution of circuit equations (1) requires the repeated solution of linear systems of the form (2). For realistic circuit simulations, all these linear systems are large-scale and have sparse coefficient matrices M. Furthermore, the matrices M are such that the linear systems (2) can be solved by means of variants of Gaussian elimination that are adapted to sparse

[1] We use the upper-case letter N for the number of components to indicate that this number is large in circuit simulation. The lower-case letter n is used to denote the iteration index in Krylov-subspace methods.

matrices. The key feature of these variants is to generate reorderings of the rows and columns of M such that the triangular factors L and U of the reordered version of M remain reasonably sparse. For general sparse matrices, such reorderings are not always possible. However, for matrices M arising in circuit simulation, sparse Gaussian elimination works amazingly well and produces triangular factors L and U that are nearly as sparse as M. In fact, all circuit simulators employ some form of sparse Gaussian elimination to solve the large-scale linear systems (2) that arise in the context of the numerical solution of (1).

For general matrix computations, the same terminology as for linear systems (2) is used. A matrix computation problem is said to be *large-scale* if it can be solved only by employing algorithms that exploit special structures of the matrices describing the problem; see, e.g., [20].

One of the most versatile tools for large-scale matrix computations are iterative methods based on Krylov subspaces. Let M be a given $N \times N$ matrix and r be a given vector of length N. For any $n = 1, 2, \ldots$, the subspace of the space of vectors of length N that is spanned by the vectors

$$r, Mr, M^2r, \ldots, M^{n-1}r \tag{3}$$

is called the *n-th Krylov subspace* (induced by M and r) and denoted by $\mathscr{K}_n(M, r)$. For many large-scale matrix computations arising in actual applications, very good approximate solutions of the large-scale problem in N-dimensional space can be obtained by solving corresponding n-dimensional problems that are obtained by means of n-th Krylov subspaces $\mathscr{K}_n(M, r)$ with $n \ll N$. However, the basis (3) used to define $\mathscr{K}_n(M, r)$ is not suitable for actual computations since the vectors (3) quickly become linearly dependent in finite-precision arithmetic as n increases. Instead, so-called Krylov-subspace methods are employed to generate more suitable bases. An important feature of these methods is that the matrix M is used only in the form of matrix-vector products with M and possibly with the transpose M^T of M. In particular, these products can be computed cheaply when M is sparse.

The two classical Krylov-subspace methods, the Lanczos algorithm [27] and the Arnoldi process [3], were introduced in the early 1950s in the context of iterative methods for systems of linear equations and eigenvalue computations. The Arnoldi process produces an orthonormal (and thus optimal) basis for $\mathscr{K}_n(M, r)$. Since the construction of such an orthonormal basis involves $(n+1)$-term recurrences of vectors of length N, the Arnoldi process requires $\mathscr{O}(n^2 N)$ operations, which makes its use problematic for very large-scale problems. The Lanczos algorithm generates a pair of bases, one for $\mathscr{K}_n(M, r)$ and one for the n-th Krylov subspace $\mathscr{K}_n(M^T, l)$ induced by M^T and a second given vector l of length N. The vectors of the two bases are constructed to be biorthogonal to each other, but neither one of the two Lanczos bases is orthonormal. As a result, the Lanczos bases are not as well-behaved in actual computations as the Arnoldi basis. However, since the construction of such biorthogonal bases can be done with three-term recurrences of vectors of length N, the Lanczos algorithm process requires only $\mathscr{O}(nN)$ operations, which allows its use for much larger problems than the Arnoldi process.

In the 6 decades since the introduction of the Lanczos algorithm and the Arnoldi process, Krylov-subspace methods have been studied extensively and have proven to be useful in many other applications besides the solution of systems of linear equations and eigenvalue computations. For example, the Lanczos algorithm was shown to be closely related to Padé approximation of transfer functions of single-input single-output time-invariant linear dynamical systems; see, e.g., Gragg's 1974 paper [24]. This so-called Lanczos-Padé connection is the basis for the PVL algorithm described in Sect. 2.2.

1.3 The Special Case of Circuit Interconnect Analysis

Given the success of sparse Gaussian elimination in solving the linear systems arising in the context of general circuit equations (1), there never was a need for even considering the use of Krylov-subspace methods for solving these linear systems. Nevertheless, in the early 1990s, Krylov-subspace methods turned out to be very efficient tools for tackling the special case of circuit equations (1) that arise in circuit interconnect analysis.

Integrated circuits contain tiny on-chip "wires" to connect transistors and other components to each other. This on-chip wiring is called the circuit *interconnect*. As the number of transistors on a single chip evolved from tens of transistors in the late 1950s to tens of billions of transistors in 2019, the amount of interconnect increased accordingly. A state-of-the-art chip in 2019 contains interconnect wires with a total length of tens of miles. Interconnect analysis uses simulation to verify and correct the interconnect design of a chip in order to ensure that the on-chip wiring does not interfere with the proper functioning of the chip.

Circuit interconnect analysis employs the lumped-element approach described in Sect. 1.1 to model interconnect structures as *RCL networks* of resistors, capacitors, and inductors that correspond to small pieces of wires of the overall interconnect. Since the BCRs of all these circuit elements are linear time-invariant equations, the electrical performance of the interconnect network is described by a system of equations of the form (1) with functions f and q that are linear in x. Moreover, the main interest in interconnect analysis is the input-output behavior of the interconnect. Given input functions, such as the voltages of voltage sources and the currents of current sources, which drive the interconnect, the task is to compute certain output functions, such as the currents of the voltage sources and the voltages of the current sources. In this case, the input-output behavior of the interconnect is described by a system of linear DAEs of the form

$$
\begin{aligned}
E \frac{\mathrm{d}}{\mathrm{d}t} x &= Ax + Bu(t), \\
y(t) &= L^T x(t),
\end{aligned}
\tag{4}
$$

together with suitable initial conditions. Here, A and E are $N \times N$ matrices, B is an $N \times m$ matrix, L is an $N \times p$ matrix, u is a vector-valued function of length m, y is a vector-valued function of length p, and the unknown $x = x(t)$ is a vector-valued function of length N the entries of which are the circuit variables at time t. The m entries of $u = u(t)$ are the given input functions, and the p entries of $y = y(t)$ are the output functions of interest. The numbers $m \geq 1$ and $p \geq 1$ are small and $m, p \ll N$. In general. the matrix E is singular and thus (4) is a system of DAEs, rather than ODEs. Finally, we always assume that the *matrix pencil*

$$sE - A, \quad s \in \mathbb{C}, \tag{5}$$

is *regular*, i.e., the matrix $sE - A$ is singular only for finitely many values of $s \in \mathbb{C}$. Here, $\mathbb{C}$ denotes the set of all complex numbers. The assumption of regularity of the matrix pencil (5) is satisfied for any realistic circuit interconnect simulation, see, e.g., [17].

Systems of equations of the form (4) are called *m-input p-output linear time-invariant linear dynamical systems*. They arise in many applications and not just in circuit simulation. However, in most applications the size N of (4) is small or only of moderate size, whereas one has to deal with large-scale systems in circuit simulation. In the large-scale case, the main interest is usually in the input-output behavior $u(t) \to y(t)$ of the system (4), rather than the complete solution vector $x(t)$. In fact, for very large N, it may not even be feasible to compute x. A standard approach to tackle large-scale systems (4) is to employ model order reduction; see, e.g. [38]. The basic idea is to replace the quantities of size N in (4) by corresponding quantities of size n. More precisely, a *reduced-order model* (ROM) of (4) is a system of the form

$$\begin{aligned} E_n \frac{\mathrm{d}}{\mathrm{d}t} \widetilde{x}(t) &= A_n \widetilde{x}(t) + B_n u(t), \\ \widetilde{y}(t) &= L_n^T \widetilde{x}(t), \end{aligned} \tag{6}$$

where A_n and E_n are $n \times n$ matrices, B_n is an $n \times m$ matrix, L_n is an $n \times p$ matrix, and $n \ll N$. Note that $u = u(t)$ is the same given input function in both the original system (4) and its ROM (6). The challenge of model order reduction is to find a value $n \ll N$ and matrices A_n, E_n, B_n, and L_n such that

$$\widetilde{y}(t) \approx y(t) \quad \text{for all 'relevant' times } t. \tag{7}$$

A standard approach in model order reduction of time-invariant linear dynamical systems is to transform (4) from time domain into complex Laplace domain. Applying the Laplace transform to (4), we obtain the Laplace-domain system

$$\begin{aligned} sE\widehat{x}(s) &= A\widehat{x}(s) + B\widehat{u}(s), \\ \widehat{y}(s) &= L^T \widehat{x}(s), \end{aligned} \tag{8}$$

where $s \in \mathbb{C}$. Elimination of $\widehat{x}(s)$ from (8) results in the Laplace-domain input-output relation

$$\widehat{y}(s) = H(s)\,\widehat{u}(s), \tag{9}$$

where

$$H : \mathbb{C} \mapsto (\mathbb{C} \cup \infty)^{p \times m}, \quad H(s) := L^T \left(sE - A\right)^{-1} B. \tag{10}$$

The function (10) is called the *transfer function* of the time-invariant linear dynamical system (4). We remark that H is a $(p \times m)$-matrix-valued rational function with potential poles at the finitely many values of $s \in \mathbb{C}$ for which the matrix $sE - A$ is singular. Analogously, the Laplace-domain input-output relation of the ROM (6) is given by

$$\widehat{\widetilde{y}}(s) = H_n(s)\,\widehat{u}(s), \tag{11}$$

where

$$H_n : \mathbb{C} \mapsto (\mathbb{C} \cup \infty)^{p \times m}, \quad H_n(s) := L_n^T \left(sE_n - A_n\right)^{-1} B_n, \tag{12}$$

is the transfer function of the ROM (6). Finally, in view of (9) and (11), the desired approximation property (7) in time domain translates into the approximation property

$$H_n(s) \;\approx\; H(s) \quad \text{for all `relevant' values of } s \in \mathbb{C} \tag{13}$$

in Laplace domain.

In control theory, the problem of constructing good approximations H_n of H in (13) has been studied extensively and many powerful methods have been developed. However, only few of these approaches are feasible in the large-scale case. The approach that is relevant for circuit interconnect analysis is *moment matching*. It is based on selecting a suitable expansion point $s_0 \in \mathbb{C}$ and then constructing H_n such that the Taylor series of $H_n(s)$ and $H(s)$ about s_0 agree in as many of their leading Taylor coefficients as possible.

1.4 Outline

In the first three decades of integrated circuits, it was sufficient to use a simple metric, the Elmore delay [7], to capture the effects of interconnect. Around 1990, the complexity of integrated circuits had reached the point where this simple metric was no longer accurate enough. Building on the concept of Elmore delay, *asymptotic waveform evaluation* (AWE) [32, 34] was proposed. Unfortunately, the initial excitement over AWE was followed by the disappointment that actual implementations of the method did not perform as expected. The remedy for these numerical problems was the *Padé via Lanczos* (PVL) algorithm [8, 9], which is based on the classical Lanczos algorithm [27]. In Sect. 2, we describe AWE and the PVL algorithm for the case $m = p = 1$ of single-input single-output linear time-invariant linear dynamical systems (4). The success of PVL quickly led to the question of how to extend the PVL algorithm to the case of general m-input p-output systems. Surprisingly, a corresponding extension of the classical Lanczos algorithm for general $m, p \geq 1$ did not exist at that time, and so a new such Krylov-subspace method, the *band Lanc-*

zos method, was developed. We describe the underlying concept of block Krylov subspaces for multiple starting vectors in Sect. 3 and the band Lanczos method itself in Sect. 4. An important issue in interconnect simulation is to preserve crucial properties, such as passivity and reciprocity, of the interconnect network model in the ROMs that are constructed via suitable Krylov-subspace methods. In Sect. 5, we discuss the problem of structure preservation in reduced-order interconnect models and the construction of such structure-preserving models by means of explicit projections. In Sect. 6, we describe a new Krylov-subspace method, the *band Arnoldi process*, that was developed to facilitate reliable implementations of such projection approaches. Finally, in Sect. 7, we mention some open problems and make some concluding remarks.

2 From AWE to the PVL Algorithm

In this section, we consider only circuit interconnect models with single input and single output functions. The system of DAEs describing such models is given by (4) with $m = p = 1$. Since B and L in (4) are vectors in this case, we use b and l instead of upper-case letters. The system of DAEs is thus of the form

$$\begin{aligned} E\frac{\mathrm{d}}{\mathrm{d}t}x &= Ax + bu(t), \\ y(t) &= l^T x(t), \end{aligned} \tag{14}$$

and its transfer function is given by

$$H : \mathbb{C} \mapsto (\mathbb{C} \cup \infty), \quad H(s) := l^T \left(sE - A\right)^{-1} b. \tag{15}$$

Note that H is a scalar rational function.

2.1 Elmore Delay and AWE

Until the late 1980s, it was sufficient to model circuit interconnect as *RC networks*, i.e., networks that contain only resistors and capacitors, but no inductors, and the Elmore delay was used as a simple metric for such RC networks. In this approach, the RC network model of the interconnect is replaced by a simple reduced model that contains only a single resistor with resistance R and a single capacitor with capacitance C. The product of R and C is the actual Elmore delay. This whole process can be viewed as the construction of an approximation of the form

$$H_1(s) = \frac{a_1}{s - b_1} \tag{16}$$

to the transfer function H of the RC network model. Here a_1 and b_1 are real parameters that are determined such that the leading two Taylor coefficients of the Taylor series of $H_1(s)$ and $H(s)$ about the expansion point $s_0 = 0$ match:

$$H_1(s) = H(s) + \mathcal{O}(s^2).$$

The values of R and C are readily obtained from a_1 and b_1; see, e.g., [18].

AWE generalizes the simple approximation (16) to rational functions of the form

$$H_n(s) = \frac{a_1}{s - b_1} + \frac{a_2}{s - b_2} + \cdots + \frac{a_n}{s - b_n}, \tag{17}$$

where the $2n$ parameters $a_1, a_2, \ldots, a_n, b_1, b_2, \ldots, b_n$ are determined such that the leading $2n$ Taylor coefficients of the Taylor series of $H_n(s)$ and $H(s)$ about some suitable expansion point $s_0 \in \mathbb{C}$ match:

$$H_n(s) = H(s) + \mathcal{O}\big((s - s_0)^{2n}\big). \tag{18}$$

In principle, any $s_0 \in \mathbb{C}$, except for the finitely many poles of H, can be chosen as expansion point in (18). For interconnect models, all poles of H have negative real parts and thus any s_0 with nonnegative real part is a safe choice. Since real values of s_0 are preferable in order to avoid complex arithmetic, expansion points $s_0 \geq 0$ are chosen in practice.

In theory, by increasing n in (17) until a sufficiently accurate approximation H_n of H is obtained, AWE should be able to easily handle much more complex interconnect models than the Elmore delay. However, in practice, the accuracy of H_n tends to stagnate already at modest values of n. The reason for this behavior is not the defining property (18) of H_n, but the algorithm that is used in AWE to compute H_n. AWE first explicitly generates the leading $2n$ coefficients (the so-called *moments*) of the Taylor series of $H(s)$ about the expansion point s_0 and then constructs the values of the $2n$ parameters in (17) such that H_n has the same $2n$ moments as H. Unfortunately, the computation of the moments is extremely sensitive to numerical round-off error and is viable only for very small values of n. For a detailed discussion of the numerical issues of AWE, we refer the reader to [9].

We remark that a function H_n defined by (17) and (18) is called a *Padé approximant* of H; see, e.g., [4]. The PVL algorithm generates the same Padé approximant H_n as AWE, but does so without computing the moments.

2.2 PVL Algorithm

The basis of the PVL algorithm is the connection between Padé approximants H_n to transfer functions H of the form (15) and the Lanczos algorithm.

Recall from Sect. 1.2 that the Lanczos algorithm involves an $N \times N$ matrix M and two vectors r and l of length N, and thus we rewrite (15) as follows:

$$H(s) = l^T \left(sE - A\right)^{-1} b = l^T \left(I + (s - s_0)\,M\right)^{-1} r,$$
$$\text{where} \quad M := \left(s_0 E - A\right)^{-1} E, \quad r := \left(s_0 E - A\right)^{-1} b, \tag{19}$$

and I denotes the identity matrix of the same size as M.

Running n iterations of the Lanczos algorithm (with M, r, and l from (19)) generates a pair of biorthogonal bases for the subspaces $\mathscr{K}_n(M,r)$ and $\mathscr{K}_n(M^T,l)$. The scalars in the three-term recurrences used to construct these bases is all that is needed to obtain H_n. More precisely,

$$H_n = (l^T r)\, e_1^T \left(I + (s - s_0)\,T_n\right)^{-1} e_1, \tag{20}$$

where e_1 denotes the first unit vector of length n and

$$T_n = \begin{bmatrix} \alpha_1 & \beta_2 & 0 & \cdots & 0 \\ \rho_2 & \alpha_2 & \beta_3 & \ddots & \vdots \\ 0 & \rho_3 & \ddots & \ddots & 0 \\ \vdots & \ddots & \ddots & \ddots & \beta_n \\ 0 & \cdots & 0 & \rho_n & \alpha_n \end{bmatrix} \tag{21}$$

is an $n \times n$ tridiagonal matrix whose entries are computed during the first n iterations of the following algorithm.

Algorithm (Lanczos algorithm)
Set $\widehat{v}_1 = r$, $\widehat{w}_1 = l$, $v_0 = w_0 = 0$, and $\delta_0 = 1$.
For $n = 1, 2, \ldots$, do:

1) Compute $\rho_n = \|\widehat{v}_n\|_2$ and $\eta_n = \|\widehat{w}_n\|_2$.
   If $\rho_n = 0$ or $\eta_n = 0$, set $n = n - 1$ and stop.
   Otherwise, set $v_n = \widehat{v}_n / \rho_n$ and $w_n = \widehat{w}_n / \eta_n$.
2) Compute $\delta_n = w_n^T v_n$.
   If $\delta_n = 0$, stop: look-ahead would be needed to continue.
3) Compute $\widehat{v}_{n+1} = M v_n$.
   Set $\beta_n = \eta_n \delta_n / \delta_{n-1}$ and $\widehat{v}_{n+1} = \widehat{v}_{n+1} - v_{n-1} \beta_n$.
4) Compute $\alpha_n = w_n^T \widehat{v}_{n+1} / \delta_n$.
   Set $\widehat{v}_{n+1} = \widehat{v}_{n+1} - v_n \alpha_n$.
5) Compute $\widehat{w}_{n+1} = M^T w_n$.
   Set $\gamma_n = \rho_n \delta_n / \delta_{n-1}$ and $\widehat{w}_{n+1} = \widehat{w}_{n+1} - w_n \alpha_n - w_{n-1} \gamma_n$.

For details and properties of the Lanczos algorithm, the reader is referred to [20, Sect. 64.5] or [37, Sect. 7.1]. Next, we list some facts about Algorithm 2.1 that are relevant for its use in the PVL algorithm and its extension to the band Lanczos method in Sect. 4:

1. In exact arithmetic, the algorithm terminates after finitely many iterations. Since $\mathscr{K}_n(M,r)$ and $\mathscr{K}_n(M^T,l)$ are subspaces of N-dimensional space, their dimensions cannot exceed N. As a result, the check in step 1) is satisfied for some

$n \leq N+1$. If $\rho_n = 0$, then $\mathscr{K}_{n-1}(M,r)$ has reached its maximum dimension $n-1$. If $\eta_n = 0$, then $\mathscr{K}_{n-1}(M^T,l)$ has reached its maximum dimension $n-1$.

2. In general, the algorithm may stop prematurely due to $\delta_n = 0$ in step 2). Such an event is called an *exact breakdown*. In practice, one also needs to stop if $\delta_n \neq 0$, but $|\delta_n|$ is 'close' to 0. Such an event is called a *near-breakdown*. Exact breakdowns and near-breakdowns can be avoided altogether by employing so-called 'look-ahead' strategies; see [22] and the references given there. The resulting *look-ahead* Lanczos algorithm is necessarily quite a bit more involved than Algorithm 2.1. To keep the exposition simple, we only discuss the Lanczos algorithm and the band Lanczos method without look-ahead.
3. The vectors $v_1, v_2, \ldots, v_n$ form a basis of $\mathscr{K}_n(M,r)$, and the vectors $w_1, w_2, \ldots, w_n$ form a basis of $\mathscr{K}_n(M^T,l)$. In exact arithmetic, the two bases are biorthogonal to each other. Using the notation

$$V_n := \begin{bmatrix} v_1 & v_2 & \cdots & v_n \end{bmatrix} \quad \text{and} \quad W_n := \begin{bmatrix} w_1 & w_2 & \cdots & w_n \end{bmatrix}, \tag{22}$$

the biorthogonality of the two bases can be stated compactly as follows:

$$W_n^T V_n = \Delta_n := \operatorname{diag}(\delta_1, \delta_2, \ldots, \delta_n) = \begin{bmatrix} \delta_1 & 0 & \cdots & 0 \\ 0 & \delta_2 & \ddots & \vdots \\ \vdots & \ddots & \ddots & 0 \\ 0 & \cdots & 0 & \delta_n \end{bmatrix}. \tag{23}$$

The recurrences to that are used in to generate the vectors $v_1, v_2, \ldots, v_n, \widehat{v}_{n+1}$ and $w_1, w_2, \ldots, w_n, \widehat{w}_{n+1}$ can be stated compactly as follows:

$$\begin{aligned} MV_n &= V_n T_n + \begin{bmatrix} 0 & 0 & \cdots & 0 & \widehat{v}_{n+1} \end{bmatrix}, \\ M^T W_n &= W_n \widetilde{T}_n + \begin{bmatrix} 0 & 0 & \cdots & 0 & \widehat{w}_{n+1} \end{bmatrix}. \end{aligned} \tag{24}$$

Here T_n is the tridiagonal matrix (21) and $\widetilde{T}_n$ is the tridiagonal matrix given by

$$\widetilde{T}_n = \begin{bmatrix} \alpha_1 & \gamma_2 & 0 & \cdots & 0 \\ \eta_2 & \alpha_2 & \gamma_3 & \ddots & \vdots \\ 0 & \eta_3 & \ddots & \ddots & 0 \\ \vdots & \ddots & \ddots & \ddots & \gamma_n \\ 0 & \cdots & 0 & \eta_n & \alpha_n \end{bmatrix}.$$

These two tridiagonal matrices are related as follows:

$$\Delta_n T_n = \widetilde{T}_n^T \Delta_n. \tag{25}$$

4. By multiplying the first equation in (24) from the left by W_n^T and by using (23) and $W_n^T \widehat{v}_{n+1} = 0$, we obtain the expression

$$T_n = \left(W_n^T V_n\right)^{-1} W_n^T M V_n = \Delta_n^{-1} W_n^T M V_n \tag{26}$$

for T_n. Since the columns of V_n and W_n are biorthogonal bases of $\mathscr{K}_n(M,r)$ and $\mathscr{K}_n(M^T,l)$, the relation (26) means that the $n \times n$ matrix T_n is the *oblique projection* of the $N \times N$ matrix M onto the subspace $\mathscr{K}_n(M,r)$ and orthogonally to the subspace $\mathscr{K}_n(M^T,l)$.

5. For the PVL algorithm, only the tridiagonal matrix T_n is needed. Its entries are generated as scalar coefficients of the three-term recurrences that are used to produce the two biorthogonal bases. In order to run these recurrences, only the 6 vectors $v_{n-1}, v_n, \widehat{v}_{n+1}, w_{n-1}, w_n, \widehat{w}_{n+1}$ need to be stored at any stage of Algorithm 2.1.
6. Each iteration of Algorithm 2.1 requires one matrix-vector product with M and one with M^T. For the PVL algorithm, M is of the form $M = \left(s_0 E - A\right)^{-1} E$, where A and E are large-scale sparse matrices. To compute the matrix-vector products efficiently in this case, one employs sparse Gaussian elimination to precompute a sparse LU factorization of the matrix $s_0 E - A$. Each matrix-vector product with M or M^T can then be computed cheaply via one multiplication with a sparse matrix and two sparse triangular solves.

2.3 An Example

The following example, which is taken from [8, 9], illustrates the numerical differences between AWE and the PVL algorithm. The circuit simulated here is a voltage filter, where the frequency range of interest is $1 \le \omega \le 10^{10}$. In Fig. 1(a) we show

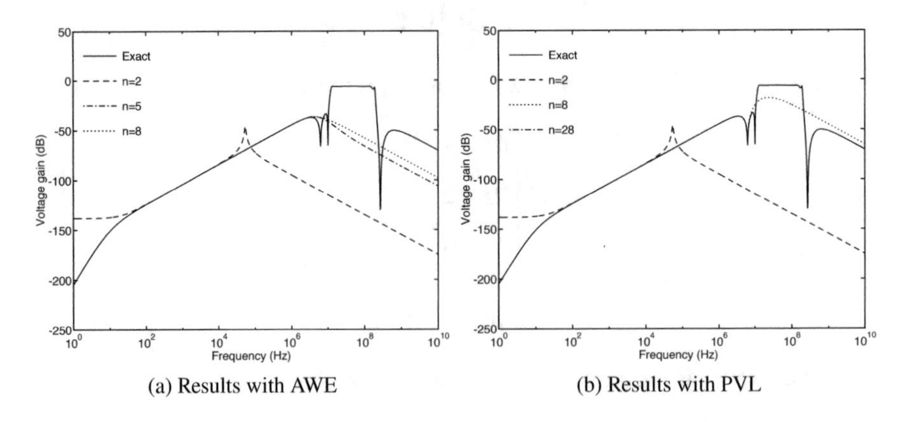

(a) Results with AWE (b) Results with PVL

Fig. 1: Simulation of a voltage filter

the exact function $|H(\mathrm{i}\,\omega)|$ and the approximations $|H_n(\mathrm{i}\,\omega)|$ generated by AWE for $n = 2, 5, 8$. Note that H_8 has clearly not yet converged to H. It turns out that the H_n's practically do not change anymore for $n \geq 8$, and so AWE never converges in this example. In Fig. 1(b) we show the exact function $|H(\mathrm{i}\,\omega)|$ and the approximations $|H_n(\mathrm{i}\,\omega)|$ generated by the PVL algorithm for $n = 2, 8, 28$. Note that the results for $n = 8$ (the dotted curves) in Fig. 1(a) and Fig. 1(b) are vastly different, although they both correspond to the same function H_8. Furthermore, note that the algorithm PVL converges, with the computed Padé approximant H_{28} being practically identical to H.

3 Krylov Subspaces with Multiple Starting Vectors

While the PVL algorithm remedies the numerical issues of AWE, it can be used only for the special case of single-input single-output systems (14). Since circuit interconnect models have multiple inputs and outputs in general, after its introduction in 1994, it quickly became clear that the PVL algorithm needed to be extended to m-input p-output systems (4). To motivate the type of Krylov-subspace method that is needed for such an extension, we first rewrite the transfer function (10) of (4) as follows:

$$\begin{aligned} H(s) &= L^T\left(sE - A\right)^{-1}B = L^T\left(I + (s - s_0)M\right)^{-1}R, \\ \text{where}\quad M &:= \left(s_0E - A\right)^{-1}E \quad\text{and}\quad R := \left(s_0E - A\right)^{-1}B. \end{aligned} \tag{27}$$

Note that M is an $N \times N$ matrix, R is an $N \times m$ matrix, L is an $N \times p$ matrix, and H is a $(p \times m)$-matrix-valued function. Instead of starting vectors r and l in the PVL algorithm, we now have blocks R and L of *multiple starting vectors*. A Lanczos-type algorithm that produces Padé approximants H_n of H via a suitable adaption of the PVL formula (4) to the m-input p-output case needs to be able to handle such multiple starting vectors.

In this section, we describe the concept of block Krylov subspaces for multiple starting vectors and briefly review the *block Lanczos method*.

3.1 Block Krylov Subspaces

Recall that for a single starting vector r, the n-th Krylov subspace $\mathscr{K}_n(M,r)$ is defined as the n-dimensional subspace spanned by the vectors (3). Here, $1 \leq n \leq n_{\max}$ and $n_{\max}$ denotes the maximum value of n such that the vectors (3) are still linearly independent.

For a block R of m starting vectors, we have $N \times m$ matrices M^iR, $i = 0, 1, \ldots,$ instead of vectors (3). To properly define Krylov subspaces $\mathscr{K}_n(M,R)$ in this case, we put the first N of these matrices into a single $N \times mN$ *right block Krylov matrix* as follows:

$$\begin{bmatrix} R & MR & M^2R & \cdots & M^{N-1}R \end{bmatrix}. \tag{28}$$

Next, we scan the mN columns of this matrix from left to right and delete any column that is linearly dependent on columns to its left. The result of this operation is the matrix

$$\begin{bmatrix} R_1 & MR_2 & M^2R_3 & \cdots & M^{i_{\max}-1}R_{i_{\max}} \end{bmatrix} \tag{29}$$

the columns of which are all linearly independent. This process of detecting and deleting the linearly dependent columns of the matrix (28) is called *exact deflation*. Note that a column of the form $M^i r$ being linearly dependent on columns to its left in (28) implies that any column $M^j r$, $j = i, i+1, \ldots,$ is linearly dependent on columns to its right. Therefore, in (29), for each $i = 1, 2, \ldots, i_{\max}$, the matrix R_i is a submatrix of R_{i-1}, where, for $i = 1$, we set $R_0 = R$. Denoting by m_i the number of columns of R_i, the number of columns of the matrix (29) is given by

$$n_{\max}^{(\mathrm{R})} := m_1 + m_2 + \cdots + m_{i_{\max}}. \tag{30}$$

By construction, the matrix (29) has full column rank $n_{\max}^{(\mathrm{R})}$.

For $n = 0, 1, \ldots, n_{\max}^{(\mathrm{R})}$, the n-dimensional subspace of the space of vectors of length N that is spanned by the first n columns of the matrix (29) is called the *n-th block Krylov subspace* (induced by M and R) and denoted by $\mathscr{K}_n(M,R)$.

To define $\mathscr{K}_n(M^T, L)$, where L is a block of p starting vectors, we proceed analogously. Applying the process of exact deflation to the *left block Krylov matrix*

$$\begin{bmatrix} L & M^T L & (M^T)^2 L & \cdots & (M^T)^{N-1} L \end{bmatrix},$$

we obtain an $N \times n_{\max}^{(\mathrm{L})}$ matrix of the form

$$\begin{bmatrix} L_1 & M^T L_2 & (M^T)^2 L_3 & \cdots & (M^T)^{j_{\max}-1} L_{j_{\max}} \end{bmatrix}, \tag{31}$$

where each L_j is a submatrix of L_{j-1}. The matrix (31) has full column rank

$$n_{\max}^{(\mathrm{L})} := p_1 + p_2 + \cdots + p_{j_{\max}}, \tag{32}$$

where p_j denotes the number of columns of L_j.

For $n = 0, 1, \ldots, n_{\max}^{(\mathrm{L})}$, the n-dimensional subspace of the space of vectors of length N that is spanned by the first n columns of the matrix (31) is called the *n-th block Krylov subspace* (induced by M^T and L) and denoted by $\mathscr{K}_n(M^T, L)$.

To distinguish the two types of block Krylov subspaces, we refer to $\mathscr{K}_n(M,R)$ and $\mathscr{K}_n(M^T, L)$ as *right* and *left* block Krylov subspaces, respectively.

3.2 Block Lanczos Method

In 1994, the problem of extending the Lanczos process for single to multiple starting vectors was not new, and a number of algorithms had been proposed. With the exception of Ruhe's band variant [36] of the symmetric Lanczos algorithm, all existing algorithms at that time are based on a block-wise construction of basis vectors for the underlying block Krylov subspaces.

For symmetric matrices $M = M^T$ and starting vectors $r = l$, the right and left Krylov subspaces $\mathcal{K}_n(M,r)$ and $\mathcal{K}_n(M^T,l)$ are identical, and the general Lanczos algorithm simplifies to the *symmetric* Lanczos algorithm. The first block extensions of the Lanczos algorithm were developed for this special case [5, 6, 23, 39]. Since the matrices M in (27) are nonsymmetric in general, block variants of the symmetric Lanczos algorithm cannot be used to extend the PVL algorithm to m-input p-output systems (4).

For the general case, Kim and Craig [25, 26] were the first to develop a block version of the classical Lanczos algorithm. Their block Lanczos method requires that $m = p$ and is essentially a variant of Algorithm 2.1, where vectors are replaced by blocks of m vectors, scalars are replaced by $m \times m$ matrices, and division by a scalar is replaced by multiplication with the inverse of an $m \times m$ matrix. The $m \times m$ matrices are chosen such that the generated blocks of basis vectors for the right and left block Krylov subspaces are block-biorthogonal to each other. Clearly, such a block approach cannot be extended to the case $m \neq p$, as this would involve 'inverses' of nonsquare matrices. Furthermore, even for the special case $m = p$, the block Lanczos method requires that the sizes for the right and left blocks of basis vectors remain the same throughout the run of the algorithm. As a result, necessary deflations to handle linearly dependent blocks can only be performed if these linear dependencies occur simultaneously in the right and left blocks. However, this is not the case in general.

In order to extend the PVL algorithm to general m-input p-output systems of the form (4), a new Lanczos-type method was needed to overcome the limitations of the block Lanczos method. Such a procedure needs to be able to handle the general case $m, p \geq 1$ and include an efficient deflation procedure. The band Lanczos method, which we describe in Sect. 4, is such a procedure. The key to the development of the band Lanczos method was the insight to construct the basis vectors of the right and left block Krylov subspaces to be vectorwise biorthogonal to each other, instead of the blockwise biorthogonality that is used in the block Lanczos method.

4 A New Approch: the Band Lanczos Method

As in Sect. 3, we assume that M is an $N \times N$ matrix, R is an $N \times m$ matrix, and L is an $N \times N$ matrix. We use the notation

$$R = \begin{bmatrix} r_1 & r_2 & \cdots & r_m \end{bmatrix} \quad \text{and} \quad L = \begin{bmatrix} l_1 & l_2 & \cdots & l_p \end{bmatrix} \tag{33}$$

for the columns of R and L.

4.1 Defining Properties

Like Algorithm 2.1, the band Lanczos method generates two sets of *right* and *left Lanczos vectors*

$$v_1, v_2, \ldots, v_n \quad \text{and} \quad w_1, w_2, \ldots, w_n \tag{34}$$

that are constructed to be (vectorwise) biorthogonal to each other. If only exact deflations are performed in the method, the vectors (34) form bases of the right and left block Krylov subspaces $\mathscr{K}_n(M,R)$ and $\mathscr{K}_n(M^T,L)$. Using the notation from (22) and (23), the biorthogonality of the vectors (34) can be stated compactly as

$$W_n^T V_n = \Delta_n := \operatorname{diag}\big(\delta_1, \delta_2, \ldots, \delta_n\big). \tag{35}$$

At any stage of the band Lanczos method, there are *right* and *left candidate vectors*

$$\widehat{v}_{n+1}, \widehat{v}_{n+2}, \ldots, \widehat{v}_{n+m_c} \quad \text{and} \quad \widehat{w}_{n+1}, \widehat{w}_{n+2}, \ldots, \widehat{w}_{n+p_c} \tag{36}$$

for the Lanczos vectors $v_{n+1}, v_{n+2}, \ldots, v_{n+m_c}$ and $w_{n+1}, w_{n+2}, \ldots, w_{n+p_c}$ to be generated in the following iterations. The vectors (36) are constructed to be biorthogonal to the Lanczos vectors (34):

$$W_n^T \widehat{v}_{n+j} = 0,\ j = 1,2,\ldots,m_c, \quad \text{and} \quad V_n^T \widehat{w}_{n+k} = 0,\ k = 1,2,\ldots,p_c. \tag{37}$$

At the start of the algorithm, the right and left candidate vectors are initialized as the columns of R and L in (33), $m_c = m$, and $p_c = p$.

The candidate vectors (36) allow for an easy way to check for necessary deflations. The next exact deflation in the right block Krylov matrix (28) occurs if, and only if, $\widehat{v}_{n+1} = 0$. The next exact deflation in the left block Krylov matrix (31) occurs if, and only if, $\widehat{w}_{n+1} = 0$. In practice, one also needs to perform deflations when these vectors are 'close' to zero vectors. In an actual algorithm, we check if

$$\|\widehat{v}_{n+1}\|_2 \le \mathtt{dftol}_{\mathrm{v}} \quad \text{or} \quad \|\widehat{w}_{n+1}\|_2 \le \mathtt{dftol}_{\mathrm{w}}, \tag{38}$$

where $\mathtt{dftol}_{\mathrm{v}}, \mathtt{dftol}_{\mathrm{w}}$ are suitably small *deflation tolerances*. If the first check in (38) is true, $\widehat{v}_{n+1}$ is labeled a *deflated* right vector, the indices of $\widehat{v}_{n+2}, \ldots, \widehat{v}_{n+m_c}$ are shifted by -1, and m_c is reduced by 1. If the second check in (38) is true, $\widehat{w}_{n+1}$ is labeled a deflated left vector, the indices of $\widehat{w}_{n+2}, \ldots, \widehat{w}_{n+p_c}$ are shifted by -1, and p_c is reduced by 1. We refer to this process as *deflation* in general, and as exact deflation when both deflation tolerances in (38) are set to 0. Note that $m - m_c$ and $p - p_c$ is the number of deflations of right and left vectors, respectively, that have occurred so far.

Similar to the relations (24) for Algorithm 2.1, the recurrences that are used to generate the vectors (34) and (36) can be stated compactly as follows:

$$\begin{aligned} MV_n &= V_n T_n + \begin{bmatrix} 0 & 0 & \cdots & 0 & \widehat{v}_{n+1} & \widehat{v}_{n+2} & \cdots & \widehat{v}_{n+m_c} \end{bmatrix} + V_n^{(\mathrm{dl})}, \\ M^T W_n &= W_n \widetilde{T}_n + \begin{bmatrix} 0 & 0 & \cdots & 0 & \widehat{w}_{n+1} & \widehat{w}_{n+2} & \cdots & \widehat{w}_{n+p_c} \end{bmatrix} + W_n^{(\mathrm{dl})}. \end{aligned} \tag{39}$$

The matrices $V_n^{(\mathrm{dl})}$ and $W_n^{(\mathrm{dl})}$ contain mostly zero columns, together with the $m-m_\mathrm{c}$ right and $p-p_\mathrm{c}$ left deflated vectors, respectively. In particular, $V_n^{(\mathrm{dl})}$ and $W_n^{(\mathrm{dl})}$ are zero matrices if no deflations have occurred so far or if only exact deflations are performed. The matrices T_n and $\widetilde{T}_n$ contain the scalar coefficients of the recurrences that are used to generate the Lanczos vectors and the candidate vectors. Since these recurrences involve at most $m_\mathrm{c}+p_\mathrm{c}+1$ terms, the matrices T_n and $\widetilde{T}_n$ are 'essentially' banded. More precisely, T_n has lower bandwidth $m_\mathrm{c}+1$ and upper bandwidth $p_\mathrm{c}+1$, where the lower bandwidth is reduced by 1 every time a right vector is deflated and the upper bandwidth is reduced by 1 every time a left vector is deflated. In addition, each deflation of a left vector causes T_n to have nonzero elements in a fixed row outside and to the right of the banded part. Analogously, $\widetilde{T}_n$ has lower bandwidth $p_\mathrm{c}+1$ and upper bandwidth $m_\mathrm{c}+1$, where the lower bandwidth is reduced by 1 every time a left vector is deflated, and the upper bandwidth is reduced by 1 every time a right vector is deflated. In addition, each deflation of a right vector causes $\widetilde{T}_n$ to have nonzero elements in a fixed row outside and to the right of the banded part.

Recall that the tridiagonal matrices from Algorithm 2.1 are connected via the relation (25). The banded parts of T_n and $\widetilde{T}_n$ in (39) are related in a similar way:

$$\Delta_n T_n^{(\mathrm{pr})} = \left(\widetilde{T}_n^{(\mathrm{pr})}\right)^T \Delta_n, \tag{40}$$

where

$$T_n^{(\mathrm{pr})} := T_n + \Delta_n^{-1} W_n^T V_n^{(\mathrm{dl})} \quad \text{and} \quad \widetilde{T}_n^{(\mathrm{pr})} := \widetilde{T}_n + \Delta_n^{-1} V_n^T W_n^{(\mathrm{dl})}. \tag{41}$$

Note that the matrix $\Delta_n^{-1} W_n^T V_n^{(\mathrm{dl})}$ and $\Delta_n^{-1} V_n^T W_n^{(\mathrm{dl})}$ has nonzero entries only below the banded part of T_n and $\widetilde{T}_n$ and in the columns corresponding to the $m-m_\mathrm{c}$ deflated right vectors and $p-p_\mathrm{c}$ deflated left vectors, respectively.

By multiplying the first equation in (39) from the left by W_n^T and using (35), (37), and (41), we obtain the expression

$$T_n^{(\mathrm{pr})} = \left(W_n^T V_n\right)^{-1} W_n^T M V_n = \Delta_n^{-1} W_n^T M V_n \tag{42}$$

for $T_n^{(\mathrm{pr})}$. The relation (42) means that the $n \times n$ matrix $T_n^{(\mathrm{pr})}$ is the oblique projection of the $N \times N$ matrix M onto the subspace spanned by $v_1, v_2, \ldots, v_n$ and orthogonally to the subspace spanned by $w_1, w_2, \ldots, w_n$. If only exact deflations are performed, then these vectors span the right and left block Krylov subspaces and thus $T_n^{(\mathrm{pr})}$ is the oblique projection of M onto $\mathcal{K}_n(M,R)$ and orthogonally to $\mathcal{K}_n(M^T,L)$.

4.2 Reduced-Order Models and Matrix Padé Approximants

In this subsection, we discuss the use of the band Lanczos method in model order reduction of m-input p-output systems (4). For this application, M, R, and L are the matrices from the representation (27) of the transfer function H of (4). In addition to the oblique projection (42), $T_n^{(\mathrm{pr})}$, of M onto the subspaces generated by n iterations of the band Lanczos method, we also need the one-sided oblique projections of R and L corresponding to (42). These projections are defined as follows:

$$\rho_n^{(\mathrm{pr})} := \Delta_n^{-1} W_n^T R \quad \text{and} \quad \eta_n^{(\mathrm{pr})} := \Delta_n^{-1} V_n^T L.$$

Using the quantities $T_n^{(\mathrm{pr})}$, $\rho_n^{(\mathrm{pr})}$, $\eta_n^{(\mathrm{pr})}$, and Δ_n, we define the approximation

$$H_n(s) = \left(\eta_n^{(\mathrm{pr})}\right)^T \Delta_n \left(I + (s - s_0)\, T_n^{(\mathrm{pr})}\right)^{-1} \rho_n^{(\mathrm{pr})} \tag{43}$$

of H. An actual reduced-order model (ROM) (6) that corresponds to H_n is readily obtained by comparing the representation (12) of the ROM transfer function with (43) and defining the matrices in (6) as follows:

$$A_n := I - s_0 T_n^{(\mathrm{pr})}, \quad E_n := T_n^{(\mathrm{pr})}, \quad B_n := \rho_n^{(\mathrm{pr})}, \quad \text{and} \quad L_n := \Delta_n \eta_n^{(\mathrm{pr})}.$$

We remark that (43) generalizes the PVL formula (20) for the single-input single-output case to the general multiple-input multiple-output case. In fact, for $m = p = 1$, the band Lanczos method reduces to Algorithm 2.1 and formula (43) reduces to (20).

The ROM transfer function (12), H_n, is called an *n-th matrix Padé approximant* (about the expansion point s_0) of the transfer function H of (4) if

$$H_n(s) = H(s) + \mathcal{O}\left((s - s_0)^{q(n)}\right),$$

where $q(n)$ is as large as possible. The ROM transfer function (43) generated via the band Lanczos method is an n-th matrix Padé approximant of H provided that only exact deflations are performed. To properly state this result, recall the definitions of $n_{\max}^{(\mathrm{R})}$ and $n_{\max}^{(\mathrm{L})}$ in (30) and (32). In addition, we define $j(n)$ and $k(n)$ as the largest values of j and k such that

$$m_1 + m_2 + \cdots + m_j \le n \quad \text{and} \quad p_1 + p_2 + \cdots + p_k \le n.$$

Theorem 4.1 *Let* $\max\{\, m_1, p_1 \,\} \le n \le \min\{\, n_{\max}^{(R)}, n_{\max}^{(L)} \,\}$. *If only exact deflations are performed in the band Lanczos method, then the function* (43), H_n, *is an n-th matrix Padé approximant of the function* (27), H, *and*

$$H_n(s) = H(s) + \mathcal{O}\left((s - s_0)^{j(n)+k(n)}\right).$$

A proof of Theorem 4.1 is given in [11].

4.3 An Actual Algorithm

The first simple version of the band Lanczos method appeared in the 1995 paper [10]. The algorithm in [10] has no built-in deflation procedure, and the computation of the Lanczos vectors is arranged such that rectangular $n \times (n+m)$ and $n \times (n+p)$ matrices instead of the $n \times n$ matrices $T_n^{(\mathrm{pr})}$ and $\widetilde{T}_n^{(\mathrm{pr})}$ are generated. After that, in the joint work [1] with Aliaga, Boley, and Hernández, we developed a complete version of this algorithm that included a proper deflation procedure and a look-ahead strategy to deal with potential breakdowns.

Working on actual code for the band Lanczos method, it became clear that arranging the computations so that square matrices $T_n^{(\mathrm{pr})}$ and $\widetilde{T}_n^{(\mathrm{pr})}$ are produced is preferable. A first version of this rearranged band Lanczos method appeared in [12]. The connection (40) of the matrices $T_n^{(\mathrm{pr})}$ and $\widetilde{T}_n^{(\mathrm{pr})}$ is exploited to explicitly compute only half of the entries of these matrices. An improved version of this algorithm was included in the survey paper [14] on Krylov-subspace methods for model order reduction. The following algorithm is essentially the version from [14]. The quantities $T_n^{(\mathrm{pr})}$, $\rho_n^{(\mathrm{pr})}$, $\eta_n^{(\mathrm{pr})}$, and Δ_n, which are needed to form the ROM corresponding to (43), are generated as outputs of this algorithm.

Algorithm (Band Lanczos algorithm)
For $k = 1, 2, \ldots, m$, set $\widehat{v}_k = r_k$.
For $k = 1, 2, \ldots, p$, set $\widehat{w}_k = l_k$.
Set $m_{\mathrm{c}} = m$, $p_{\mathrm{c}} = p$, and $\mathscr{I}_{\mathrm{v}} = \mathscr{I}_{\mathrm{w}} = \emptyset$.
For $n = 1, 2, \ldots,$ until convergence or $m_{\mathrm{c}} = 0$ or $p_{\mathrm{c}} = 0$ or $\delta_n = 0$ do:

1) Compute $t_{n,n-m_{\mathrm{c}}} = \|\widehat{v}_n\|_2$.
   Decide if $\widehat{v}_n$ should be deflated. If yes, do the following:

   a) Set $\widehat{v}_{n-m_{\mathrm{c}}}^{(\mathrm{dl})} = \widehat{v}_n$ and store this deflated vector. Set $\mathscr{I}_{\mathrm{v}} = \mathscr{I}_{\mathrm{v}} \cup \{n - m_{\mathrm{c}}\}$.
   b) Set $m_{\mathrm{c}} = m_{\mathrm{c}} - 1$. If $m_{\mathrm{c}} = 0$, set $n = n - 1$ and stop.
   c) For $k = n, n+1, \ldots, n + m_{\mathrm{c}} - 1$, set $\widehat{v}_k = \widehat{v}_{k+1}$.
   d) Repeat all of step 1).

2) Compute $\widetilde{t}_{n,n-p_{\mathrm{c}}} = \|\widehat{w}_n\|_2$.
   Decide if $\widehat{w}_n$ should be deflated. If yes, do the following:

   a) Set $\widehat{w}_{n-p_{\mathrm{c}}}^{(\mathrm{dl})} = \widehat{w}_n$ and store this deflated vector. Set $\mathscr{I}_{\mathrm{w}} = \mathscr{I}_{\mathrm{w}} \cup \{n - p_{\mathrm{c}}\}$.
   b) Set $p_{\mathrm{c}} = p_{\mathrm{c}} - 1$. If $p_{\mathrm{c}} = 0$, set $n = n - 1$ and stop.
   c) For $k = n, n+1, \ldots, n + p_{\mathrm{c}} - 1$, set $\widehat{w}_k = \widehat{w}_{k+1}$.
   d) Repeat all of step 2).

3) Set $v_n = \widehat{v}_n / t_{n,n-m_{\mathrm{c}}}$ and $w_n = \widehat{w}_n / \widetilde{t}_{n,n-p_{\mathrm{c}}}$.
4) Compute $\delta_n = w_n^T v_n$.
   If $\delta_n = 0$, stop: look-ahead would be needed to continue.
5) For $k = n+1, n+2, \ldots, n + m_{\mathrm{c}} - 1$, do:
   Compute $t_{n,k-m_{\mathrm{c}}} = w_n^T \widehat{v}_k / \delta_n$ and set $\widehat{v}_k = \widehat{v}_k - v_n t_{n,k-m_{\mathrm{c}}}$.

6) For $k = n+1, n+2, \dots, n+p_c - 1$, do:
   Compute $\widetilde{t}_{n,k-p_c} = \widehat{w}_k^T v_n / \delta_n$ and set $\widehat{w}_k = \widehat{w}_k - w_n \widetilde{t}_{n,k-p_c}$.
7) Compute $\widehat{v}_{n+m_c} = M v_n$.
8) a) For $k \in \mathscr{I}_w$ (in ascending order), do:
   Compute $\widetilde{\sigma} = \left(\widehat{w}_k^{(dl)}\right)^T v_n$ and set $\widetilde{t}_{n,k} = \widetilde{\sigma} / \delta_n$.
   If $k > 0$, set $t_{k,n} = \widetilde{\sigma} / \delta_k$ and $\widehat{v}_{n+m_c} = \widehat{v}_{n+m_c} - v_k t_{k,n}$.
   b) Set $k_v = \max\{1, n - p_c\}$.
   c) For $k = k_v, k_v + 1, \dots, n-1$, do:
   Set $t_{k,n} = \widetilde{t}_{n,k} \delta_n / \delta_k$ and $\widehat{v}_{n+m_c} = \widehat{v}_{n+m_c} - v_k t_{k,n}$.
   d) Compute $t_{n,n} = w_n^T \widehat{v}_{n+m_{mc}} / \delta_n$ and set $\widehat{v}_{n+m_{mc}} = \widehat{v}_{n+m_{mc}} - v_n t_{n,n}$.
9) Compute $\widehat{w}_{n+p_c} = M^T w_n$.
10) a) For $k \in \mathscr{I}_v$ (in ascending order), do:
   Compute $\sigma = w_n^T \widehat{v}_k^{(dl)}$ and set $t_{n,k} = \sigma / \delta_n$.
   If $k > 0$, set $\widetilde{t}_{k,n} = \sigma / \delta_k$ and $\widehat{w}_{n+p_c} = \widehat{w}_{n+p_c} - w_k \widetilde{t}_{k,n}$.
   b) Set $k_w = \max\{1, n - m_c\}$.
   c) For $k = k_w, k_w + 1, \dots, n-1$, do:
   Set $\widetilde{t}_{k,n} = t_{n,k} \delta_n / \delta_k$ and $\widehat{w}_{n+p_c} = \widehat{w}_{n+p_c} - w_k \widetilde{t}_{k,n}$.
   d) Set $\widetilde{t}_{n,n} = t_{n,n}$ and $\widehat{w}_{n+p_c} = \widehat{w}_{n+p_c} - w_n \widetilde{t}_{n,n}$.
11) Set $T_n^{(pr)} = \left[t_{i,k}\right]_{i,k=1,2,\dots,n}$ and $\Delta_n = \operatorname{diag}\left(\delta_1, \delta_2, \dots, \delta_n\right)$.
   Set $k_\rho = m + \min\{0, n - m_c\}$ and $\rho_n^{(pr)} = \left[t_{i,k-m}\right]_{i=1,2,\dots,n;k=1,2,\dots,k_\rho}$.
   Set $k_\eta = p + \min\{0, n - p_c\}$ and $\eta_n^{(pr)} = \left[\widetilde{t}_{i,k-p}\right]_{i=1,2,\dots,n;k=1,2,\dots,k_\eta}$.
12) Check if n is large enough. If yes, stop.

5 Structure Preservation

An important class of interconnect models are RCL networks with only independent voltage and current sources. Such models are described by DAEs of the form (4) where $m = p$. Moreover, the equations in (4) can be formulated such that

$$B = L, \quad E = E^T \succeq 0, \quad A + A^T \preceq 0 \tag{44}$$

and the matrices A, E, and B have certain block structures; see, e.g., [17, 19]. Here, the notation "$\succeq$" and "$\preceq$" means that a matrix is *symmetric positive semidefinite* and *symmetric negative semidefinite*, respectively. In this section, we consider the problem of model order reduction of DAEs (4) for this class of RCL networks.

An important property of RCL networks is *passivity*, which means that such networks do not generate energy. In fact, the matrix properties (44) imply passivity. It is desirable and for some applications crucial that ROMs of RCL networks are also passive. One of the disadvantages of Lanczos-based approaches is that the resulting ROMs are not guaranteed to be passive for general RCL networks.

A simple approach to generate passive ROMs is based on *explicit projection* of the matrices in (4). Let V_n be a real $N \times n$ matrix with full column rank n. Setting

$$A_n := V_n^T A V_n, \quad E_n := V_n^T A V_n, \quad B_n := V_n^T B, \quad \text{and} \quad L_n := B_n, \tag{45}$$

one obtains a ROM (6) with matrices that satisfy the same conditions (44) as the matrices of (4). In particular, this ROM is passive.

By combining the projection approach with block Krylov subspaces, the transfer function H_n of the ROM defined by (45) satisfies a *Padé-type approximation* property. To this end, we choose a suitable expansion point $s_0 \geq 0$ and rewrite the transfer function of (4) (with $B = L$) as follows:

$$\begin{aligned} H(s) &= B^T \left(sE - A\right)^{-1} B = B^T \left(I + (s - s_0) M\right)^{-1} R, \\ \text{where} \quad M &:= \left(s_0 E - A\right)^{-1} E \quad \text{and} \quad R := \left(s_0 E - A\right)^{-1} B. \end{aligned} \tag{46}$$

If we choose the matrix V_n such that its range[2] contains the $\widehat{n}$-the block Krylov subspace $\mathscr{K}_{\widehat{n}}(M,R)$ for some $\widehat{n} \leq n$, then the ROM transfer function H_n is an n-th matrix Padé-type approximant of H. As in Theorem 4.1, $n_{\max}^{(R)}$ is the integer defined in (30) and $j(\widehat{n})$ denotes the largest value of j such that $m_1 + m_2 + \cdots + m_j \leq \widehat{n}$.

Theorem 5.1 *Let V_n be an $N \times n$ matrix with full column rank n, and assume that*

$$\mathscr{K}_{\widehat{n}}(M,R) \subseteq \operatorname{range}(V_n) \tag{47}$$

for some $m_1 \leq \widehat{n} \leq n_{\max}^{(R)}$. Then, the ROM transfer function H_n is an n-th matrix Padé-type approximant of the transfer function (46)*, H, and*

$$H_n(s) = H(s) + \mathscr{O}\left((s - s_0)^{j(\widehat{n})}\right). \tag{48}$$

A proof of Theorem 5.1 is given in [16].

The first ROM algorithm based on explicit projection was PRIMA [29, 30]. It employs a simple block variant of the Arnoldi process without deflation to generate an orthonormal basis for $\mathscr{K}_n(M,R)$, and uses these basis vectors as the columns of the projection matrix V_n. Note that for PRIMA, we have $\widehat{n} = n$ and equality of the two subspaces in (47). While the ROMs generated by PRIMA are passive by construction, they do not preserve any of the other properties of RCL networks, such as reciprocity and the block structure of the matrices A, E, and B. We also remark that for a robust implementation of PRIMA, a variant of the Arnoldi process with a proper built-in deflation procedure needs to be used. The band Arnoldi process discussed in Sect. 6 is such a variant.

SRIM [15, 19] was introduced as an improvement of PRIMA that in addition to passivity, preserves both reciprocity and the block structure of the matrices A, E, and B. SPRIM employs the band Arnoldi process to first generate an orthonormal

---

[2] The *range*, denoted by $\operatorname{range}(M)$, of a matrix M is defined as the subspace spanned by the columns of M.

basis for $\mathscr{K}_{\widehat{n}}(M,R)$. Let $V_{\widehat{n}}$ denote the $N \times \widehat{n}$ matrix the columns of which are these basis vectors. Instead of using $V_{\widehat{n}}$ as the projection matrix, $V_{\widehat{n}}$ is turned into an $N \times n$ matrix V_n for some $\widehat{n} < n \leq 2\widehat{n}$ such that (47) is satisfied and the projected matrices (45) of the SPRIM ROMs preserve both reciprocity and the block structure of A, E,and B; see [17, 19] for details of the construction of V_n. Finally, we remark that the block structure of the SPRIM ROMs imply a higher accuracy than the corresponding PRIMA ROMs. More precisely, as shown in [16], the SPRIM transfer function H_n is an n-th matrix Padé-type approximant of H that matches $2j(\widehat{n})$ moments instead of $j(\widehat{n})$ moments in (48).

The following example, which is taken from [19], illustrates the higher accuracy of SPRIM. The example is a RCL network with $m = p = 16$ and $N = 1841$ that models the pin package of a chip. The expansion point $s_0 = 2\pi \times 10^{10}$ was used. For this example, $\widehat{n} = 128$ was needed for the SPRIM transfer function H_n to converge to the exact transfer function H. Fig. 2 depicts the absolute values of the $(8,1)$-component of the transfer functions. Note that for $\widehat{n} = 128$ PRIMA has not converged yet.

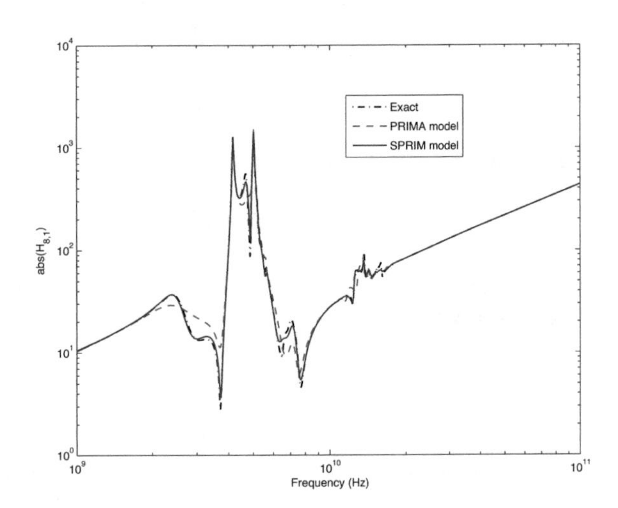

Fig. 2 Package example, $(8,1)$-component of transfer functions

6 Band Arnoldi Process

In this section, we state an algorithm for the band Arnoldi process applied to an $N \times N$ matrix M and an $N \times m$ matrix R. This is essentially the algorithm that first appeared in [14]. Since M and R are complex matrices for some applications of the band Arnoldi process, we use the *complex conjugate transpose* $v^H := \overline{v}^T$ instead of the transpose v^T in the statement of the algorithm.

The notation is similar to the one we used in Sect. 4. The algorithm produces *Arnoldi vectors* and *candidate vectors*

$$v_1.v_2,\dots,v_n \quad \text{and} \quad v_{n+1},v_{n+2},\dots,v_{n+m_c}, \tag{49}$$

where the Arnoldi vectors are constructed to be orthonormal to each other and the candidate vectors are constructed to be orthogonal to the Arnoldi vectors. Using the notation from (22), the orthogonality of the vectors (49) can be stated compactly as

$$V_n^H V_n = I, \quad \text{and} \quad V_n^H v_{n+j} = 0, \; j = 1,2,\dots,m_c, \tag{50}$$

where I denotes the $n \times n$ identity matrix. The recurrences that are employed to generate the vectors (49) can be stated compactly as follows:

$$MV_n = V_n H_n + \begin{bmatrix} 0 & 0 & \cdots & 0 & \widehat{v}_{n+1} & \widehat{v}_{n+2} & \cdots & \widehat{v}_{n+m_c} \end{bmatrix} + V_n^{(\mathrm{dl})}. \tag{51}$$

Similarly to Algorithm 4.2, the following algorithm produces the matrices

$$H_n^{(\mathrm{pr})} := H_n + V_n^H V_n^{(\mathrm{dl})}, \quad \rho_n^{(\mathrm{pr})} := V_n^H R \tag{52}$$

as outputs. We remark that by multiplying (51) from the left by V_n^H and using (50) and (52), it follows that $H_n^{(\mathrm{pr})} = V_n^H M V_n$. In particular, $H_n^{(\mathrm{pr})}$ and $\rho_n^{(\mathrm{pr})}$ are the orthogonal projections of M and R onto the subspace spanned by the Arnoldi vectors $v_1.v_2,\dots,v_n$. If only exact deflations are performed in the algorithm or if no deflations occur, then these vectors span the n-th block Krylov subspace $\mathscr{K}_n(M,R)$.

Algorithm (Band Arnoldi process)
For $k = 1,2,\dots,m$, set $\widehat{v}_k = r_k$.
Set $m_c = m$ and $\mathscr{I} = \emptyset$.
For $n = 1,2,\dots,$ until convergence or $m_c = 0$ do:

1) Compute $h_{n,n-m_c} = \|\widehat{v}_n\|_2$.
   Decide if $\widehat{v}_n$ should be deflated. If yes, do the following:
   a) Set $\widehat{v}_{n-m_c}^{(\mathrm{dl})} = \widehat{v}_n$ and store this deflated vector. Set $\mathscr{I} = \mathscr{I} \cup \{\, n - m_c \,\}$.
   b) Set $m_c = m_c - 1$. If $m_c = 0$, set $n = n - 1$ and stop.
   c) For $k = n, n+1, \dots, n+m_c-1$, set $\widehat{v}_k = \widehat{v}_{k+1}$.
   d) Repeat all of step 1).
2) Set $v_n = \widehat{v}_n / h_{n,n-m_c}$.
3) For $k = n+1, n+2, \dots, n+m_c-1$, do:
   Compute $h_{n,k-m_c} = v_n^H \widehat{v}_k$ and set $\widehat{v}_k = \widehat{v}_k - v_n h_{n,k-m_c}$.
4) Compute $\widehat{v}_{n+m_c} = M v_n$.
5) For $k = 1,2,\dots,n$, do:
   Compute $h_{k,n} = v_k^H \widehat{v}_{n+m_c}$ and set $\widehat{v}_{n+m_c} = \widehat{v}_{n+m_c} - v_k h_{k,n}$.
   Compute $h_{n,k} = v_n^H \widehat{v}_k^{(\mathrm{dl})}$ if $k \in \mathscr{I}$, and set $h_{n,k} = 0$ if $k \notin \mathscr{I}$.
6) Set $H_n^{(\mathrm{pr})} = \begin{bmatrix} h_{i,k} \end{bmatrix}_{i,k=1,2,\dots,n}$.
   Set $k_\rho = m + \min\{\, 0, n - m_c \,\}$ and $\rho_n^{(\mathrm{pr})} = \begin{bmatrix} h_{i,k-m} \end{bmatrix}_{i=1,2,\dots,n;\,k=1,2,\dots,k_\rho}$.
7) Check if n is large enough. If yes, stop.

7 Concluding Remarks

In this chapter, we gave an account of how the need to efficiently and accurately simulate the effects of on-chip wiring of integrated circuits has led to the development of new band versions of the Lanczos algorithm and the Arnoldi process for multiple starting vectors. We stress that the applications of these new band Krylov-subspace methods are not restricted to electronic circuit simulation. In fact, they can be employed wherever there is a need for model order reduction of large-scale time-invariant linear dynamical systems with multiple inputs and outputs. Such applications include structural analysis, microelectromechanical systems, transport networks, and computational acoustics. Multiple starting vectors also arise in the context of matrix functions. The use of band Krylov-subspace methods in the efficient evaluation of matrix functions has yet to be explored.

Just as the classical Lanczos algorithm is related to formally orthogonal polynomials (FOPs), there is a connection of the band Lanczos method to matrix-valued FOPs. Some of the underlying relations were derived in [13], but as the recent paper [2] indicates, the connection to matrix-valued FOPs needs to be explored further.

Robust implementations of band Krylov-subspace methods, especially of the band Lanczos method, are not as straightforward as implementations of the classical Krylov-subspace methods. In order to facilitate the use of these band algorithms, the author has produced the software package BANDITS [21], which provides Matlab implementations of various band Krylov-subspace methods.

The band Lanczos method described in this chapter (and implemented in BANDITS) does not include a look-ahead procedure to deal with potential breakdowns. While a version of the band Lanczos method with look-ahead is described in [1], this algorithm was never implemented in an actual code. Since Algorithm 4.2 is preferable to the one in [1], a version of Algorithm 4.2 with look-ahead should be developed and implemented in a production-quality code.

References

[1] Aliaga, J.I., Boley, D.L., Freund, R.W., Hernández, V.: A Lanczos-type method for multiple starting vectors. Math. Comp. **69**(232), 1577–1601 (2000)

[2] Alqahtani, H., Reichel, L.: Multiple orthogonal polynomials applied to matrix function evaluation. BIT Numer. Math. **58**, 835–849 (2018)

[3] Arnoldi, W.E.: The principle of minimized iterations in the solution of the matrix eigenvalue problem. Quart. Appl. Math. **9**, 17–29 (1951)

[4] Baker, Jr., G.A., Graves-Morris, P.: Padé Approximants, second edn. Cambridge University Press, New York, New York (1996)

[5] Cullum, J.K., Donath, W.E.: A block Lanczos algorithm for computing the q algebraically largest eigenvalues and a corresponding eigenspace for large,

sparse symmetric matrices. In: Proc. 1974 IEEE Conference on Decision and Control, pp. 505–509. IEEE Press, New York, New York (1974)
[6] Cullum, J.K., Willoughby, R.A.: Lanczos Algorithms for Large Symmetric Eigenvalue Computations, Volume 1, Theory. Birkhäuser, Basel, Switzerland (1985)
[7] Elmore, W.C.: The transient response of damped linear networks with particular regard to wideband amplifiers. J. Appl. Phys. **19**(1), 55–63 (1948)
[8] Feldmann, P., Freund, R.W.: Efficient linear circuit analysis by Padé approximation via the Lanczos process. In: Proceedings of EURO-DAC '94 with EURO-VHDL '94, pp. 170–175. IEEE Computer Society Press, Los Alamitos, California (1994)
[9] Feldmann, P., Freund, R.W.: Efficient linear circuit analysis by Padé approximation via the Lanczos process. IEEE Trans. Computer-Aided Design **14**, 639–649 (1995)
[10] Feldmann, P., Freund, R.W.: Reduced-order modeling of large linear subcircuits via a block Lanczos algorithm. In: Proc. 32nd ACM/IEEE Design Automation Conference, pp. 474–479. ACM, New York, New York (1995)
[11] Freund, R.W.: Computation of matrix Padé approximations of transfer functions via a Lanczos-type process. In: C. Chui, L. Schumaker (eds.) Approximation Theory VIII, Vol. 1: Approximation and Interpolation, pp. 215–222. World Scientific Publishing Co., Inc., Singapore (1995)
[12] Freund, R.W.: Band Lanczos method (Section 4.6). In: Z. Bai, J. Demmel, J. Dongarra, A. Ruhe, H. van der Vorst (eds.) Templates for the Solution of Algebraic Eigenvalue Problems: A Practical Guide, pp. 80–88. SIAM Publications, Philadelphia, Pennsylvania (2000)
[13] Freund, R.W.: Computation of matrix-valued formally orthogonal polynomials and applications. J. Comput. Appl. Math. **127**(1–2), 173–199 (2001)
[14] Freund, R.W.: Model reduction methods based on Krylov subspaces. Acta Numerica **12**, 267–319 (2003)
[15] Freund, R.W.: SPRIM: structure-preserving reduced-order interconnect macromodeling. In: Tech. Dig. 2004 IEEE/ACM International Conference on Computer-Aided Design, pp. 80–87. IEEE Computer Society Press, Los Alamitos, California (2004)
[16] Freund, R.W.: On Padé-type model order reduction of J-Hermitian linear dynamical systems. Linear Algebra Appl. **429**, 2451–2464 (2008)
[17] Freund, R.W.: Structure-preserving model order reduction of RCL circuit equations. In: W. Schilders, H. van der Vorst, J. Rommes (eds.) Model Order Reduction: Theory, Research Aspects and Applications, Mathematics in Industry, vol. 13, pp. 49–73. Springer-Verlag, Berlin/Heidelberg, Germany (2008)
[18] Freund, R.W.: Recent advances in structure-preserving model order reduction. In: P. Li, L. Silveira, P. Feldmann (eds.) Simulation and Verification of Electronic and Biological Systems, chap. 3, pp. 43–70. Springer-Verlag, Dordrecht/Heidelberg/London/New York (2011)
[19] Freund, R.W.: The SPRIM algorithm for structure-preserving order reduction of general RCL circuits. In: P. Benner, M. Hinze, E.J.W. ter Maten (eds.) Model

Reduction for Circuit Simulation, Lecture Notes in Electrical Engineering, vol. 74, chap. 2, pp. 25–52. Springer-Verlag, Dordrecht/Heidelberg/London/New York (2011)

[20] Freund, R.W.: Large-scale matrix computations. In: L. Hogben (ed.) Handbook of Linear Algebra, second edn., chap. 64. Chapman & Hall/CRC, Boca Raton (2013)

[21] Freund, R.W.: BANDITS: a Matlab package of band Krylov subspace iterations. Software package (2019). Available online at `https://www.math.ucdavis.edu/~freund/BANDITS/`

[22] Freund, R.W., Gutknecht, M.H., Nachtigal, N.M.: An implementation of the look-ahead Lanczos algorithm for non-Hermitian matrices. SIAM J. Sci. Comput. **14**, 137–158 (1993)

[23] Golub, G.H., Underwood, R.: The block Lanczos method for computing eigenvalues. In: J.R. Rice (ed.) Mathematical Software III, pp. 361–377. Academic Press, New York, New York (1977)

[24] Gragg, W.B.: Matrix interpretations and applications of the continued fraction algorithm. Rocky Mountain J. Math. **4**, 213–225 (1974)

[25] Kim, H.M., Craig, Jr., R.R.: Structural dynamics analysis using an unsymmetric block Lanczos algorithm. Internat. J. Numer. Methods Engrg. **26**, 2305–2318 (1988)

[26] Kim, H.M., Craig, Jr., R.R.: Computational enhancement of an unsymmetric block Lanczos algorithm. Internat. J. Numer. Methods Engrg. **30**, 1083–1089 (1990)

[27] Lanczos, C.: An iteration method for the solution of the eigenvalue problem of linear differential and integral operators. J. Res. Nat. Bur. Standards **45**, 255–282 (1950)

[28] Nagel, L.W.: SPICE2: A computer program to simulate semiconductor circuits. Ph.D. thesis, EECS Department, University of California, Berkeley (1975)

[29] Odabasioglu, A., Celik, M., Pileggi, L.T.: PRIMA: passive reduced-order interconnect macromodeling algorithm. In: Tech. Dig. 1997 IEEE/ACM International Conference on Computer-Aided Design, pp. 58–65. IEEE Computer Society Press, Los Alamitos, California (1997)

[30] Odabasioglu, A., Celik, M., Pileggi, L.T.: PRIMA: passive reduced-order interconnect macromodeling algorithm. IEEE Trans. Computer-Aided Design **17**(8), 645–654 (1998)

[31] Pederson, D.O.: A historical review of circuit simulation. IEEE Trans. Circuits Syst. **31**(1), 103–111 (1984)

[32] Pillage, L.T., Rohrer, R.A.: Asymptotic waveform evaluation for timing analysis. IEEE Trans. Computer-Aided Design **9**, 352–366 (1990)

[33] Pillage, L.T., Rohrer, R.A., Visweswariah, C.: Electronic Circuit and System Simulation Methods. McGraw-Hill, Inc., New York, New York (1995)

[34] Raghavan, V., Rohrer, R.A., Pillage, L.T., Lee, J.Y., Bracken, J.E., Alaybeyi, M.M.: AWE–inspired. In: Proc. IEEE Custom Integrated Circuits Conference, pp. 18.1.1–18.1.8. IEEE, Piscataway, New Jersey (1993)

[35] Rohrer, R.A.: Circuit simulation — the early years. IEEE Circuits Devices Mag. **8**(3), 32–37 (1992)
[36] Ruhe, A.: Implementation aspects of band Lanczos algorithms for computation of eigenvalues of large sparse symmetric matrices. Math. Comp. **33**(146), 680–687 (1979)
[37] Saad, Y.: Iterative Methods for Sparse Linear Systems, second edn. SIAM Publications, Philadelphia, Pennsylvania (2003)
[38] Schilders, W., van der Vorst, H., Rommes, J. (eds.): Model Order Reduction: Theory, Research Aspects and Applications, Mathematics in Industry, vol. 13. Springer-Verlag, Berlin/Heidelberg, Germany (2008)
[39] Underwood, R.: An iterative block Lanczos method for the solution of large sparse symmetric eigenproblems. Ph.D. thesis, Computer Science Department, Stanford University, Stanford, California (1975)

Modular Time Integration of Coupled Problems in System Dynamics

Martin Arnold

Abstract In industrial design processes, the system dynamics of complex engineering structures is modelled by a network approach that results in differential-algebraic model equations. For this problem class, well-established modular simulation techniques like multi-rate or multi-method approaches, co-simulation or waveform relaxation may suffer from exponential instability. With a novel framework for the stability and convergence analysis of modular time integration methods for differential-algebraic systems the sources of numerical instability could be identified and eliminated. Stabilized modular methods have been developed for such diverse fields of application like multibody system dynamics and circuit simulation. This mathematical research has strongly influenced the design of an industrial interface standard for co-simulation in system dynamics.

1 Introduction

Modular time integration methods allow the efficient numerical solution of coupled differential equations using different tailored discretization schemes for different subsystems. Furthermore, the time step size may be adjusted to the specific dynamical behaviour of solution components or groups of solution components.

Until now, these methods have not yet achieved the maturity and robustness being typical of monolithic simulation techniques in the field of nonlinear system dynamics. The present paper addresses one specific stability problem that was observed about 25 years ago in various fields of application whenever a network approach was used to setup the model equations in differential-algebraic form.

New mathematical tools have been developed to analyse the stability and convergence of modular time integration methods for this class of problems and to stabilize

Martin Arnold
Martin Luther University Halle-Wittenberg, Institute of Mathematics, 06099 Halle (Saale), Germany, e-mail: martin.arnold@mathematik.uni-halle.de

M. Günther, W. Schilders (eds.), *Novel Mathematics Inspired by Industrial Challenges*,
Mathematics in Industry 38, https://doi.org/10.1007/978-3-030-96173-2_3

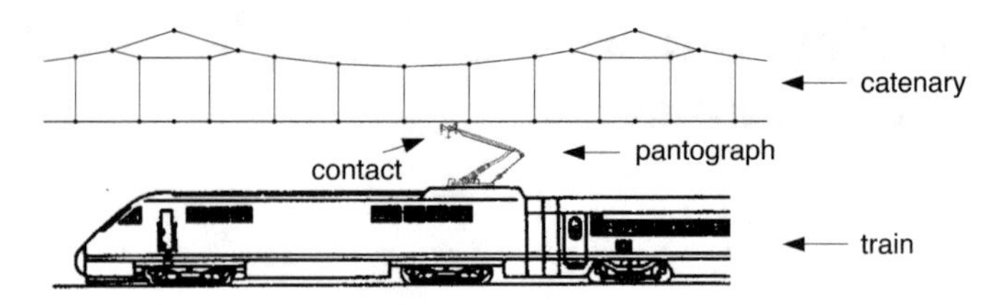

Fig. 1: System pantograph–catenary, see [20, Fig. 1].

existing schemes by appropriate easy-to-use algorithmic modifications. To focus on the numerical problems being caused by the coupling of methods or coupling of simulation tools, all error bounds are derived in function spaces neglecting the details of time integration in the individual subsystems.

Throughout the paper, the dynamical interaction of pantograph and catenary in railway technology is used as non-trivial application scenario. A contractivity condition is found to be essential for the numerical stability and convergence of the modular approach. The condition can always be satisfied by a preconditioning technique that makes use of the Jacobians of all coupling terms.

2 Model based simulation of pantograph-catenary interaction

The interaction of pantograph and catenary is one of the most challenging problems in the design of modern high speed trains. The pantograph guarantees electrical power supply. Its head moves with high speed along the overhead line and causes oscillations in the overhead equipment that is known as *catenary* in railway engineering, see Fig. 1.

The oscillations in the overhead equipment result in wear and damage in catenary and pantograph head. The problem is really challenging for long-distance trains being composed of two coupled half-trains with individual pantographs since the first pantograph excites the catenary resulting in high loads on the pantograph of the second half-train.

In the late 1990's DLR German Aerospace Center was involved in a large scale industrial project aiming at substantial technical improvements of the pantograph–catenary system for the German high-speed railway network. Control engineers came up with the idea to control the pantograph pressure in order to minimize the excitation of the catenary. Because of the high voltages in overhead line and pantograph head these novel control strategies could not be validated by a prototype implementation with measurements in the real technical system but had to rely on simulation based verification and optimization.

Pantograph and pantograph head are sophisticated engineering systems that are analyzed most conveniently by the methods and software tools of multibody system dynamics [17, 20, 22, 27]. On the other hand, the catenary with highly flexible components of more than 1,000 m length is far beyond the capabilities of any standard multibody system simulation software [19].

The Vehicle System Dynamics group at DLR German Aerospace Center had contributed in the 1980's to the development of multibody system simulation software for the design of the first generation of German high-speed trains (ICE 1). For the dynamical simulation of the interaction of pantograph and catenary, DLR's multibody system simulation package SIMPACK [21] was coupled in a *co-simulation* framework to the in-house simulation tool PrOSA of Deutsche Bahn AG for the dynamical simulation of catenaries [27]. In a first attempt, this modular approach failed completely: Even for very frequent data exchange between both simulation tools, the co-simulation suffered from exponential instability [28].

This strange numerical effect was not expected since co-simulation had been used successfully before in automotive applications to couple sophisticated tyre models to multibody system models of passenger cars. In railway engineering, the dynamical interaction of multibody systems and large elastic structures had been studied before by modular simulation of vehicle-guideway interaction [13].

For a detailed analysis of the numerical stability problems in the simulation of pantograph and catenary we started with a precise definition of the mathematical model [26] and defined a non-trivial benchmark scenario for future numerical tests [8]. After space discretization of all flexible system components [25] the coupled model equations get the form

$$\mathbf{M}_{\mathrm{p}}(\boldsymbol{q}_{\mathrm{p}})\ddot{\boldsymbol{q}}_{\mathrm{p}} = \boldsymbol{f}_{\mathrm{p}}(t,\boldsymbol{q}_{\mathrm{p}},\dot{\boldsymbol{q}}_{\mathrm{p}}) - \mathbf{G}_{\mathrm{p}}^{\top}(t,\boldsymbol{q}_{\mathrm{p}},\boldsymbol{q}_{\mathrm{c}})\boldsymbol{\lambda}\,, \tag{1a}$$

$$\mathbf{M}_{\mathrm{c}}(\boldsymbol{q}_{\mathrm{c}})\ddot{\boldsymbol{q}}_{\mathrm{c}} = \boldsymbol{f}_{\mathrm{c}}(t,\boldsymbol{q}_{\mathrm{c}},\dot{\boldsymbol{q}}_{\mathrm{c}}) - \mathbf{G}_{\mathrm{c}}^{\top}(t,\boldsymbol{q}_{\mathrm{p}},\boldsymbol{q}_{\mathrm{c}})\boldsymbol{\lambda}\,, \tag{1b}$$

$$\mathbf{0} = \boldsymbol{g}(t,\boldsymbol{q}_{\mathrm{p}},\boldsymbol{q}_{\mathrm{c}}) \tag{1c}$$

with the holonomic constraints (1c) representing the contact condition(s) between pantograph head(s) and catenary. These constraint equations are coupled by constraint forces with Lagrange multipliers $\boldsymbol{\lambda}$ and constraint Jacobians

$$\mathbf{G}_{\mathrm{p}}(t,\boldsymbol{q}_{\mathrm{p}},\boldsymbol{q}_{\mathrm{c}}) := \frac{\partial \boldsymbol{g}}{\partial \boldsymbol{q}_{\mathrm{p}}}(t,\boldsymbol{q}_{\mathrm{p}},\boldsymbol{q}_{\mathrm{c}})\,,\;\; \mathbf{G}_{\mathrm{c}}(t,\boldsymbol{q}_{\mathrm{p}},\boldsymbol{q}_{\mathrm{c}}) := \frac{\partial \boldsymbol{g}}{\partial \boldsymbol{q}_{\mathrm{c}}}(t,\boldsymbol{q}_{\mathrm{p}},\boldsymbol{q}_{\mathrm{c}})$$

to the dynamical equations of pantograph ("p") and catenary ("c"). The mass matrices $\mathbf{M}_{\mathrm{p}}$, $\mathbf{M}_{\mathrm{c}}$ of both subsystems are supposed to be symmetric positive definite. The vectors $\boldsymbol{f}_{\mathrm{p}}$, $\boldsymbol{f}_{\mathrm{c}}$ collect all internal force terms of the two subsystems.[1] For time integration, the constraints (1c) are substituted by their second time derivatives

$$\begin{aligned}\mathbf{0} &= \frac{\mathrm{d}^2}{\mathrm{d}t^2}\boldsymbol{g}\big(t,\boldsymbol{q}_{\mathrm{p}}(t),\boldsymbol{q}_{\mathrm{c}}(t)\big)\\ &= \mathbf{G}_{\mathrm{p}}(t,\boldsymbol{q}_{\mathrm{p}},\boldsymbol{q}_{\mathrm{c}})\ddot{\boldsymbol{q}}_{\mathrm{p}} + \mathbf{G}_{\mathrm{c}}(t,\boldsymbol{q}_{\mathrm{p}},\boldsymbol{q}_{\mathrm{c}})\ddot{\boldsymbol{q}}_{\mathrm{c}} + \boldsymbol{g}^{(II)}(t,\boldsymbol{q}_{\mathrm{p}},\boldsymbol{q}_{\mathrm{c}},\dot{\boldsymbol{q}}_{\mathrm{p}},\dot{\boldsymbol{q}}_{\mathrm{c}})\,,\end{aligned} \tag{1d}$$

[1] For more sophisticated pantograph models the model equations (1a) of this subsystem have to be substituted by a differential-algebraic equation with internal constraints [27].

i.e., by the hidden constraints at the level of acceleration coordinates [4, 16]. Standard projection techniques help to avoid the undesired drift-off effect.

We consider an initial value problem on a finite time interval $[t_0, t_e]$ with consistent initial values $\boldsymbol{q}_{p,0}$, $\boldsymbol{q}_{c,0}$, $\boldsymbol{\lambda}_0$ at $t = t_0$ and approximate the solution of (1) by continuously differentiable functions $\widetilde{q}_p(t), \widetilde{q}_c(t)$ and a piecewise continuous function $\widetilde{\lambda}(t)$ proceeding in macro steps $T_m \to T_{m+1} = T_m + H$ of size H from the consistent initial values at $T_0 := t_0$ to the end of the time interval. Following the classical force-displacement coupling paradigm for the coupled simulation of mechanical systems, there are basically two approaches for a modular simulation of pantograph–catenary interaction:

Catenary first

a) Fix $\widetilde{\boldsymbol{q}}_{p,m} = \widetilde{\boldsymbol{q}}_p(T_m)$ and the Lagrange multipliers $\widetilde{\boldsymbol{\lambda}}_m = \widetilde{\boldsymbol{\lambda}}(T_m)$ for a sufficiently small macro step $T_m \to T_{m+1} := T_m + H$.

b) Get the response of the elastic structure solving

$$\mathbf{M}_c(\widetilde{\boldsymbol{q}}_c)\ddot{\widetilde{\boldsymbol{q}}}_c = \boldsymbol{f}_c(t, \widetilde{\boldsymbol{q}}_c, \dot{\widetilde{\boldsymbol{q}}}_c) - \mathbf{G}_c^\top(t, \widetilde{\boldsymbol{q}}_{p,m}, \widetilde{\boldsymbol{q}}_c)\widetilde{\boldsymbol{\lambda}}_m\,, \tag{2a}$$

on $[T_m, T_{m+1}]$ with initial values $\widetilde{\boldsymbol{q}}_c(T_m)$, $\dot{\widetilde{\boldsymbol{q}}}_c(T_m)$ being known from the previous macro step.

c) Now, the state variables $\widetilde{\boldsymbol{q}}_c(t)$, $\dot{\widetilde{\boldsymbol{q}}}_c(t)$, $(t \in [T_m, T_{m+1}])$, are known and the equations of motion for the pantograph may be solved taking into account the contact condition (1c) in its differentiated form (1d):

$$\mathbf{M}_p(\widetilde{\boldsymbol{q}}_p)\ddot{\widetilde{\boldsymbol{q}}}_p = \boldsymbol{f}_p(t, \widetilde{\boldsymbol{q}}_p, \dot{\widetilde{\boldsymbol{q}}}_p) - \mathbf{G}_p^\top(t, \widetilde{\boldsymbol{q}}_p, \widetilde{\boldsymbol{q}}_c)\widetilde{\boldsymbol{\lambda}}\,, \tag{2b}$$

$$\mathbf{0} = \mathbf{G}_p(t, \widetilde{\boldsymbol{q}}_p, \widetilde{\boldsymbol{q}}_c)\ddot{\widetilde{\boldsymbol{q}}}_p + \mathbf{G}_c(t, \widetilde{\boldsymbol{q}}_p, \widetilde{\boldsymbol{q}}_c)\ddot{\widetilde{\boldsymbol{q}}}_c + \boldsymbol{g}^{(II)}(t, \widetilde{\boldsymbol{q}}_p, \widetilde{\boldsymbol{q}}_c, \dot{\widetilde{\boldsymbol{q}}}_p, \dot{\widetilde{\boldsymbol{q}}}_c)\,. \tag{2c}$$

As before, the initial values $\widetilde{\boldsymbol{q}}_p(T_m)$, $\dot{\widetilde{\boldsymbol{q}}}_p(T_m)$ for position and velocity coordinates are known from the previous macro step. The solution of (2b,c) defines $\widetilde{\boldsymbol{q}}_p(t)$, $\dot{\widetilde{\boldsymbol{q}}}_p(t)$ and $\widetilde{\boldsymbol{\lambda}}(t)$ for all $t \in (T_m, T_{m+1}]$.

d) Now, the macro step $T_m \to T_{m+1} = T_m + H$ is complete and the modular integration may be continued with macro step $T_{m+1} \to T_{m+2}$.

Pantograph first

a) Fix $\widetilde{\boldsymbol{q}}_{c,m} = \widetilde{\boldsymbol{q}}_c(T_m)$ and the Lagrange multipliers $\widetilde{\boldsymbol{\lambda}}_m = \widetilde{\boldsymbol{\lambda}}(T_m)$ for a sufficiently small macro step $T_m \to T_{m+1} := T_m + H$.

b) Solve the equations of motion for the pantograph

$$\mathbf{M}_p(\widetilde{\boldsymbol{q}}_p)\ddot{\widetilde{\boldsymbol{q}}}_p = \boldsymbol{f}_p(t, \widetilde{\boldsymbol{q}}_p, \dot{\widetilde{\boldsymbol{q}}}_p) - \mathbf{G}_p^\top(t, \widetilde{\boldsymbol{q}}_p, \widetilde{\boldsymbol{q}}_{c,m})\widetilde{\boldsymbol{\lambda}}_m\,, \tag{3a}$$

on $[T_m, T_{m+1}]$ with initial values $\widetilde{\boldsymbol{q}}_p(T_m)$, $\dot{\widetilde{\boldsymbol{q}}}_p(T_m)$ being known from the previous macro step.

c) Now, the state variables $\widetilde{\boldsymbol{q}}_{\rm p}(t)$, $\dot{\widetilde{\boldsymbol{q}}}_{\rm p}(t)$, ($t \in [T_m, T_{m+1}]$), are known and the equations of motion for the catenary may be solved taking into account the contact condition (1c) in its differentiated form (1d):

$$\mathbf{M}_{\rm c}(\widetilde{\boldsymbol{q}}_{\rm c})\ddot{\widetilde{\boldsymbol{q}}}_{\rm c} = \boldsymbol{f}_{\rm c}(t,\widetilde{\boldsymbol{q}}_{\rm c},\dot{\widetilde{\boldsymbol{q}}}_{\rm c}) - \mathbf{G}_{\rm c}^\top(t,\widetilde{\boldsymbol{q}}_{\rm p},\widetilde{\boldsymbol{q}}_{\rm c})\widetilde{\boldsymbol{\lambda}}\,, \tag{3b}$$

$$\mathbf{0} = \mathbf{G}_{\rm p}(t,\widetilde{\boldsymbol{q}}_{\rm p},\widetilde{\boldsymbol{q}}_{\rm c})\ddot{\widetilde{\boldsymbol{q}}}_{\rm p} + \mathbf{G}_{\rm c}(t,\widetilde{\boldsymbol{q}}_{\rm p},\widetilde{\boldsymbol{q}}_{\rm c})\ddot{\widetilde{\boldsymbol{q}}}_{\rm c} + \boldsymbol{g}^{(II)}(t,\widetilde{\boldsymbol{q}}_{\rm p},\widetilde{\boldsymbol{q}}_{\rm c},\dot{\widetilde{\boldsymbol{q}}}_{\rm p},\dot{\widetilde{\boldsymbol{q}}}_{\rm c})\,. \tag{3c}$$

As before, the initial values $\widetilde{\boldsymbol{q}}_{\rm c}(T_m)$, $\dot{\widetilde{\boldsymbol{q}}}_{\rm c}(T_m)$ for position and velocity coordinates are known from the previous macro step. The solution of (3b,c) defines $\widetilde{\boldsymbol{q}}_{\rm c}(t)$, $\dot{\widetilde{\boldsymbol{q}}}_{\rm c}(t)$ and $\widetilde{\boldsymbol{\lambda}}(t)$ for all $t \in (T_m, T_{m+1}]$.

d) Now, the macro step $T_m \to T_{m+1} = T_m + H$ is complete and the modular integration may be continued with macro step $T_{m+1} \to T_{m+2}$.

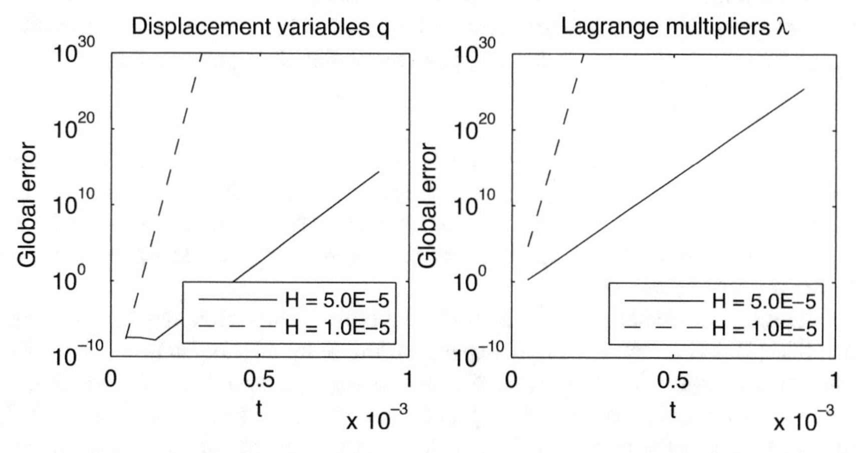

Fig. 2: **Catenary first** approach in the modular simulation of a benchmark problem for pantograph-catenary interaction [8]: Global errors in the displacement variables $\boldsymbol{q}_{\rm p}$, $\boldsymbol{q}_{\rm c}$ (left plot: $\|\boldsymbol{e}_m^{\boldsymbol{q}}\|$) and in the Lagrange multipliers $\boldsymbol{\lambda}$ (right plot: $\|\boldsymbol{e}_m^{\boldsymbol{\lambda}}\|$).

Both modular simulation techniques (2) and (3) are expected to define numerical solutions $\big(\widetilde{\boldsymbol{q}}_{\rm p}(t), \widetilde{\boldsymbol{q}}_{\rm c}(t), \widetilde{\boldsymbol{\lambda}}(t)\big)$ that approximate the solution $\big(\boldsymbol{q}_{\rm p}(t), \boldsymbol{q}_{\rm c}(t), \boldsymbol{\lambda}(t)\big)$ of (1) with increasing accuracy if the macro step size H is decreased. However, the numerical test results in Fig. 2 illustrate that the **Catenary first** solution strategy (2) fails spectacularly in the application to a standard benchmark problem in this field. The norm of the global errors

$$\boldsymbol{e}_m^{\boldsymbol{q}} := \begin{pmatrix} \boldsymbol{q}_{\rm p}(T_m) - \widetilde{\boldsymbol{q}}_{\rm p}(T_m) \\ \boldsymbol{q}_{\rm c}(T_m) - \widetilde{\boldsymbol{q}}_{\rm c}(T_m) \end{pmatrix}, \quad \boldsymbol{e}_m^{\boldsymbol{\lambda}} := \boldsymbol{\lambda}(T_m) - \widetilde{\boldsymbol{\lambda}}(T_m) = \boldsymbol{\lambda}(T_m) - \widetilde{\boldsymbol{\lambda}}_m$$

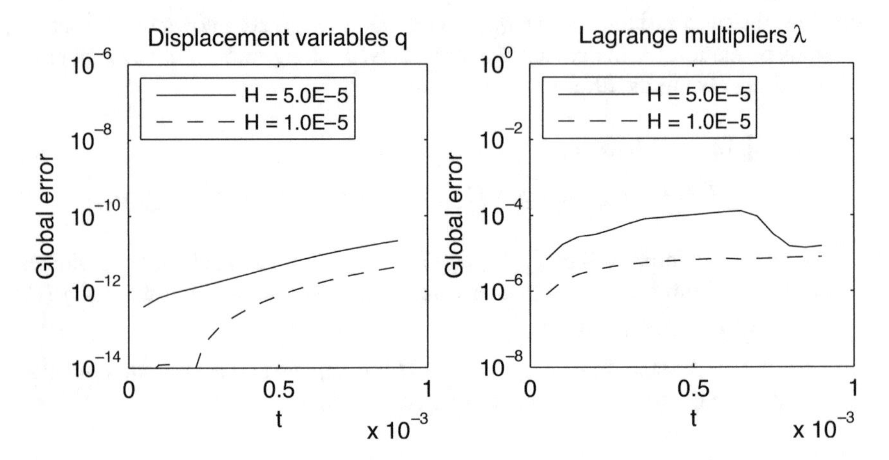

Fig. 3: **Pantograph first** approach in the modular simulation of a benchmark problem for pantograph-catenary interaction [8]: Global errors in the displacement variables $\boldsymbol{q}_{\mathrm{p}}, \boldsymbol{q}_{\mathrm{c}}$ (left plot: $\|\boldsymbol{e}_m^{\boldsymbol{q}}\|$) and in the Lagrange multipliers $\boldsymbol{\lambda}$ (right plot: $\|\boldsymbol{e}_m^{\boldsymbol{\lambda}}\|$).

grows like a geometric series α^m with $\alpha > 1$. Furthermore, the errors grow even more rapidly if the macro step size is decreased from $H = 50\,\mu s$ to $H = 10\,\mu s$. (In the numerical tests, the local initial value problems (2a) and (2b,c) are solved with high accuracy to focus on the discretization errors being caused by the modular **Catenary first** approach.)

Stability and convergence of Gauss-Seidel type modular time integration methods like (2) and (3) may strongly depend on the order of subsystems. This fact is illustrated by numerical test results for the **Pantograph first** approach, see Fig. 3: For $H = 50\,\mu\mathrm{s}$, the global errors in $\boldsymbol{q}$ and $\boldsymbol{\lambda}$ remain in the size of 10^{-10} and 10^{-4}, respectively, and the errors are further decreased reducing the macro step size to $H = 10\,\mu\mathrm{s}$. A more refined error analysis shows that the **Pantograph first** approach provides a numerical solution $\big(\widetilde{\boldsymbol{q}}_{\mathrm{p}}(t), \widetilde{\boldsymbol{q}}_{\mathrm{c}}(t), \widetilde{\boldsymbol{\lambda}}(t)\big)$ with errors of size $\mathscr{O}(H)$ in the supremum norm on $[t_0, t_{\mathrm{e}}]$, i.e., we observe numerically first order convergence for this benchmark problem.

Engineering application

From the viewpoint of engineering application, the stability problems of modular time integration could simply be fixed using the **Pantograph first** approach (3), see [28]. The modular algorithm was implemented in a developer version of the industrial SIMPACK – PrOSA tool coupling and tested for a sophisticated multibody system pantograph model with 56 degrees of freedom, see [27, Section 5], and a 4th order finite difference approximation of a beam model for a catenary of 1,200 m length, see [19].

In a classical monolithic time integration of the coupled system (1), the time step size is restricted by a CFL condition $\Delta_t \leq c\Delta_x^2$ for the semi-discretized beam model representing the catenary [19], see also [8]. In the modular approach (3), this time step restriction applies only to the catenary subsystem (3b,c). The accuracy of the simulation results was still acceptable if up to eight times larger step sizes were used for the communication between both subsystems ($H = 8\Delta_t$) and for the time integration of the multibody system pantograph model in SIMPACK by a user-defined fixed step size BDF2 solver. Taking into account the overhead for the interprocess communication of both simulation tools [27], an overall speed-up by a factor of 6.1 could be achieved in practically relevant simulation scenarios [29], see also [2].

Mathematical aspects of modular simulation

From the mathematical viewpoint, the unexpected systematic failure of modular integration in the **Catenary first** approach was rather disappointing and further analysis was necessary

- to identify qualitatively potential sources of numerical instability in modular time integration,
- to fix such stability problems by modified coupling strategies,
- to study quantitatively the convergence of numerically stable modular time integration schemes in the sense of local and global error estimates and
- to verify the practical relevance of such theoretical investigations by numerical tests for nontrivial benchmark problems.

This research was guided by the classical perturbation theory for initial value problems of ordinary differential equations (ODEs) with Lipschitz continuous right hand side. For the application to differential-algebraic equations (DAEs) like the coupled system (1), this perturbation theory had to be combined with the analysis of the coupled error propagation in differential and algebraic solution components using contractivity conditions to guarantee numerical stability and convergence.

3 Modular time integration: The ODE case

The numerically observed first order convergence of the **Pantograph first** approach (3) fits to engineering intuition since continuously differentiable inputs $\boldsymbol{u}(t)$ in ODE initial value problems

$$\dot{\boldsymbol{x}}(t) = \boldsymbol{f}\big(\boldsymbol{x}(t), \boldsymbol{u}(t)\big)\,,\ \big(t \in [t_0, t_{\mathrm{e}}]\big)\,,\ \boldsymbol{x}(t_0) = \boldsymbol{x}_0 \tag{4}$$

may safely be substituted by sampled data if the sampling time $H > 0$ is sufficiently small:

$$\widetilde{\boldsymbol{u}} : [t_0, t_e] \to \mathbb{R}^{n_{\boldsymbol{u}}} , \ \widetilde{\boldsymbol{u}}\big|_{[T_m, T_{m+1})} := \boldsymbol{u}(T_m) \tag{5}$$

with $T_m := t_0 + mH$. The sampled input results in an approximation $\widetilde{\boldsymbol{x}}(t)$ with

$$\dot{\widetilde{\boldsymbol{x}}}(t) = \boldsymbol{f}\big(\widetilde{\boldsymbol{x}}(t), \widetilde{\boldsymbol{u}}(t)\big) , \ (t \in [t_0, t_e]) , \ \widetilde{\boldsymbol{x}}(t_0) = \boldsymbol{x}_0 \tag{6}$$

that has on $[T_m, T_{m+1}]$ a global error of the form

$$\begin{aligned} \boldsymbol{x}(\tau) - \widetilde{\boldsymbol{x}}(\tau) &= \boldsymbol{x}(T_m) - \widetilde{\boldsymbol{x}}(T_m) + \int_{T_m}^{\tau} \Big(\dot{\boldsymbol{x}}(s) - \dot{\widetilde{\boldsymbol{x}}}(s) \Big) \, \mathrm{d}s \\ &= \boldsymbol{x}(T_m) - \widetilde{\boldsymbol{x}}(T_m) + \int_{T_m}^{\tau} \Big(\boldsymbol{f}\big(\boldsymbol{x}(s), \boldsymbol{u}(s)\big) - \boldsymbol{f}\big(\widetilde{\boldsymbol{x}}(s), \boldsymbol{u}(T_m)\big) \Big) \, \mathrm{d}s \, . \end{aligned}$$

If the right hand side $\boldsymbol{f}$ satisfies Lipschitz conditions w.r.t. $\boldsymbol{x}$ and $\boldsymbol{u}$ with constants $L_{\boldsymbol{x}}$, $L_{\boldsymbol{u}}$, respectively, then $\|\boldsymbol{x}(\tau) - \widetilde{\boldsymbol{x}}(\tau)\|$, $(\tau \in [T_m, T_{m+1}])$, is bounded by the continuously differentiable function

$$\psi_m(\tau) := \|\boldsymbol{x}(T_m) - \widetilde{\boldsymbol{x}}(T_m)\| + \int_{T_m}^{\tau} L_{\boldsymbol{x}} \|\boldsymbol{x}(s) - \widetilde{\boldsymbol{x}}(s)\| \, \mathrm{d}s + L_{\boldsymbol{u}} (\tau - T_m) \boldsymbol{\Delta}_m^{\boldsymbol{u}}$$

with the sampling error $\boldsymbol{\Delta}_m^{\boldsymbol{u}} := \max_{s \in [T_m, T_{m+1}]} \|\boldsymbol{u}(s) - \boldsymbol{u}(T_m)\|$ that is of size $\mathscr{O}(H)$ for any continuously differentiable input function $\boldsymbol{u}(t)$. This upper bound ψ_m satisfies

$$\frac{\mathrm{d}}{\mathrm{d}\tau} \psi_m(\tau) = L_{\boldsymbol{x}} \|\boldsymbol{x}(\tau) - \widetilde{\boldsymbol{x}}(\tau)\| + L_{\boldsymbol{u}} \boldsymbol{\Delta}_m^{\boldsymbol{u}} \le L_{\boldsymbol{x}} \psi_m(\tau) + L_{\boldsymbol{u}} \boldsymbol{\Delta}_m^{\boldsymbol{u}} ,$$

$$\frac{\mathrm{d}}{\mathrm{d}\tau} \big(\mathrm{e}^{L_{\boldsymbol{x}}(t-\tau)} \psi_m(\tau) \big) = \mathrm{e}^{L_{\boldsymbol{x}}(t-\tau)} \Big(-L_{\boldsymbol{x}} \psi_m(\tau) + \frac{\mathrm{d}}{\mathrm{d}\tau} \psi_m(\tau) \Big) \le \mathrm{e}^{L_{\boldsymbol{x}}(t-\tau)} L_{\boldsymbol{u}} \boldsymbol{\Delta}_m^{\boldsymbol{u}} ,$$

i.e.,

$$\begin{aligned} \|\boldsymbol{x}(t) - \widetilde{\boldsymbol{x}}(t)\| \le \psi_m(t) &= \mathrm{e}^{L_{\boldsymbol{x}}(t-t)} \psi_m(t) \\ &= \mathrm{e}^{L_{\boldsymbol{x}}(t-T_m)} \psi_m(T_m) + \int_{T_m}^{t} \frac{\mathrm{d}}{\mathrm{d}\tau} \big(\mathrm{e}^{L_{\boldsymbol{x}}(t-\tau)} \psi_m(\tau) \big) \, \mathrm{d}\tau \\ &\le \mathrm{e}^{L_{\boldsymbol{x}}(t-T_m)} \psi_m(T_m) + \int_{T_m}^{t} \mathrm{e}^{L_{\boldsymbol{x}}(t-\tau)} L_{\boldsymbol{u}} \boldsymbol{\Delta}_m^{\boldsymbol{u}} \, \mathrm{d}\tau \\ &= \mathrm{e}^{L_{\boldsymbol{x}}(t-T_m)} \|\boldsymbol{x}(T_m) - \widetilde{\boldsymbol{x}}(T_m)\| + L_{\boldsymbol{u}} \frac{\mathrm{e}^{L_{\boldsymbol{x}}(t-T_m)} - 1}{L_{\boldsymbol{x}}} \boldsymbol{\Delta}_m^{\boldsymbol{u}} \end{aligned}$$

and

$$\|\boldsymbol{x}(T_{m+1}) - \widetilde{\boldsymbol{x}}(T_{m+1})\| \le \mathrm{e}^{L_{\boldsymbol{x}}(T_{m+1}-T_m)} \|\boldsymbol{x}(T_m) - \widetilde{\boldsymbol{x}}(T_m)\| + 2 L_{\boldsymbol{u}} H \boldsymbol{\Delta}_m^{\boldsymbol{u}}$$

if $H \in (0, H_0]$ with $H_0 > 0$ being defined such that $\mathrm{e}^{L_{\boldsymbol{x}} H_0} \le 1 + 2 L_{\boldsymbol{x}} H_0$. A simple induction argument shows for all m with $0 \le m \le n$ the estimate

$$\|\boldsymbol{x}(T_n) - \widetilde{\boldsymbol{x}}(T_n)\| \le \mathrm{e}^{L_{\boldsymbol{x}}(T_n - T_m)} \|\boldsymbol{x}(T_m) - \widetilde{\boldsymbol{x}}(T_m)\| + 2 L_{\boldsymbol{u}} \int_{T_m}^{T_n} \mathrm{e}^{L_{\boldsymbol{x}}(T_n - \tau)} \, \mathrm{d}\tau \max_{m \le r < n} \boldsymbol{\Delta}_r^{\boldsymbol{u}} ,$$

see [5, Theorem 4.11] which proves first order convergence since $\boldsymbol{\Delta}_r^{\boldsymbol{u}} = \mathscr{O}(H)$ and $\boldsymbol{x}(t_0) - \widetilde{\boldsymbol{x}}(t_0) = \boldsymbol{x}_0 - \boldsymbol{x}_0 = \boldsymbol{0}$:

Theorem 3.1 *If the right hand side of the initial value problem (4) satisfies Lipschitz conditions w.r.t.* $\boldsymbol{x}$ *and* $\boldsymbol{u}$ *and the input function* $\boldsymbol{u}(t)$ *is continuously differentiable then the piecewise constant approximation of* $\boldsymbol{u}$ *according to (5) results in an approximate solution* $\widetilde{\boldsymbol{x}}(t)$ *with*

$$\|\boldsymbol{x}(t) - \widetilde{\boldsymbol{x}}(t)\| \leq CH\,,\ (t \in [t_0, t_e])\,, \tag{7}$$

provided that $H > 0$ *is sufficiently small. The constant* C *in (7) is independent of* H *but depends in general on the Lipschitz constants of* $\boldsymbol{f}$ *as well as on the length of the time interval* $[t_0, t_e]$ *and on bounds on* $\dot{\boldsymbol{u}}(t)$.

The convergence result in Theorem 3.1 combines the classical perturbation theory for ODE initial value problems, see, e.g., [31], with error bounds that are known from the convergence analysis of ODE one step methods [15]. The analysis of numerical errors being caused by the time-discrete sampling (5) was simplified substantially neglecting all details of time integration for the perturbed system (6).

In [6], it was shown how to generalize this approach to inputs $\boldsymbol{u}(t)$ that depend on the solution $\boldsymbol{x}(t)$ itself like in block-structured coupled systems

$$\left.\begin{aligned} \dot{\boldsymbol{x}}_j(t) &= \boldsymbol{f}_j\big(\boldsymbol{x}_j(t), \boldsymbol{u}_j(t), \boldsymbol{u}_{\mathrm{ex}}(t)\big) \\ \boldsymbol{y}_j(t) &= \boldsymbol{g}_j\big(\boldsymbol{x}_j(t)\big) \end{aligned}\right\},\ (1 \leq j \leq N)\,, \tag{8a}$$

with $\boldsymbol{x}_j$, $\boldsymbol{u}_j$, $\boldsymbol{y}_j$ denoting the state, input and output vectors and some external input $\boldsymbol{u}_{\mathrm{ex}}(t)$. The $N \geq 2$ subsystems in (8a) are coupled by input-output relations

$$\boldsymbol{u}_j(t) = \boldsymbol{c}_j\big(\boldsymbol{y}_1(t), \ldots, \boldsymbol{y}_{j-1}(t), \boldsymbol{y}_{j+1}(t), \ldots, \boldsymbol{y}_r(t)\big)\,,\ (j = 1, \ldots, N) \tag{8b}$$

resulting in

$$\left.\begin{aligned} \dot{\boldsymbol{x}}(t) &= \boldsymbol{f}\big(\boldsymbol{x}(t), \boldsymbol{u}(t), \boldsymbol{u}_{\mathrm{ex}}(t)\big) \\ \boldsymbol{u}(t) &= \boldsymbol{c}\big(\boldsymbol{g}\big(\boldsymbol{x}(t)\big)\big) \end{aligned}\right\} \tag{9}$$

with super-vectors $\boldsymbol{u}$, $\boldsymbol{x}$, $\boldsymbol{y}$ that summarize the vectors $\boldsymbol{u}_j$, $\boldsymbol{x}_j$, $\boldsymbol{y}_j$, $(1 \leq j \leq N)$ in compact form. The error analysis is not limited to the piecewise constant approximation (5) with $\boldsymbol{u}(T_m) = \boldsymbol{c}\big(\boldsymbol{g}\big(\boldsymbol{x}(T_m)\big)\big)$ but may be extended straightforwardly to linear, quadratic or higher order approximations that may be represented by an extrapolation operator

$$\boldsymbol{\Phi}_m : \mathscr{C}(T_{m-k}, T_m] \to \mathscr{C}(T_m, T_{m+1}]\,,\ \widetilde{\boldsymbol{u}}\big|_{(T_{m-k}, T_m]} \mapsto \boldsymbol{\Phi}_m[\widetilde{\boldsymbol{u}}] \tag{10}$$

with an appropriate initialization scheme for the first k macro steps. If all inputs $\boldsymbol{u}(t)$ in (9) are substituted by their extrapolation $\boldsymbol{\Phi}_m[\boldsymbol{u}](t)$ then this coupled system is for $t \in [T_m, T_{m+1}]$ decomposed into N separate subsystems

$$\dot{\widetilde{\boldsymbol{x}}}_j(t) = \boldsymbol{f}_j\big(\widetilde{\boldsymbol{x}}(t), \boldsymbol{\Phi}_{m,j}[\widetilde{\boldsymbol{u}}](t), \boldsymbol{u}_{\mathrm{ex}}(t)\big)\,,\ (1 \leq j \leq N)\,,$$

that may be solved independently in a modular time integration approach. As before, the approximation of $\boldsymbol{u}$ results in an approximation error $\boldsymbol{x}(t) - \widetilde{\boldsymbol{x}}(t)$ that may be bounded in terms of the *extrapolation error*

$$\boldsymbol{\Delta}_m^{\boldsymbol{u}} := \max_{s \in [T_m, T_{m+1}]} \| \boldsymbol{u}(s) - \boldsymbol{\Phi}_m[\boldsymbol{u}](s) \| . \tag{11}$$

(A typical example are extrapolation operators $\boldsymbol{\Phi}_m$ that are defined by interpolation polynomials of degree $\leq k$. Then, the extrapolation error is of size $\mathscr{O}(H^{k+1})$ if $\boldsymbol{u}$ is sufficiently often continuously differentiable.)

Theorem 3.2 *If the functions* $\boldsymbol{f}$, $\boldsymbol{c}$ *and* $\boldsymbol{g}$ *in the initial value problem (9) satisfy Lipschitz conditions w.r.t.* $\boldsymbol{x}$ *and* $\boldsymbol{u}$ *and the input function* $\boldsymbol{u}(t)$ *is substituted by its piecewise defined extrapolation* $\boldsymbol{\Phi}_m[\boldsymbol{u}](t)$ *then there is a constant* $C > 0$ *such that the approximate solution* $\widetilde{\boldsymbol{x}}(t)$ *satisfies*

$$\| \boldsymbol{x}(t) - \widetilde{\boldsymbol{x}}(t) \| \leq C \max_{\{m : T_0 + mH \leq t\}} \boldsymbol{\Delta}_m^{\boldsymbol{u}} , \; (t \in [t_0, t_{\mathrm{e}}]) ,$$

provided that $H > 0$ *is sufficiently small.*

This convergence result may be generalized to subsystems (8a) with *direct feed-through* being represented by output equations

$$\boldsymbol{y}_j(t) = \boldsymbol{g}_j\big(\boldsymbol{x}_j(t), \boldsymbol{u}_j(t)\big)$$

as long as the overall coupled system is free of algebraic loops, see [6] for a detailed discussion. This problem class is highly relevant from the practical viewpoint and includes also force-displacement couplings of mechanical systems by spring-damper systems. The coupling of subsystems by constraints like the contact condition (1c) in the pantograph–catenary model is, however, substantially more complex and requires mathematical tools beyond classical ODE theory.

4 Modular time integration: The DAE case

For constrained systems, the divergence of modular time integration methods may be observed for rather simple test problems. Consider a coupled problem (1) with constant mass matrices $\mathbf{M}_{\mathrm{p}}$, $\mathbf{M}_{\mathrm{c}}$ and linear constraints

$$\mathbf{0} = \boldsymbol{g}(t, \boldsymbol{q}_{\mathrm{p}}, \boldsymbol{q}_{\mathrm{c}}) := \mathbf{G}_{\mathrm{p}} \boldsymbol{q}_{\mathrm{p}} + \mathbf{G}_{\mathrm{c}} \boldsymbol{q}_{\mathrm{c}} \tag{12}$$

as an academic test problem without any practical relevance that is, however, helpful to understand the basic mechanism of error propagation in the algebraic variables $\boldsymbol{\lambda}$. Here, the **Catenary first** approach defines $\ddot{\widetilde{\boldsymbol{q}}}_{\mathrm{c}}$ by (2a) as

$$\ddot{\widetilde{\boldsymbol{q}}}_{\mathrm{c}} = -\mathbf{M}_{\mathrm{c}}^{-1} \mathbf{G}_{\mathrm{c}}^{\top} \widetilde{\boldsymbol{\lambda}}_m .$$

Multiplying (2b) by $\left(\mathbf{G}_\mathrm{p}\mathbf{M}_\mathrm{p}^{-1}\mathbf{G}_\mathrm{p}^\top\right)^{-1}\mathbf{G}_\mathrm{p}\mathbf{M}_\mathrm{p}^{-1}$, we get

$$\widetilde{\boldsymbol{\lambda}}(t) = -\left(\mathbf{G}_\mathrm{p}\mathbf{M}_\mathrm{p}^{-1}\mathbf{G}_\mathrm{p}^\top\right)^{-1}\mathbf{G}_\mathrm{c}\mathbf{M}_\mathrm{c}^{-1}\mathbf{G}_\mathrm{c}^\top\,\widetilde{\boldsymbol{\lambda}}_m \tag{13}$$

since

$$\mathbf{G}_\mathrm{p}\,\ddot{\widetilde{\boldsymbol{q}}}_\mathrm{p} = -\mathbf{G}_\mathrm{c}\,\ddot{\widetilde{\boldsymbol{q}}}_\mathrm{c} = \mathbf{G}_\mathrm{c}\mathbf{M}_\mathrm{c}^{-1}\mathbf{G}_\mathrm{c}^\top\,\widetilde{\boldsymbol{\lambda}}_m\,,$$

see (2c) and (12). The explicit expression (13) shows

$$\widetilde{\boldsymbol{\lambda}}_n = \left(-\left(\mathbf{G}_\mathrm{p}\mathbf{M}_\mathrm{p}^{-1}\mathbf{G}_\mathrm{p}^\top\right)^{-1}\mathbf{G}_\mathrm{c}\mathbf{M}_\mathrm{c}^{-1}\mathbf{G}_\mathrm{c}^\top\right)^n\widetilde{\boldsymbol{\lambda}}_0$$

and $(\widetilde{\boldsymbol{\lambda}}_n)_{n\geq 0}$ will diverge unless all eigenvalues of $\left(\mathbf{G}_\mathrm{p}\mathbf{M}_\mathrm{p}^{-1}\mathbf{G}_\mathrm{p}^\top\right)^{-1}\mathbf{G}_\mathrm{c}\mathbf{M}_\mathrm{c}^{-1}\mathbf{G}_\mathrm{c}^\top$ are inside the unit circle. In the benchmark problem that was used for the numerical tests in Section 2, the spectral radius of this matrix product is, however, larger than 30 which explains the numerical instability of the **Catenary first** approach, see Fig. 2.

For the convergence analysis of **Catenary first** and **Pantograph first** approaches (2) and (3), a contractivity constant is introduced by

$$\alpha := \max_{t\in[t_0,t_\mathrm{e}]}\left\|\left[\left(\mathbf{G}_\mathrm{p}\mathbf{M}_\mathrm{p}^{-1}\mathbf{G}_\mathrm{p}^\top\right)^{-1}\mathbf{G}_\mathrm{c}\mathbf{M}_\mathrm{c}^{-1}\mathbf{G}_\mathrm{c}^\top\right]\left(\boldsymbol{q}_\mathrm{p}(t),\boldsymbol{q}_\mathrm{c}(t)\right)\right\| \tag{14a}$$

for the **Catenary first** approach and by

$$\alpha := \max_{t\in[t_0,t_\mathrm{e}]}\left\|\left[\left(\mathbf{G}_\mathrm{c}\mathbf{M}_\mathrm{c}^{-1}\mathbf{G}_\mathrm{c}^\top\right)^{-1}\mathbf{G}_\mathrm{p}\mathbf{M}_\mathrm{p}^{-1}\mathbf{G}_\mathrm{p}^\top\right]\left(\boldsymbol{q}_\mathrm{p}(t),\boldsymbol{q}_\mathrm{c}(t)\right)\right\| \tag{14b}$$

for the **Pantograph first** approach. Furthermore, the analysis is extended to higher order extrapolation operators resulting in

$$\mathbf{M}_\mathrm{p}(\widetilde{\boldsymbol{q}}_\mathrm{p})\,\ddot{\widetilde{\boldsymbol{q}}}_\mathrm{p} = \boldsymbol{f}_\mathrm{p}(t,\widetilde{\boldsymbol{q}}_\mathrm{p},\dot{\widetilde{\boldsymbol{q}}}_\mathrm{p}) - \mathbf{G}_\mathrm{p}^\top(t,\widetilde{\boldsymbol{q}}_\mathrm{p},\widetilde{\boldsymbol{q}}_\mathrm{c})\widetilde{\boldsymbol{\lambda}}\,, \tag{15a}$$

$$\mathbf{M}_\mathrm{c}(\widetilde{\boldsymbol{q}}_\mathrm{c})\,\ddot{\widetilde{\boldsymbol{q}}}_\mathrm{c} = \boldsymbol{f}_\mathrm{c}(t,\widetilde{\boldsymbol{q}}_\mathrm{c},\dot{\widetilde{\boldsymbol{q}}}_\mathrm{c}) - \mathbf{G}_\mathrm{c}^\top(t,\boldsymbol{\Phi}_m^{\boldsymbol{q}}[\widetilde{\boldsymbol{q}}_\mathrm{p}],\widetilde{\boldsymbol{q}}_\mathrm{c})\,\boldsymbol{\Phi}_m^{\boldsymbol{\lambda}}[\widetilde{\boldsymbol{\lambda}}]\,, \tag{15b}$$

$$\mathbf{0} = \mathbf{G}_\mathrm{p}(t,\widetilde{\boldsymbol{q}}_\mathrm{p},\widetilde{\boldsymbol{q}}_\mathrm{c})\,\ddot{\widetilde{\boldsymbol{q}}}_\mathrm{p} + \mathbf{G}_\mathrm{c}(t,\widetilde{\boldsymbol{q}}_\mathrm{p},\widetilde{\boldsymbol{q}}_\mathrm{c})\,\ddot{\widetilde{\boldsymbol{q}}}_\mathrm{c} + \boldsymbol{g}^{(II)}(t,\widetilde{\boldsymbol{q}}_\mathrm{p},\widetilde{\boldsymbol{q}}_\mathrm{c},\dot{\widetilde{\boldsymbol{q}}}_\mathrm{p},\dot{\widetilde{\boldsymbol{q}}}_\mathrm{c})\,. \tag{15c}$$

for the **Catenary first** approach and in

$$\mathbf{M}_\mathrm{p}(\widetilde{\boldsymbol{q}}_\mathrm{p})\,\ddot{\widetilde{\boldsymbol{q}}}_\mathrm{p} = \boldsymbol{f}_\mathrm{p}(t,\widetilde{\boldsymbol{q}}_\mathrm{p},\dot{\widetilde{\boldsymbol{q}}}_\mathrm{p}) - \mathbf{G}_\mathrm{p}^\top(t,\widetilde{\boldsymbol{q}}_\mathrm{p},\boldsymbol{\Phi}_m^{\boldsymbol{q}}[\widetilde{\boldsymbol{q}}_\mathrm{c}])\,\boldsymbol{\Phi}_m^{\boldsymbol{\lambda}}[\widetilde{\boldsymbol{\lambda}}]\,, \tag{16a}$$

$$\mathbf{M}_\mathrm{c}(\widetilde{\boldsymbol{q}}_\mathrm{c})\,\ddot{\widetilde{\boldsymbol{q}}}_\mathrm{c} = \boldsymbol{f}_\mathrm{c}(t,\widetilde{\boldsymbol{q}}_\mathrm{c},\dot{\widetilde{\boldsymbol{q}}}_\mathrm{c}) - \mathbf{G}_\mathrm{c}^\top(t,\widetilde{\boldsymbol{q}}_\mathrm{p},\widetilde{\boldsymbol{q}}_\mathrm{c})\widetilde{\boldsymbol{\lambda}}\,, \tag{16b}$$

$$\mathbf{0} = \mathbf{G}_\mathrm{p}(t,\widetilde{\boldsymbol{q}}_\mathrm{p},\widetilde{\boldsymbol{q}}_\mathrm{c})\,\ddot{\widetilde{\boldsymbol{q}}}_\mathrm{p} + \mathbf{G}_\mathrm{c}(t,\widetilde{\boldsymbol{q}}_\mathrm{p},\widetilde{\boldsymbol{q}}_\mathrm{c})\,\ddot{\widetilde{\boldsymbol{q}}}_\mathrm{c} + \boldsymbol{g}^{(II)}(t,\widetilde{\boldsymbol{q}}_\mathrm{p},\widetilde{\boldsymbol{q}}_\mathrm{c},\dot{\widetilde{\boldsymbol{q}}}_\mathrm{p},\dot{\widetilde{\boldsymbol{q}}}_\mathrm{c})\,. \tag{16c}$$

for the **Pantograph first** approach.

Theorem 4.1 *Consider coupled model equations (1) with functions $\boldsymbol{f}_\mathrm{p}$, $\boldsymbol{f}_\mathrm{c}$, $\boldsymbol{g}$, $\mathbf{M}_\mathrm{p}$, $\mathbf{M}_\mathrm{c}$ being sufficiently often continuously differentiable and consider furthermore piecewise defined extrapolations $\boldsymbol{\Phi}_m^{\boldsymbol{q}}$, $\boldsymbol{\Phi}_m^{\boldsymbol{\lambda}}$ with operators that map from $\mathscr{C}(T_{m-k},T_m]$ to $\mathscr{C}(T_m,T_{m+1}]$ and satisfy a Lipschitz condition with constant $L_{\boldsymbol{\Phi}}$. If the contractivity* *constant α being defined in (14a) and (14b), respectively, satisfies*

$$\alpha < 1\,, \;\; L_{\boldsymbol{\Phi}}\alpha < 1/k \tag{17}$$

then there is a constant $C > 0$ such that the error of the modular method is bounded by

$$\|\boldsymbol{q}_{\mathrm{p}}(t) - \widetilde{\boldsymbol{q}}_{\mathrm{p}}(t)\| + \|\boldsymbol{q}_{\mathrm{c}}(t) - \widetilde{\boldsymbol{q}}_{\mathrm{c}}(t)\| + \|\boldsymbol{\lambda}(t) - \widetilde{\boldsymbol{\lambda}}(t)\| \leq C \max_{\{m:T_0+mH\leq t\}} \boldsymbol{\Delta}_m\,,$$

($t \in [t_0, t_{\mathrm{e}}]$), provided that $H > 0$ is sufficiently small. Here, $\boldsymbol{\Delta}_m$ denotes the extrapolation error in components $\boldsymbol{\lambda}$ and in $\boldsymbol{q}_{\mathrm{p}}$ and $\boldsymbol{q}_{\mathrm{c}}$, respectively, see (11).

Proof The proof of this convergence result is based on the analysis of the coupled error propagation in differential and algebraic solution components. The contractivity condition (17) allows to consider the error propagation in the differential solution components similar to the error analysis in the ODE case, see Section 3, and to use a contractivity argument [12] to guarantee a stable error propagation in the algebraic solution components. For modular time integration methods, such error analysis has been given for the first time in [7]; the extension to extrapolation operators with $k > 1$ is discussed in [3]. □

Preconditioning

In the simulation of pantograph–catenary interaction, the contractivity condition (17) is satisfied in the **Pantograph first** approach but violated if we start with the catenary. Guided by the convergence analysis, the modular method may be modified to enforce numerical stability and convergence in any case.

Starting with the catenary, the **Stabilized catenary first** approach is obtained from (15) substituting in the dynamical equations (15a) of the subsystem pantograph the numerical solution $\widetilde{\boldsymbol{\lambda}}$ by a suitable linear combination of $\widetilde{\boldsymbol{\lambda}}$ and $\boldsymbol{\Phi}_m^{\boldsymbol{\lambda}}[\widetilde{\boldsymbol{\lambda}}]$:

$$\mathbf{M}_{\mathrm{p}}(\widetilde{\boldsymbol{q}}_{\mathrm{p}})\ddot{\widetilde{\boldsymbol{q}}}_{\mathrm{p}} = \boldsymbol{f}_{\mathrm{p}}(t,\widetilde{\boldsymbol{q}}_{\mathrm{p}},\dot{\widetilde{\boldsymbol{q}}}_{\mathrm{p}}) - \mathbf{G}_{\mathrm{p}}^{\top}(t,\widetilde{\boldsymbol{q}}_{\mathrm{p}},\widetilde{\boldsymbol{q}}_{\mathrm{c}})\Big(\big(\mathbf{I}-\mathbf{A}(t)\big)\widetilde{\boldsymbol{\lambda}} + \mathbf{A}(t)\,\boldsymbol{\Phi}_m^{\boldsymbol{\lambda}}[\widetilde{\boldsymbol{\lambda}}]\Big)\,, \tag{18a}$$

$$\mathbf{M}_{\mathrm{c}}(\widetilde{\boldsymbol{q}}_{\mathrm{c}})\ddot{\widetilde{\boldsymbol{q}}}_{\mathrm{c}} = \boldsymbol{f}_{\mathrm{c}}(t,\widetilde{\boldsymbol{q}}_{\mathrm{c}},\dot{\widetilde{\boldsymbol{q}}}_{\mathrm{c}}) - \mathbf{G}_{\mathrm{c}}^{\top}(t,\boldsymbol{\Phi}_m^{\boldsymbol{q}}[\widetilde{\boldsymbol{q}}_{\mathrm{p}}],\widetilde{\boldsymbol{q}}_{\mathrm{c}})\,\boldsymbol{\Phi}_m^{\boldsymbol{\lambda}}[\widetilde{\boldsymbol{\lambda}}]\,, \tag{18b}$$

$$\mathbf{0} = \mathbf{G}_{\mathrm{p}}(t,\widetilde{\boldsymbol{q}}_{\mathrm{p}},\widetilde{\boldsymbol{q}}_{\mathrm{c}})\ddot{\widetilde{\boldsymbol{q}}}_{\mathrm{p}} + \mathbf{G}_{\mathrm{c}}(t,\widetilde{\boldsymbol{q}}_{\mathrm{p}},\widetilde{\boldsymbol{q}}_{\mathrm{c}})\ddot{\widetilde{\boldsymbol{q}}}_{\mathrm{c}} + \boldsymbol{g}^{(II)}(t,\widetilde{\boldsymbol{q}}_{\mathrm{p}},\widetilde{\boldsymbol{q}}_{\mathrm{c}},\dot{\widetilde{\boldsymbol{q}}}_{\mathrm{p}},\dot{\widetilde{\boldsymbol{q}}}_{\mathrm{c}}) \tag{18c}$$

with

$$\mathbf{A}(t) := -\big[\big(\mathbf{G}_{\mathrm{p}}\mathbf{M}_{\mathrm{p}}^{-1}\mathbf{G}_{\mathrm{p}}^{\top}\big)^{-1}\mathbf{G}_{\mathrm{c}}\mathbf{M}_{\mathrm{c}}^{-1}\mathbf{G}_{\mathrm{c}}^{\top}\big]\big(\boldsymbol{q}_{\mathrm{p}}(t),\boldsymbol{q}_{\mathrm{c}}(t)\big)\,. \tag{18d}$$

This *preconditioning* [1, 7] may be implemented efficiently and allows to avoid the contractivity condition (17) in Theorem 4.1. The theoretical convergence result is again nicely illustrated by numerical test results for the pantograph–catenary benchmark problem, see Fig. 4. As in Figs. 2 and 3, the most simple extrapolation scheme was used resulting in $\boldsymbol{\Phi}_m^{\boldsymbol{\lambda}}[\widetilde{\boldsymbol{\lambda}}] = \widetilde{\boldsymbol{\lambda}}_m$ etc. The global errors remain very small

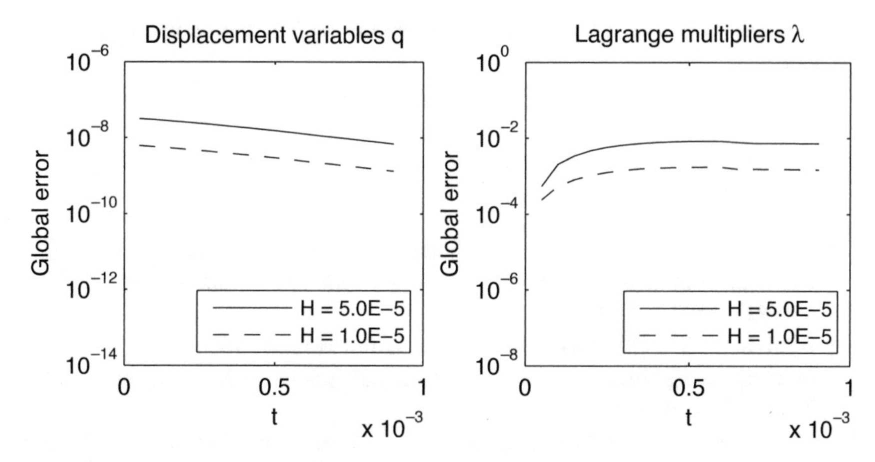

Fig. 4: **Stabilized catenary first** approach in the modular simulation of a benchmark problem for pantograph-catenary interaction [8]: Global errors in the displacement variables $\boldsymbol{q}_{\mathrm{p}}$, $\boldsymbol{q}_{\mathrm{c}}$ (left plot: $\|\boldsymbol{e}_m^{\boldsymbol{q}}\|$) and in the Lagrange multipliers $\boldsymbol{\lambda}$ (right plot: $\|\boldsymbol{e}_m^{\boldsymbol{\lambda}}\|$).

in all solution components and are approximately reduced by a factor of five if the macro step size is reduced from $H = 50\,\mu$s to $H = 10\,\mu$s, i.e., first order convergence may be observed numerically.

Related work

In the present paper, we discussed modular time integration methods for coupled problems in nonlinear system dynamics guided by an application scenario from computational mechanics. Similar mechanisms of error propagation and numerical stability are observed for problems from electrical circuit simulation [7]. In that field of application, waveform relaxation or dynamic iteration found special interest [7, 9, 10].

Matthies et al. [18] discuss coupling strategies for the modular simulation of fluid-structure interaction and observe similar stability problems as the ones in the **Catenary first** approach of Fig. 2. Part of our stability and convergence analysis can be extended to such coupled partial differential equations as well.

From the algorithmic viewpoint, Gauss-Seidel type iteration schemes are attractive for sequential computations and for coupled systems with a small number of subsystems. Alternatives include Jacobi type iteration [9] and overlapping iteration schemes [23]. A novel approach that is tailored to the coupling of beam like structures to multibody system models has been proposed by Schneider et al. [24].

The stability analysis of modular time integration methods in system dynamics underlines the central role of Jacobian matrices for a numerically stable time integration of coupled systems. The recently published industrial interface standard *FMI for Model Exchange and Co-Simulation* [14] provides interfaces to get subsystem Jacobians for advanced numerically stable master algorithms in co-simulation [11, 30].

References

[1] M. Arnold. A pre-conditioned method for the dynamical simulation of coupled mechanical multibody systems. *Z. Angew. Math. Mech.*, Supplement 3 to Vol. 80:S817–S818, 2000.

[2] M. Arnold. Simulator coupling – A straightforward approach to the simulation of coupled technical systems? 72thAnnual Scientific Meeting of GAMM, Zurich, February 2001.

[3] M. Arnold. Stability of sequential modular time integration methods for coupled multibody system models. *J. Comput. Nonlinear Dynam.*, 5:031003, 2010.

[4] M. Arnold. DAE aspects of multibody system dynamics. In A. Ilchmann and T. Reis, editors, *Surveys in Differential-Algebraic Equations IV*, Differential-Algebraic Equations Forum, pages 41–106. Springer International Publishing, Cham, 2017.

[5] M. Arnold, A. Cardona, and O. Brüls. A Lie algebra approach to Lie group time integration of constrained systems. In P. Betsch, editor, *Structure-Preserving Integrators in Nonlinear Structural Dynamics and Flexible Multibody Dynamics*, volume 565 of *CISM Courses and Lectures*, pages 91–158. Springer International Publishing, Cham, 2016.

[6] M. Arnold, C. Clauß, and T. Schierz. Error analysis and error estimates for co-simulation in FMI for Model Exchange and Co-Simulation v2.0. *Archive of Mechanical Engineering*, LX:75–94, 2013.

[7] M. Arnold and M. Günther. Preconditioned dynamic iteration for coupled differential-algebraic systems. *BIT Numerical Mathematics*, 41:1–25, 2001.

[8] M. Arnold and B. Simeon. Pantograph and catenary dynamics: a benchmark problem and its numerical solution. *Applied Numerical Mathematics*, 34:345–362, 2000.

[9] A. Bartel, M. Brunk, M. Günther, and S. Schöps. Dynamic iteration for coupled problems of electric circuits and distributed devices. *SIAM J. Sci. Comput.*, 35:B315–B335, 2013.

[10] A. Bartel, M. Brunk, and S. Schöps. On the convergence rate of dynamic iteration for coupled problems with multiple subsystems. *Journal of Computational and Applied Mathematics*, 262:14–24, 2014.

[11] T. Blochwitz, M. Otter, J. Åkesson, M. Arnold, C. Clauß, H. Elmqvist, M. Friedrich, A. Junghanns, J. Mauss, D. Neumerkel, H. Olsson, and A. Viel. Functional Mockup Interface 2.0: The standard for tool independent exchange of simulation models. In M. Otter and D. Zimmer, editors, *Proc. of the* 9th *International Modelica Conference, September 3–5, 2012*, Munich, Germany, 2012.

[12] P. Deuflhard, E. Hairer, and J. Zugck. One–step and extrapolation methods for differential–algebraic systems. *Numer. Math.*, 51:501–516, 1987.

[13] W. Duffek. Ein Fahrbahnmodell zur Simulation der dynamischen Wechselwirkung zwischen Fahrzeug und Fahrweg. Technical Report IB 515–91–18, DLR, D-5000 Köln 90, 1991.

[14] FMI. The Functional Mockup Interface. `https://www.fmi-standard.org/`.

[15] E. Hairer, S.P. Nørsett, and G. Wanner. *Solving Ordinary Differential Equations. I. Nonstiff Problems*. Springer–Verlag, Berlin Heidelberg New York, 2nd edition, 1993.

[16] E. Hairer and G. Wanner. *Solving Ordinary Differential Equations. II. Stiff and Differential-Algebraic Problems*. Springer–Verlag, Berlin Heidelberg New York, 2nd edition, 1996.

[17] W. Kortüm, W.O. Schiehlen, and M. Arnold. Software tools: From multibody system analysis to vehicle system dynamics. In H. Aref and J.W. Phillips, editors, *Mechanics for a New Millennium*, pages 225–238, Dordrecht, 2001. Kluwer Academic Publishers.

[18] H.G. Matthies, R. Niekamp, and J. Steindorf. Algorithms for strong coupling procedures. *Computer Methods in Applied Mechanics and Engineering*, 195:2028–2049, 2006.

[19] G. Poetsch. *Untersuchung und Verbesserung numerischer Verfahren zur Simulation von Stromabnehmer–Kettenwerk–Systemen*. PhD thesis, Universität Paderborn, Heinz–Nixdorf–Institut, 2000.

[20] G. Poetsch, J. Evans, R. Meisinger, W. Kortüm, W. Baldauf, A. Veitl, and J. Wallaschek. Pantograph/catenary dynamics and control. *Vehicle System Dynamics*, 28:159–195, 1997.

[21] W. Rulka. SIMPACK – A computer program for simulation of large–motion multibody systems. In W.O. Schiehlen, editor, *Multibody Systems Handbook*. Springer–Verlag, Berlin Heidelberg New York, 1990.

[22] W.O. Schiehlen, editor. *Multibody Systems Handbook*. Springer–Verlag, Berlin Heidelberg New York, 1990.

[23] T. Schierz and M. Arnold. Stabilized overlapping modular time integration of coupled differential-algebraic equations. *Applied Numerical Mathematics*, 62:1491–1502, 2012.

[24] F. Schneider, M. Burger, M. Arnold, and B. Simeon. A new approach for force-displacement co-simulation using kinematic coupling constraints. *ZAMM - Journal of Applied Mathematics and Mechanics / Zeitschrift für Angewandte Mathematik und Mechanik*, 97:1147–1166, 2017.

[25] B. Simeon. *Computational Flexible Multibody Dynamics: A Differential-Algebraic Approach*. Differential-Algebraic Equations Forum. Springer–Verlag, Berlin Heidelberg, 2013.

[26] B. Simeon and M. Arnold. Coupling DAE's and PDE's for simulating the interaction of pantograph and catenary. *Mathematical and Computer Modelling of Dynamical Systems*, 6:129–144, 2000.

[27] A. Veitl. *Integrierter Entwurf innovativer Stromabnehmer*. Fortschritt-Berichte VDI Reihe 12, Nr. 449. VDI–Verlag, Düsseldorf, 2001.

[28] A. Veitl and M. Arnold. Coupled simulation of multibody systems and elastic structures. In J.A.C. Ambrósio and W.O. Schiehlen, editors, *Advances in Computational Multibody Dynamics*, pages 635–644, IDMEC/IST Lisbon, Portugal, 1999.

[29] A. Veitl and M. Arnold. Joint work in an internal project on numerical aspects of simulator coupling SIMPACK – PrOSA. DLR German Aerospace Center, Vehicle System Dynamics Group Oberpfaffenhofen, 1999.

[30] A. Viel. Implementing stabilized co-simulation of strongly coupled systems using the Functional Mock-up Interface 2.0. Proc. of the 10th International Modelica Conference, March 10–12, 2014, Lund (Sweden), 2014.

[31] W. Walter. *Ordinary Differential Equations*. Number 182 in Graduate Texts in Mathematics. Springer, 1998.

Differential-Algebraic Equations and Beyond: From Smooth to Nonsmooth Constrained Dynamical Systems

Jan Kleinert and Bernd Simeon

Abstract In the 1970s of the last century, the progress in powerful simulation software for mechanical multibody systems and for electrical circuits led to a new class of models that is characterized by differential equations and algebraic constraints. These Differential-Algebraic Equations (DAEs) became soon a hot topic, and the methodology that has emerged since then represents now a general field in applied and computational mathematics, with various new applications in science and engineering. Taking a historical approach, the present article introduces a summarizing view at DAEs, with emphasis on numerical aspects and without aiming for completeness. Recent numerical methods for nonsmooth dynamical systems subject to unilateral contact and friction illustrate the topicality of this development.

1 Introduction

Differential-Algebraic Equations (DAEs) are a prominent example for application-driven research that leads to new concepts and methodology in mathematics. Until the early 1980s of the last century, this topic was widely unknown but the introduction of powerful simulation software in electrical and mechanical engineering created a strong demand for the analysis and numerical solution of dynamical systems with constraints. Mathematicians all over the world then started to work on DAEs, which resulted in an avalanche of research over the following decades. Meanwhile, most issues have been resolved and sophisticated numerical methods have been found, but despite this maturity, the field is constantly expanding due to the ongoing trend to model complex phenomena in science and engineering by means of differential

Jan Kleinert, Hochschule Bonn-Rhein-Sieg, Grantham-Allee 20, D-53757 Sankt Augustin, e-mail: jan.kleinert@h-brs.de and German Aerospace Center, Simulation and Software Technology, Linder Höhe, D-51147 Cologne, e-mail: jan.kleinert@dlr.de · Bernd Simeon, Felix-Klein-Zentrum, TU Kaiserslautern, D-67663 Kaiserslautern, Germany, e-mail: simeon@mathematik.uni-kl.de

M. Günther, W. Schilders (eds.), *Novel Mathematics Inspired by Industrial Challenges*, Mathematics in Industry 38, https://doi.org/10.1007/978-3-030-96173-2_4

equations and additional constraints that stem from network structures, boundary and coupling conditions or physical conservation properties.

In this paper, we strive for a survey of this development and even more also discuss recent extensions towards dynamical systems that are nonsmooth. The latter topic is quite timely and arises, e.g., in the modeling of granular material. Clearly, our approach is exemplary in nature and does not aim at completeness. There is, moreover, a strong bias on mechanical systems and the simulation of granular material. In the section on nonsmooth phenomena, the mathematical tools to deal with discontinuous phenomena is presented, while the intricate physical impact and friction models behind the nonsmooth equations are not discussed. Those readers who would like to know more about the topic of DAEs and the rich oeuvre that has accumulated over the years are referred to the monographs of Brenan, Campbell & Petzold [24], Griepentrog & März [60], Hairer & Wanner [69], Kunkel & Mehrmann [88], Lamour, März & Tischendorf [92], and to the survey of Rabier & Rheinboldt [142]. Refer to the textbooks by Acary and Brogliato [4, 27] and references therein for a more elaborate literature survey on nonsmooth dynamical systems.

The paper is organized as follows. Section 2 outlines the emergence of DAEs and shows how it was influenced by applications from electrical and mechanical engineering problems. Section 3 then highlights some major results and numerical methods that were developed in the DAE context. Some extensions to classical DAEs for partial and stochastic differential equations are discussed in Section 4. Section 5 discusses nonsmooth dynamical systems. An overview of the most important theoretical concepts is given in Section 5.1, the equations of motion for a unilaterally constrained mechanical system are motivated in Section 5.2 and finally, Section 5.3 provides some insights into numerical methods and recent developments in the field of nonsmooth dynamical systems.

2 Differential-algebraic equations

In this section, we describe how the DAEs became a hot topic in the 1980s and 1990s and then go further back in time to the works of Kirchhoff [80] and Lagrange [91] who introduced differential equations with constraints in order to model electrical circuits and mechanical systems.

2.1 How the topic of DAEs emerged

In the beginning of the 1980s, the term 'DAE' was widely unknown in mathematics. But this changed rapidly, due to an increasing demand in several engineering fields but also due to the pioneering work of Bill Gear. The first occurrence of the term *Differential-Algebraic Equation* can be found in the title of Gear's paper *Simultaneous numerical solution of differential-algebraic equations* [54] from 1971, and in the

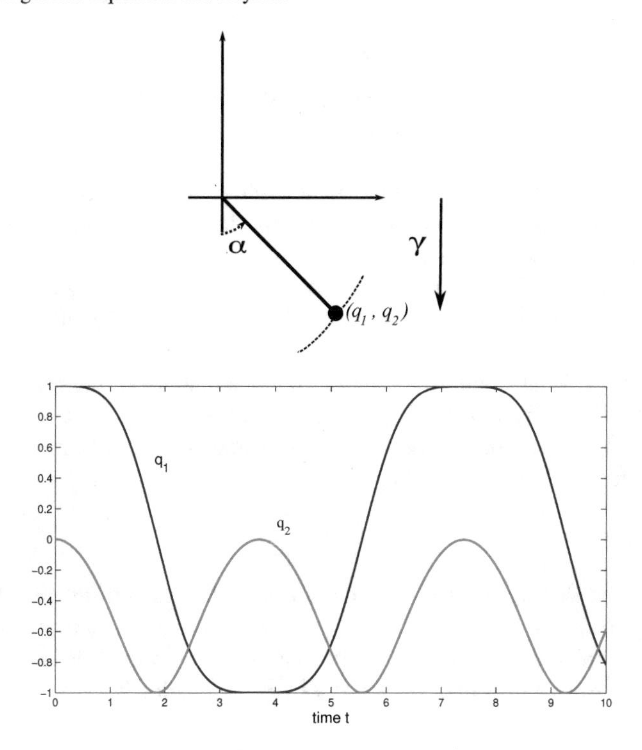

Fig. 1: The mathematical pendulum.

same year his famous book *Numerical Initial Value Problems in Ordinary Differential Equations* [53] appeared where he already considers examples from electrical circuit analysis in the form

$$E\dot{x} = \phi(x,t) \tag{1}$$

with possibly singular capacitance matrix $E \in \mathbb{R}^{n_x \times n_x}$ and right hand side function ϕ. For a regular matrix E, (1) can be easily converted into a system of ordinary differential equations (ODEs). Otherwise, however, a DAE arises that calls for new approaches, both in theory and in numerical analysis.

Two application fields, namely electrical circuit analysis and constrained mechanical systems, have been the major driving forces for the development of DAEs. Below, this statement will be made more explicit by looking at the corresponding modeling concepts. Bill Gear had the farsightedness to perceive very early the importance of these modeling approaches for today's simulation software. During an Oberwolfach workshop in 1981, he suggested to study the *mathematical pendulum in Cartesian coordinates*

$$\ddot{q}_1 = -2q_1\lambda, \tag{2a}$$
$$\ddot{q}_2 = -\gamma - 2q_2\lambda, \tag{2b}$$
$$0 = q_1^2 + q_2^2 - 1 \tag{2c}$$

that describes the motion of a mass point with coordinates (q_1, q_2) in the plane subject to a constraint. The constraint models the massless rod of length 1 that connects the mass point to a pivot placed in the origin of the coordinate system, Fig. 1. The motion of the mass point is then determined by the gravity (parameter γ) and by the constraint forces that are expressed in terms of the unknown Lagrange multiplier λ.

The DAE (2) is an example for the Lagrange equations of the first kind that we will discuss below. By introducing velocity variables, it can be easily converted to a system of first order that fits into the class of linear-implicit DAEs (1). The most general form of a DAE is a *fully implicit system*

$$F(\dot{x}, x, t) = 0 \tag{3}$$

with state variables $x(t) \in \mathbb{R}^{n_x}$ and a nonlinear, vector-valued function F of corresponding dimension. Clearly, if the $n_x \times n_x$ Jacobian $\partial F/\partial \dot{x}$ is invertible, then by the implicit function theorem, it is theoretically possible to transform (3), at least locally, to an explicit system of ODEs. If $\partial F/\partial \dot{x}$ is singular, however, (3) constitutes a DAE.

Linda Petzold, a student of Bill Gear, continued and extended his pioneering work in various directions. In particular, the development of the DASSL code (the 'Differential-Algebraic System SoLver') that she had started in the early 1980s [24, 135] set a corner stone that still persists today. DASSL is based on the Backward Differentiation Formulas (BDF), which are also popular for solving systems of stiff ODEs. The extension of the BDF methods to implicit systems (3) is intriguingly simple. One replaces the differential operator $\mathrm{d}/\mathrm{d}t$ in (3) by the difference operator

$$\rho x_{n+k} := \sum_{i=0}^{k} \alpha_i x_{n+i} = \tau \dot{x}(t_{n+k}) + \mathcal{O}(\tau^{k+1}) \tag{4}$$

where x_{n+i} stands for the discrete approximation of $x(t_{n+i})$ with stepsize τ and where the α_i, $i = 0, \ldots, k$, denote the method coefficients. Using the finite difference approximation $\rho x_{n+k}/\tau$ of the time derivative, the numerical solution of the DAE (3) then boils down to solving the nonlinear system

$$F\left(\frac{\rho x_{n+k}}{\tau}, x_{n+k}, t_{n+k}\right) = 0 \tag{5}$$

for x_{n+k} in each time step. This is exactly the underlying idea of the DASSL code, see Fig. 2.

Soon after the first release of the DASSL code, it became very popular among engineers and mathematicians. For quite some problems, however, the code would fail, which in turn triggered new research in numerical analysis in order to understand

```
      SUBROUTINE DDASSL (RES,NEQ,T,Y,YPRIME,TOUT,INFO,RTOL,ATOL,
     +   IDID,RWORK,LRW,IWORK,LIW,RPAR,IPAR,JAC)
C***BEGIN PROLOGUE  DDASSL
C***PURPOSE  This code solves a system of differential/algebraic
C            equations of the form G(T,Y,YPRIME) = 0.
```

Fig. 2: Calling sequence of the DASSL code [24, 135] that has had an enormous impact on the subject of DAEs and that is still in wide use today. The original code is written in FORTRAN77 in double precision. A recent implementation in C is part of the SUNDIALS suite of codes [76].

such phenomena. As it turned out, the notion of an *index* of the DAE (3) was the key to obtain further insight.

Gear [56, 57] introduced what we call today the *differentiation index*. This non-negative integer k is defined by

$k=0$: If $\partial F/\partial \dot{x}$ is non-singular, the index is 0.
$k>0$: Otherwise, consider the system of equations

$$\begin{aligned} F(\dot{x},x,t) &= 0,\\ \frac{\mathrm{d}}{\mathrm{d}t}F(\dot{x},x,t) = \frac{\partial}{\partial \dot{x}}F(\dot{x},x,t)x^{(2)} + \ldots &= 0, \\ &\vdots \\ \frac{\mathrm{d}^s}{\mathrm{d}t^s}F(\dot{x},x,t) = \frac{\partial}{\partial \dot{x}}F(\dot{x},x,t)x^{(s+1)} + \ldots &= 0 \end{aligned} \tag{6}$$

as a system in the separate dependent variables $\dot{x}, x^{(2)}, \ldots, x^{(s+1)}$, with x and t as independent variables. Then the index k is the smallest s for which it is possible, using algebraic manipulations only, to extract an ordinary differential equation $\dot{x} = \psi(x,t)$ (the underlying ODE) from (6).

Meanwhile other notions of an index have emerged, but despite its ambiguity with respect to the algebraic manipulations, the differentiation index is still the most popular and widespread tool to classify DAEs.

In the next chapter, other index concepts and their relation to the differential index will be addressed, and also more protagonists will enter the stage. This first section on the early days of DAEs closes now with a look at the application fields that set the ball rolling.

2.2 Electrical circuits

In 1847, Kirchhoff first published his *circuit laws* that describe the conservation properties of electrical circuits [80]. These laws consist of the current law and the

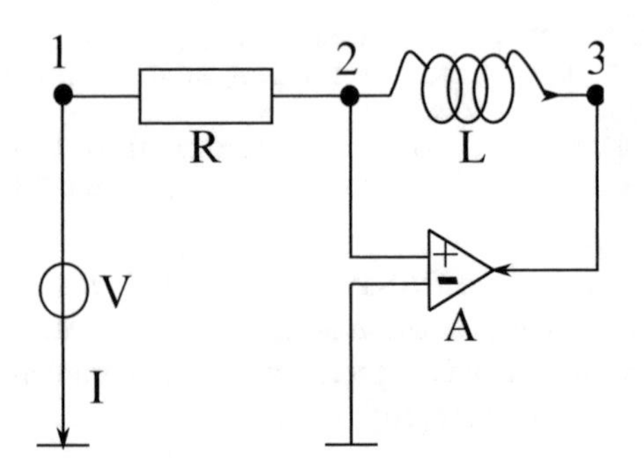

Fig. 3: Differentiator circuit.

voltage law, which both follow from Maxwell's equations of electro-dynamics. When these laws are applied to circuits with time-dependent behavior, the corresponding equations are typically given as a linear-implicit system (1). Often, the structure even turns out to be a linear constant coefficient DAE

$$E\dot{x} + Hx = c(t) \tag{7}$$

with matrices $E, H \in \mathbb{R}^{n_x \times n_x}$ and a time-dependent source term $c(t) \in \mathbb{R}^{n_x}$.

An example for such an electrical circuit is the *differentiator* [63] shown in Fig. 3. It consists of a resistance R, an inductance L, an ideal operational amplifier $A = \infty$, and a given voltage source $V(t)$. The $n_x = 6$ unknowns read here $x = (V_1, V_2, V_3, I, I_L, I_V)$ with voltages V_i and currents I, I_L, I_V. From Kirchhoff's laws and the properties of the amplifier and the inductance one obtains the relations

$$\begin{aligned} I + (V_1 - V_2)/R &= 0, \\ -(V_1 - V_2)/R + I_L &= 0, \\ -I_L + I_V &= 0, \\ V_1 &= V(t), \\ V_2 &= 0, \\ V_2 - V_3 &= L \cdot \dot{I}_L. \end{aligned}$$

This linear system has the form (7) with singular inductance matrix

$$E = \begin{pmatrix} 0 & 0 & \dots & 0 & 0 \\ \vdots & \vdots & \ddots & \vdots & \vdots \\ 0 & 0 & \dots & 0 & 0 \\ 0 & 0 & 0 & L & 0 \end{pmatrix}. \tag{8}$$

If the matrix E was regular, it could be brought to the right hand side by formal inversion, ending up in a system of ODEs. Here however, it is singular and thus we face a DAE problem.

Weierstrass and Kronecker were in Berlin at the same time as Kirchhoff, and it is quite obvious to suppose that they knew his work. [1] Weierstrass and later Kronecker were thus inspired to study such singular systems and provided an elegant theory that is still fundamental today in order to understand the specific properties of DAEs.

We assume that the *matrix pencil* $\mu E + H \in \mathbb{R}^{n_x \times n_x}[\mu]$ is regular. I.e., there exists $\mu \in \mathbb{C}$ such that the matrix $\mu E + H$ is regular. Otherwise, the pencil is singular, and (7) has either no or infinitely many solutions. This latter case has been first studied by Kronecker [87], see also [33, 51].

If $\mu E + H$ is regular, there exist nonsingular matrices U and V such that

$$UEV = \begin{pmatrix} I & 0 \\ 0 & N \end{pmatrix}, \quad UHV = \begin{pmatrix} C & 0 \\ 0 & I \end{pmatrix} \tag{9}$$

where N is a nilpotent matrix, I the identity matrix, and C a matrix that can be assumed to be in Jordan canonical form. Note that the dimensions of these square blocks in (9) are uniquely determined. The transformation (9) is called the *Weierstrass canonical form* [168]. It is a generalization of the Jordan canonical form and contains the essential structure of the linear system (7).

In the Weierstrass canonical form (9), the singularity of the DAE is represented by the nilpotent matrix N. Its degree of nilpotency, i.e., the smallest positive integer k such that $N^k = 0$, plays a key role when studying closed-form solutions of the linear system (7) and is identical to the differentiation index of (7).

To construct a solution of (7), we introduce new variables and right hand side vectors

$$V^{-1}x =: \begin{pmatrix} y \\ z \end{pmatrix}, \quad Uc =: \begin{pmatrix} \delta \\ \theta \end{pmatrix}. \tag{10}$$

Premultiplying (7) by U then leads to the *decoupled system*

$$\dot{y} + Cy = \delta\,, \tag{11a}$$

$$N\dot{z} + z = \theta\,. \tag{11b}$$

While the solution of the ODE (11a) follows by integrating and results in an expression based on the matrix exponential $\exp(-C(t-t_0))$, the equation (11b) for z can be solved recursively by differentiating. More precisely, it holds

$$N\ddot{z} + \dot{z} = \dot{\theta} \quad \Rightarrow \quad N^2\ddot{z} = -N\dot{z} + N\dot{\theta} = z - \theta + N\dot{\theta}\,.$$

Repeating the differentiation and multiplication by N, we can eventually exploit the nilpotency and get

$$0 = N^k z^{(k)} = (-1)^k z + \sum_{\ell=0}^{k-1} (-1)^{k-1-\ell} N^\ell \theta^{(\ell)}.$$

[1] The relation of the work of Weierstrass and Kronecker to Kirchhoff's circuit laws was pointed out to us by Volker Mehrmann when we met in September 2014 during a Summer School on DAEs in Elgersburg, Germany.

This implies the explicit representation

$$z = \sum_{\ell=0}^{k-1} (-1)^{\ell} N^{\ell} \theta^{(\ell)}. \tag{12}$$

The above solution procedure illustrates several crucial points about DAEs and how they differ from ODEs. Remarkably, the linear constant coefficient case displays already these points, and thus the work of Weierstrass and Kronecker represents still the fundamental of DAE theory today.

We highlight two crucial points:

(i) The solution of (7) rests on $k-1$ differentiation steps. This requires that the derivatives of certain components of θ exist up to $\ell = k-1$. Furthermore, some components of z may only be continuous but not differentiable depending on the smoothness of θ.
(ii) The components of z are directly given in terms of the right hand side data θ and its derivatives. Accordingly, the initial value $z(t_0) = z_0$ is fully determined by (12) and, in contrast to y_0, cannot be chosen arbitrarily. Initial values (y_0, z_0) where z_0 satisfies (12) are called *consistent*. The same terminology applies to the initial value x_0, which is consistent if, after the transformation (10), z_0 satisfies (12).

Today, more than 150 years after the discoveries of Kirchhoff, electrical circuit analysis remains one of the driving forces in the development of DAEs. The interplay of modeling and mathematical analysis is particularly important in this field, and the interested reader is referred to Günther & Feldmann [62] and März & Tischendorf [107] as basic works. The first simulation code that generated a model in differential-algebraic form was the SPICE package [122].

2.3 Constrained mechanical systems

Even older than the DAEs arising from Kirchhoff's laws are the Euler-Lagrange equations. They were first published in Lagrange's famous work *Mécanique analytique* [91] from 1788.

Consider a mechanical system that consists of rigid bodies interacting via springs, dampers, joints, and actuators, Fig. 4. The bodies possess a certain geometry and mass while the interconnection elements are massless. Let $q(t) \in \mathbb{R}^{n_q}$ denote a vector that comprises the coordinates for position and orientation of all bodies in the system. Revolute, translational, universal, and spherical joints are examples for bondings in such a multibody system. They may constrain the motion q and hence determine its kinematics.

If constraints are present, we express the resulting conditions on q in terms of n_λ constraint equations

$$0 = g(q). \tag{13}$$

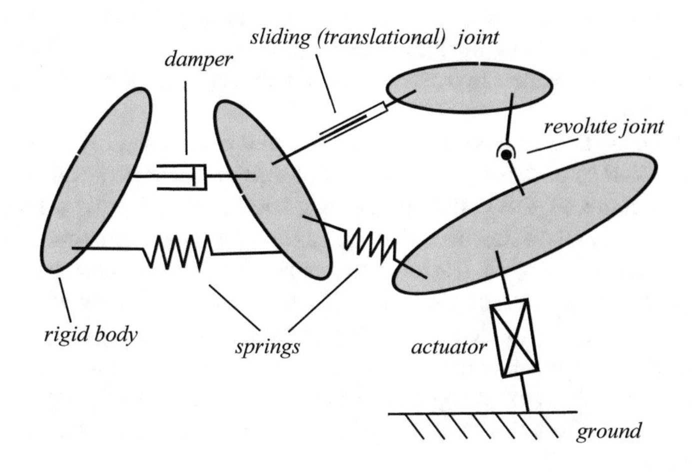

Fig. 4: Sketch of a multibody system with rigid bodies and typical interconnections.

Obviously, a meaningful model requires $n_\lambda < n_q$. The equations (13) that restrict the motion q are called *holonomic constraints*, and the rectangular matrix

$$G(q) := \frac{\partial g(q)}{\partial q} \in \mathbb{R}^{n_\lambda \times n_q}$$

is called the *constraint Jacobian*.

Using both the redundant position variables q and additional Lagrange multipliers λ to describe the dynamics leads to the *equations of constrained mechanical motion*, also called the *Lagrange equations of the first kind* or the *Euler-Lagrange equations*

$$M(q)\ddot{q} = f(q,\dot{q},t) - G(q)^T\lambda\,, \tag{14a}$$

$$0 = g(q)\,, \tag{14b}$$

where $M(q) \in \mathbb{R}^{n_q \times n_q}$ stands for the *mass matrix* and $f(q,\dot{q},t) \in \mathbb{R}^{n_q}$ for the vector of *applied and internal forces*.

The standard example for such a constrained mechanical system are the equations (2) of the mathematical pendulum. For a long time, it was common sense that the Euler-Lagrange equations should be transformed to the *state space form*, also called the *Lagrange equations of the second kind*. In case of the pendulum, this means that the Cartesian coordinates can be expressed as $q_1 = \sin\alpha$, $q_2 = -\cos\alpha$ with the angle α as minimal coordinate, Fig. 1. By inserting these relations into (2), the constraints and the Lagrange multiplier cancel, and one arrives at the second order ODE

$$\ddot{\alpha} = -\gamma \sin\alpha \tag{15}$$

as state space form.

It seems obvious that a state space form such as (15) constitutes a more appropriate and easier model than the differential-algebraic system (14), or (2), respectively, in redundant coordinates. In practice, however, the state space form suffers from serious drawbacks:

The analytical complexity of the constraint equations (13) makes it in various applications impossible to obtain a set of minimal coordinates that is valid for all configurations of the multibody system. Moreover, although we know from the theorem on implicit functions that such a set exists in a neighborhood of the current configuration, it might loose its validity when the configuration changes. This holds in particular for multibody systems with so-called *closed kinematic loops*.

Even more, the modeling of subsystems like electrical and hydraulic feedback controls, which are essential for the performance of modern mechanical systems, is limited. The differential-algebraic model, on the other hand, bypasses topological analysis and offers the choice of using a set of coordinates q that possess physical significance.

This reasoning in favor of the differential-algebraic model (14) became more and more widespread in the 1980s, driven by the development of sophisticated software packages, so-called *multibody formalisms*. One of the first packages that fully exploited this new way of modeling is due to Haug [72].

A look at the leading software tools in the field today shows a clear picture. Some of the codes generate a differential-algebraic model whenever a constraint is present, while others try to generate a state space form as long as it is convenient. But the majority of the commercial products relies on the differential-algebraic approach as the most general way to handle complex technical applications [52, 150].

The main difference between the DAEs arising from electrical circuit analysis and the DAEs that model constrained mechanical systems is the richer structure of the latter. E.g., for *conservative multibody systems*, i.e., systems where the applied forces can be written as the gradient of a potential U, the Euler-Lagrange equations (14) result from Hamilton's principle of least action

$$\int_{t_0}^{t_1} \left(T - U - g(q)^T \lambda \right) \mathrm{d}t \to \text{ stationary !} \tag{16}$$

where the kinetic energy possesses a representation as quadratic form

$$T(q,\dot{q}) = \frac{1}{2}\dot{q}^T M(q)\dot{q}\,.$$

In the least action principle (16), we observe the fundamental Lagrange multiplier technique for coupling constraints and dynamics [26]. Extensions of the multiplier technique exist in various more general settings such as dissipative systems or even inequality constraints, see Section 5.

The pendulum equations (2) are the example for a constrained mechanical system. Though they simply describe the motion of a single mass point, several key properties of the Euler-Lagrange equations can already be studied: the differential equations are of second order, the constraint equations are mostly nonlinear, and one observes a clear semi-explicit structure with differential variables q and algebraic variables λ.

The Euler-Lagrange equations are of index 3 and form the prototype for a system of higher index. Index reduction techniques are thus required and in fact, already in 1972 this issue was addressed by Baumgarte [21]. He observed that in (14), the Lagrange multipliers can be eliminated by differentiating the constraints twice. The first differentiation leads to the constraints at velocity level

$$0 = \frac{\mathrm{d}}{\mathrm{d}t} g(q) = G(q)\,\dot{q}. \tag{17}$$

A second differentiation step yields the *constraints at acceleration level*

$$0 = \frac{\mathrm{d}^2}{\mathrm{d}t^2} g(q) = G(q)\,\ddot{q} + \kappa(q,\dot{q}), \quad \kappa(q,\dot{q}) := \frac{\partial G(q)}{\partial q}(\dot{q},\dot{q}), \tag{18}$$

where the two-form κ comprises additional derivative terms. The combination of the dynamic equation

$$M(q)\,\ddot{q} = f(q,\dot{q},t) - G(q)^T\lambda$$

with (18) results in a linear system for $\ddot{q}$ and λ with the saddle point matrix

$$\begin{pmatrix} M(q) & G(q)^T \\ G(q) & 0 \end{pmatrix} \in \mathbb{R}^{(n_q+n_\lambda)\times(n_q+n_\lambda)}. \tag{19}$$

For a well-defined multibody system, this matrix is invertible in a neighborhood of the solution, and in this way, the Lagrange multiplier can be computed as a function of q and $\dot{q}$.

However, the well-known drift-off phenomenon requires additional stabilization measures, and Baumgarte came up with the idea to combine original and differentiated constraints as

$$0 = G(q)\,\ddot{q} + \kappa(q,\dot{q}) + 2\alpha G(q)\dot{q} + \beta^2 g(q) \tag{20}$$

with scalar parameters α and β. The free parameters α and β should be chosen in such a way that

$$0 = \ddot{w} + 2\alpha\dot{w} + \beta^2 w \tag{21}$$

becomes an asymptotically stable equation, with $w(t) := g(q(t))$.

From today's perspective, the crucial point in Baumgarte's approach is the choice of the parameters. Nevertheless, it was the very beginning of a long series of works that tried to reformulate the Euler-Lagrange equations in such a way that the index is lowered while still maintaining the information of all constraint equations. For a detailed analysis of this stabilization and related techniques we refer to Ascher et al. [16, 17].

Another – very early – stabilization of the Euler-Lagrange equations is due to Gear, Gupta & Leimkuhler [55]. This formulation represents still the state-of-the-art in multibody dynamics. It uses a formulation of the equations of motion as system of first order with velocity variables $v = \dot{q}$ and simultaneously enforces the constraints at velocity level (17) and the position constraints (13), where the latter are interpreted as invariants and appended by means of extra Lagrange multipliers.

In this way, one obtains an enlarged system

$$\begin{aligned} \dot{q} &= v - G(q)^T \mu\,, \\ M(q)\,\dot{v} &= f(q,v,t) - G(q)^T \lambda\,, \\ 0 &= G(q)\,v\,, \\ 0 &= g(q) \end{aligned} \tag{22}$$

with additional multipliers $\mu(t) \in \mathbb{R}^{n_\lambda}$. A straightforward calculation shows

$$0 = \frac{\mathrm{d}}{\mathrm{d}t}\, g(q) = G(q)\dot{q} = G(q)\,v - G(q)G^T(q)\mu = -G(q)G^T(q)\mu$$

and one concludes $\mu = 0$ since $G(q)$ is of full rank and hence $G(q)G^T(q)$ invertible. With the additional multipliers μ vanishing, (22) and the original equations of motion (14) coincide along any solution. Yet, the index of the *GGL formulation* (22) is 2 instead of 3. Some authors refer to (22) also as *stabilized index-2 system* [50].

The last paragraphs on stabilized formulations of the Euler-Lagrange equations demonstrate that the development of theory and numerical methods for DAEs was strongly intertwined with the mathematical models. This holds for all application fields where DAEs arise. Even more, the application fields typically provide rich structural features of the model equations that are crucial for determining the index and for the numerical treatment.

Parallel to these developments in mechanical multibody systems, general multidisciplinary industrial simulation tools for nonlinear system modelling, such as Modelica [109], became a strong driving force for the topic of DAEs and made the methodology available in various other fields. For such powerful software tools, it is often natural and straightforward to use a redundant description in combination with coupling conditions between subsystems. In the end, however, the resulting DAE might possess a high index that calls for index reduction and stabilization.

3 Major results and numerical methods

Between 1989 and 1996, both theory and numerical analysis of DAEs were booming, and many groups in mathematics and engineering started to explore this new research topic. Driven by the development of powerful simulation packages in the engineering sciences, the demand for efficient and robust integration methods was growing steadily while at the same time, it had become apparent that higher index problems require stabilization measures or appropriate reformulations.

3.1 Perturbation index and implicit Runge-Kutta methods

The groundbreaking monograph on *The Numerical Solution of Differential-Algebraic Equations by Runge-Kutta Methods* [65] by Ernst Hairer, Christian Lubich and Michel Roche presented several new method classes, a new paradigm for the construction of convergence proofs, a new index concept, and the new RADAU5 code. From then on, Hairer and Lubich played a very strong role in the further development of DAEs and corresponding numerical methods.

The *perturbation index* as defined in [65] sheds a different light on DAEs and adopts the idea of a well-posed mathematical model. While the differential index is based on successively differentiating the original DAE (3) until the obtained system can be solved for $\dot{x}$, the perturbation index measures the sensitivity of the solutions to perturbations in the equation:

The system $F(\dot{x},x,t)=0$ has perturbation index $k \geq 1$ along a solution $x(t)$ on $[t_0,t_1]$ if k is the smallest integer such that, for all functions $\widehat{x}$ having a defect

$$F(\dot{\widehat{x}},\widehat{x},t)=\delta(t)\,,$$

there exists on $[t_0,t_1]$ an estimate

$$\|\widehat{x}(t)-x(t)\| \leq c\Big(\|\widehat{x}(t_0)-x(t_0)\| + \max_{t_0\leq\xi\leq t}\|\delta(\xi)\| + \ldots + \max_{t_0\leq\xi\leq t}\|\delta^{(k-1)}(\xi)\|\Big)$$

whenever the expression on the right hand side is sufficiently small. Note that the constant c depends only on F and on the length of the interval, but not on the perturbation δ. The perturbation index is $k=0$ if

$$\|\widehat{x}(t)-x(t)\| \leq c\Big(\|\widehat{x}(t_0)-x(t_0)\| + \max_{t_0\leq\xi\leq t}\|\int_{t_0}^{\xi}\delta(\tau)\,\mathrm{d}\tau\|\Big)\,,$$

which is satisfied for ordinary differential equations.

If the perturbation index exceeds $k=1$, derivatives of the perturbation show up in the estimate and indicate a certain degree of ill-posedness. E.g., if δ contains a small high frequency term $\varepsilon \sin \omega t$ with $\varepsilon \ll 1$ and $\omega \gg 1$, the resulting derivatives will induce a severe amplification in the bound for $\widehat{x}(t)-x(t)$.

Unfortunately, the differential and the perturbation index are not equivalent in general and may even differ substantially [32].

The definition of the perturbation index is solely a prelude in [65]. As the title says, most of the monograph deals with Runge-Kutta methods, in particular implicit ones. These are extended to linear-implicit systems $E\dot{x}=\phi(x,t)$ by discretizing $\dot{x}=E^{-1}\phi(x,t)$ and assuming for a moment that the matrix E is invertible. Multiplying the resulting scheme by E, one gets the method definition

$$EX_i = Ex_0 + \tau \sum_{j=1}^{s} a_{ij}\phi(X_j, t_0 + c_j\tau), \quad i = 1,\ldots,s; \tag{23a}$$

$$x_1 = \left(1 - \sum_{i,j=1}^{s} b_i\gamma_{ij}\right) x_0 + \tau \sum_{i,j=1}^{s} b_i\gamma_{ij}X_j. \tag{23b}$$

Here, the method coefficients are denoted by $(a_{ij})_{i,j=1}^{s}$ and $b_1,\ldots,b_s$ while $(\gamma_{ij}) = (a_{ij})^{-1}$ is the inverse of the coefficient matrix, with s being the number of stages. Obviously, (23) makes sense also in the case where E is singular.

Using *stiffly accurate methods* for differential-algebraic equations is advantageous, which becomes evident if we consider the discretization of the semi-explicit system

$$\dot{y} = a(y,z), \tag{24a}$$

$$0 = b(y,z) \tag{24b}$$

with differential variables y and algebraic variables z. The method (23) then reads

$$Y_i = y_0 + \tau \sum_{j=1}^{s} a_{ij}a(Y_j, Z_j), \quad i = 1,\ldots,s, \tag{25a}$$

$$0 = b(Y_i, Z_i), \tag{25b}$$

for the internal stages and

$$y_1 = y_0 + \tau \sum_{j=1}^{s} b_j a(Y_j, Z_j), \tag{26a}$$

$$z_1 = \left(1 - \sum_{i,j=1}^{s} b_i\gamma_{ij}\right) z_0 + \tau \sum_{i,j=1}^{s} b_i\gamma_{ij}Z_j \tag{26b}$$

as update for the numerical solution after one step. For stiffly accurate methods, we have $\sum_{i,j=1}^{s} b_i\gamma_{ij} = 1$ and $y_1 = Y_s, z_1 = Z_s$. The update (26) is hence superfluous and furthermore, the constraint $0 = b(y_1, z_1)$ is satisfied by construction.

It is not the purpose of this article to dive further into the world of Runge-Kutta methods, but like in numerical ODEs, the rivalry between multistep methods and Runge-Kutta methods also characterizes the situation for DAEs. While Linda Petzold's DASSL code is the most prominent multistep implementation, the RADAU5 and RADAU codes [69, 70] represent the one-step counter parts and have also become widespread in various applications.

The competition for the best code was a major driving force in the numerical analysis of DAEs, and from time to time those in favor of multistep methods looked also at one-step methods, e.g., in [18], and vice versa.

Simultaneously to the joint work with Ernst Hairer and Michel Roche, Christian Lubich investigated a different class of discretization schemes, the *half-explicit methods* [102]. These methods are tailored for semi-explicit DAEs and discretize the differential equations explicitly while the constraint equations are enforced in an implicit fashion. As example, consider the Euler-Lagrange equations (14) with velocity constraint (17). The half-explicit Euler method as generic algorithm for the method class reads

$$\begin{aligned} q_{n+1} &= q_n + \tau v_n\,, \\ M(q_n)\,v_{n+1} &= M(q_n)\,v_n + \tau f(q_n, v_n, t_n) - \tau G(q_n)^T \lambda_n\,, \\ 0 &= G(q_{n+1})\,v_{n+1}\,. \end{aligned} \tag{27}$$

Only a linear system of the form

$$\begin{pmatrix} M(q_n) & G(q_n)^T \\ G(q_{n+1}) & 0 \end{pmatrix} \begin{pmatrix} v_{n+1} \\ \tau\lambda_n \end{pmatrix} = \begin{pmatrix} M(q_n)v_n + \tau f(q_n, v_n, t_n) \\ 0 \end{pmatrix}$$

arises here in each step. The scheme (27) forms the basis for a class of extrapolation methods [102, 104], and also for half-explicit Runge-Kutta methods as introduced in [65] and then further enhanced by Brasey & Hairer [23] and Arnold & Murua [14].

These methods have in common that only information of the velocity constraints is required. As remedy for the drift off, which grows only linearly but might still be noticeable, the following projection, which is also due to Lubich [102], can be applied: Let q_{n+1} and v_{n+1} denote the numerical solution of the system, obtained by integration from consistent values q_n and v_n. Then, the projection consists of the following steps:

$$\textit{solve} \begin{cases} 0 = M(\widetilde{q}_{n+1})(\widetilde{q}_{n+1} - q_{n+1}) + G(\widetilde{q}_{n+1})^T \mu, \\ 0 = g(\widetilde{q}_{n+1}) \end{cases} \textit{for } \widetilde{q}_{n+1}, \mu\,; \tag{28a}$$

$$\textit{solve} \begin{cases} 0 = M(\widetilde{q}_{n+1})(\widetilde{v}_{n+1} - v_{n+1}) + G(\widetilde{q}_{n+1})^T \eta, \\ 0 = G(\widetilde{q}_{n+1})\,\widetilde{v}_{n+1} \end{cases} \textit{for } \widetilde{v}_{n+1}, \eta\,. \tag{28b}$$

A simplified Newton method can be used to solve the nonlinear system (28a) while (28b) represents a linear system for $\widetilde{v}_{n+1}$ and η with similar structure.

The projection can also be employed for stabilizing the equations of motion with acceleration constraint (18) where the position and velocity constraints are invariants and not preserved by the time integration, see Eich [43] and von Schwerin [155]. Such projection methods are particularly attractive in combination with explicit ODE integrators.

3.2 DAEs and differential geometry

Already in 1984, Werner Rheinboldt had investigated DAEs from the viewpoint of differential geometry [147]. While the approaches discussed so far are mainly inspired by differential calculus and algebraic considerations, a fundamentally different aspect comes into play by his idea of *differential equations on manifolds*.

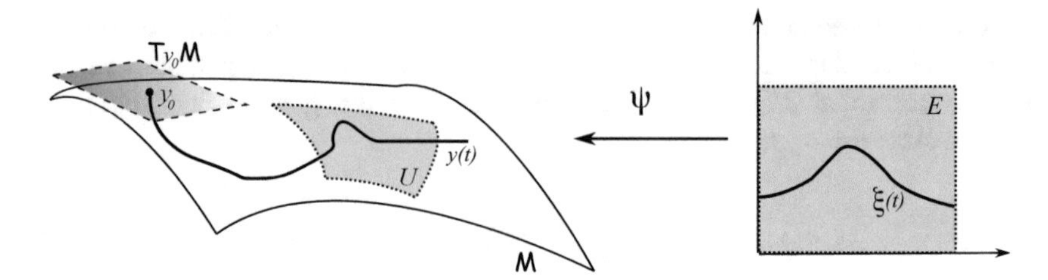

Fig. 5: Manifold $\mathscr{M}$, tangent space $\mathscr{T}_y\mathscr{M}$, and local parametrization.

Referring to [1, 15] for the theoretical underpinnings, we shortly illustrate this approach by considering the semi-explicit system

$$\dot{y} = a(y,z), \tag{29a}$$
$$0 = b(y) \tag{29b}$$

under the assumption

$$\frac{\partial b}{\partial y}(y) \cdot \frac{\partial a}{\partial z}(y,z) \in \mathbb{R}^{n_z \times n_z} \quad \text{is invertible} \tag{30}$$

in a neighborhood of the solution. Clearly, (29) is of index 2 where the constraint $0 = b(y)$, assuming sufficient differentiability, defines the manifold

$$\mathscr{M} := \{y \in \mathbb{R}^{n_y} : b(y) = 0\}. \tag{31}$$

The full rank condition (30) for the matrix product $\partial b/\partial y \cdot \partial a/\partial z$ implies that the Jacobian $B(y) = \partial b(y)/\partial y \in \mathbb{R}^{n_z \times n_y}$ possesses also full rank n_z. Hence, for fixed $y \in \mathscr{M}$, the tangent space

$$\mathscr{T}_y\mathscr{M} := \{v \in \mathbb{R}^{n_y} : B(y)v = 0\} \tag{32}$$

is the kernel of B and has the same dimension $n_y - n_z$ as the manifold $\mathscr{M}$. Fig. 5 depicts $\mathscr{M}$, $\mathscr{T}_y\mathscr{M}$, and a solution of the DAE (29), which, starting from a consistent initial value, is required to proceed on the manifold.

The differential equation on the manifold $\mathscr{M}$ that is equivalent to the DAE (29) is obtained as follows: The hidden constraint

$$0 = B(y)a(y,z)$$

can be solved for $z(y)$ according to the rank condition (30) and the implicit function theorem. Moreover, for $y \in \mathscr{M}$ it holds $a(y,z(y)) \in \mathscr{T}_y\mathscr{M}$, which defines a vector field on the manifold $\mathscr{M}$. Overall,

$$\dot{y} = a(y,z(y)) \quad \text{for } y \in \mathscr{M} \tag{33}$$

represents then a differential equation on the manifold [147].

In theory, and also computationally [148], it is possible to transform the differential equation (33) from the manifold to an ordinary differential equation in a linear space of dimension $n_y - n_z$. For this purpose, one introduces a *local parametrization*

$$\psi : E \to \mathscr{U} \tag{34}$$

where E is an open subset of $\mathbb{R}^{n_y - n_z}$ and $\mathscr{U} \subset \mathscr{M}$, see Fig. 5. Such a parametrization is not unique and holds only locally in general. It is, however, possible to extend it to a family of parametrizations such that the whole manifold is covered. For $y \in \mathscr{U}$ and local coordinates $\xi \in E$ we thus get the relations

$$y = \psi(\xi), \qquad \dot{y} = \Psi(\xi)\dot{\xi}, \quad \Psi(\xi) := \frac{\partial \psi}{\partial \xi}(\xi) \in \mathbb{R}^{n_y \times (n_y - n_z)}.$$

Premultiplying (33) by the transpose of the Jacobian $\Psi(\xi)$ of the parametrization and substituting y by $\psi(\xi)$, we arrive at

$$\Psi(\xi)^T \Psi(\xi)\dot{\xi} = \Psi(\xi)^T a(\psi(\xi), z(\psi(\xi))). \tag{35}$$

Since the Jacobian Ψ has full rank for a valid parametrization, the matrix $\Psi^T \Psi$ is invertible, and (35) constitutes the desired ordinary differential equation in the local coordinates ξ. In analogy to a mechanical system in minimal coordinates, we call (35) a *local state space form*.

The process of transforming a differential equation on a manifold to a local state space form constitutes a *push forward* operator, while the reverse mapping is called a *pull back* operator [1]. It is important to realize that the previously defined concept of an index does not appear in the theory of differential equations on manifolds. Finding hidden constraints by differentiation, however, is also crucial for the classification of DAEs from a geometric point of view.

The geometrical viewpoint was also considered very early by Sebastian Reich [143], but its full potential became clear only a couple of years later when the topic of geometric numerical integration emerged, cf. [66]. We furthermore remark that solving DAEs by structural analysis as presented in the Nedialkov-Pryce theory [125, 126, 141] can be seen as a related approach. Also, index reduction by dummy derivatives, Mattson & Söderlind [108], is in effect a practical implementation of choosing and switching between local coordinate systems.

3.3 Singularly perturbed problems and regularization

In the early days of DAEs, regularization was a quite popular means to convert the algebraic part into a differential equation. Motivated by physical examples such as stiff springs or parasitic effects in electrical circuits, a number of authors have looked into this topic. Furthermore, it is also interesting to start with a singularly perturbed

ODE, discretize it, and then to analyze the behavior of exact and numerical solution in the limit case.

To study an example for a semi-explicit system, we consider Van der Pol's equation

$$\varepsilon \ddot{q} + (q^2 - 1)\dot{q} + q = 0 \tag{36}$$

with parameter $\varepsilon > 0$. This is an oscillator equation with a nonlinear damping term that acts as a controller. For large amplitudes $q^2 > 1$, the damping term introduces dissipation into the system while for small values $q^2 < 1$, the sign changes and the damping term is replaced by an excitation, leading thus to a self-exciting oscillator. Introducing Liénhard's coordinates [68]

$$z := q, \quad y := \varepsilon \dot{z} + (z^3/3 - z),$$

we transform (36) into the first order system

$$\dot{y} = -z, \tag{37a}$$

$$\varepsilon \dot{z} = y - \frac{z^3}{3} + z. \tag{37b}$$

The case $\varepsilon \ll 1$ is of special interest. In the limit $\varepsilon = 0$, the equation (37b) turns into a constraint and we arrive at the semi-explicit system

$$\dot{y} = -z, \tag{38a}$$

$$0 = y - \frac{z^3}{3} + z. \tag{38b}$$

In other words, Van der Pol's equation (37) in Liénhard's coordinates is an example of a *singularly perturbed system* which tends to the semi-explicit DAE (38) when $\varepsilon \to 0$.

Such a close relation between a singularly perturbed system and a differential-algebraic equation is quite common and can be found in various application fields. Often, the parameter ε stands for an almost negligible physical quantity or the presence of strongly different time scales. Analyzing the *reduced system*, in this case (38), usually proves successful to gain a better understanding of the original perturbed equation [128]. In the context of regularization methods, this relation is also exploited, but in reverse direction [71]. One starts with a DAE such as (38) and replaces it by a singularly perturbed ODE, in this case (37).

In numerical analysis, the derivation and study of integration schemes via a singularly perturbed ODE has been termed the *indirect approach* [65] and lead to much additional insight [64, 101, 103], both for the differential-algebraic equation as limit case and for the stiff ODE case. A particularly interesting method class for the indirect approach are Rosenbrock methods as investigated by Rentrop, Roche & Steinebach [146].

3.4 General fully implicit DAEs

In contrast to the solution theory in the linear constant coefficient case, the treatment of fully implicit DAEs without a given internal structure is still challenging, even from today's perspective. For this purpose, Campbell [34] introduced the derivative array as key concept that carries all the information of the DAE system. The derivative array is constructed from the definition of the differential index, i.e., one considers the equations

$$\begin{aligned} F(\dot{x},x,t) &= 0, \\ \frac{\mathrm{d}}{\mathrm{d}t}F(\dot{x},x,t) = \frac{\partial}{\partial \dot{x}}F(\dot{x},x,t)\,x^{(2)} + \ldots &= 0, \\ &\vdots \\ \frac{\mathrm{d}^k}{\mathrm{d}t^k}F(\dot{x},x,t) = \frac{\partial}{\partial \dot{x}}F(\dot{x},x,t)\,x^{(k+1)} + \ldots &= 0 \end{aligned} \tag{39}$$

for a DAE of index k. Upon discretization, (39) becomes an overdetermined system that can be tackled by least squares techniques. The challenge in this procedure, however, is the in general unknown index k and its determination.

Algorithms based on the derivative array are a powerful means for general unstructured DAE systems, and this holds even for the linear constant coefficient case since the computation of the Weierstrass form or the Drazin inverse are very sensitive to small perturbations and thus problematic in finite precision arithmetic. For the derivative array, in contrast, so-called staircase algorithms have been developed that rely on orthogonal matrix multiplications and are much more stable [22].

3.5 Constrained Hamiltonian systems

In the conservative case, the Lagrange equations (14) of constrained mechanical motion can be reformulated by the transformation to Hamilton's canonical equations. This leads to a mathematical model that is typical for *molecular dynamics* simulations. Again, constraints come into play in this application field, and the time discretizations need to cope with the index and stability issues.

We define the Lagrange function

$$L(q,\dot{q}) := T(q,\dot{q}) - U(q) \tag{40}$$

as the difference of kinetic and potential energy. The conjugate momenta $p(t) \in \mathbb{R}^{n_q}$ are then given by

$$p := \frac{\partial}{\partial \dot{q}}L(q,\dot{q}) = M(q)\dot{q}, \tag{41}$$

and for the Hamiltonian we set

$$H := p^T \dot{q} - L(q, \dot{q}). \tag{42}$$

Since the velocity $\dot{q}$ can be expressed as $\dot{q}(p,q)$ due to (41) if the mass matrix is invertible, we view the Hamiltonian as a function $H = H(p,q)$. Moreover, we observe that H is the total energy of the system because

$$H = p^T M(q)^{-1} p - \frac{1}{2} \dot{q}^T M(q) \dot{q} + U(q) = T + U.$$

Using the least action principle (16) in the new coordinates p and q and applying the Lagrange multiplier technique as above in the presence of constraints, we can express the equations of motion as

$$\begin{aligned} \dot{q} &= \frac{\partial}{\partial p} H(p,q), \\ \dot{p} &= -\frac{\partial}{\partial q} H(p,q) - G(q)^T \lambda, \\ 0 &= g(q). \end{aligned} \tag{43}$$

The *Hamiltonian equations* (43) possess a rich mathematical structure that should be preserved by numerical methods. In the 1990s, this problem class led to the new field of *geometric integration*, see the monograph by Hairer, Lubich & Wanner [66] for an extensive exposition. For brevity, we simply mention the *SHAKE* scheme as one of the established methods. In case of a Hamiltonian $H = \frac{1}{2} p^T M^{-1} p + U(q)$ with constant mass matrix, it reads

$$\begin{aligned} q_{n+1} - 2q_n + q_{n-1} &= -\tau^2 M^{-1} (\nabla U(q_n) + G(q_n)^T \lambda_n), \\ 0 &= g(q_{n+1}). \end{aligned} \tag{44}$$

In each time step thus a nonlinear system for q_{n+1} and λ_n needs to be solved, which is closely related to the projection step (28a).

4 Beyond classical DAEs

At the end of the last century, new topics emerged that were closely related to DAEs but also included important aspects from other fields. In particular, the topic of Partial Differential Algebraic Equations (PDAEs) became attractive by then, driven by time-dependent partial differential equations that were treated by the method of lines and that featured additional constraints.

4.1 Navier-Stokes incompressible

It requires convincing examples to demonstrate the benefits of a differential-algebraic viewpoint in the PDE context. One such example is sketched next.

A classical example for a PDAE is given by the Navier-Stokes equations

$$\dot{u} + (u \cdot \nabla)u + \tfrac{1}{\rho}\nabla p = \nu \Delta u + l, \tag{45a}$$

$$0 = \nabla \cdot u \tag{45b}$$

for the velocity field $u(x,t)$ and the pressure $p(x,t)$ in a d-dimensional domain Ω, with mass density ρ, viscosity ν, and source term $l(x,t)$. The second equation (45b) models the incompressibility of the fluid and defines a constraint for the velocity field. For simplification, the convection term $(u \cdot \nabla)u$ in (45a) can be omitted, which makes the overall problem linear and more amenable for the analysis. In an abstract notation, the resulting Stokes problem then reads

$$\dot{u} + \mathscr{A} u + \mathscr{B}' p = l, \tag{46a}$$

$$\mathscr{B} u = 0, \tag{46b}$$

with differential operators $\mathscr{A}$ and $\mathscr{B}$ expressing the Laplacian and the divergence, respectively. The notation $\mathscr{B}'$ stands for the conjugate operator of $\mathscr{B}$, which here is the gradient.

The discretization, e.g., by a Galerkin-projection

$$u(x,t) \doteq N(x) q(t), \qquad p(x,t) \doteq Q(x) \lambda(t)$$

with ansatz functions N and Q in some finite element spaces, transforms the infinite-dimensional PDAE (46) to the DAE

$$M\dot{q} + Aq + B^T \lambda = l, \tag{47a}$$

$$Bq = 0. \tag{47b}$$

While the mass matrix M and stiffness matrix A are symmetric positive definite and symmetric positive semi-definite, respectively, and easy to handle, the constraint matrix B is generated by mixing the discretizations for the velocity field and the pressure. It is well-known in mixed finite elements [25] that a bad choice for the discretization will either result in a rank-deficient matrix B or in a situation where the smallest singular value of B is approaching zero for a decreasing mesh size. This means that the DAE (47) may become singular or almost singular due to the spatial discretization. The famous LBB-condition by Ladyshenskaja, Bab*v*uska, and Brezzi [25] gives a means to classify the discretization pairs for u and p. If the matrix B has full rank, the index of the DAE (47) is $k = 2$.

To summarize, PDEs with constraints such as the Navier-Stokes equations often feature a rich structure that should be exploited, and building on the available PDE methodology reveals interesting cross-connections with the differential-algebraic viewpoint. In this context, the abstract formulation (46) as *transient saddle point problem* defines a rather broad problem class where many application fields can be subsumed [158].

By combining the state-of-the-art in DAEs with advanced PDE methodology and numerics, powerful algorithms can then be developed that break new ground. Time-space adaptivity for PDAEs is one such topic where many different aspects are put together in order to set up numerical schemes with sophisticated error control. The work by Lang [94] defines a cornerstone in this field. In electrical circuit simulation, the inclusion of heating effects or semi-conductors results also in PDAE models where ODEs, DAEs, and PDEs are coupled via network approaches, see, e.g., [8, 61].

4.2 Stochastic DAEs

In the spring of 2006, Oberwolfach offered again the showcase for the latest developments in DAEs – 25 years after the workshop where Bill Gear had first investigated the mathematical pendulum (2) in Cartesian coordinates. The organizers were Stephen Campbell, Roswitha März, Linda Petzold, and Peter Rentrop. Among the participants from all over the world was Bill Gear himself, and during the week it became evident that DAEs were now well-established in many fields.

The same year, the book by Kunkel & Mehrmann [88] appeared, which shed new light on topics such as boundary value problems in differential-algebraic equations and the numerical treatment of fully implicit systems (3).

Most talks at the meeting addressed the field of PDAEs, but among the other prominent topics were also optimization and optimal control problems with constraints described by DAEs, see, e.g., [31, 85] and model order reduction for descriptor systems [144]. An emerging topic at the time were *stochastic differential-algebraic equations* or SDAEs in short. Since many models in science and engineering contain uncertain quantities, it is natural to extend the methodology for DAEs by corresponding random terms. This could either be a parameter or coefficient that is only known approximately or even an extra diffusion term in the differential equation that is expressed in terms of a Wiener process. For the constant coefficient system (7), such a diffusion term leads to the linear SDAE

$$E\mathrm{d}x(t) + Hx(t) = c(t) + C\mathrm{d}W(t) \tag{48}$$

with a Wiener process W in $\mathbb{R}^{n_x}$ and a square matrix C. For work in this field and applications in electrical circuit analysis we refer to [75, 169].

5 Nonsmooth dynamical systems

Differential equations for dynamical systems become particularly challenging when nonsmooth functions must be considered in one way or the other. Economical and financial mathematics have been a driving force for nonsmooth dynamical systems, where finding equilibria in market situations can be solved using variational inequalities that are closely related to nonlinear programming and convex optimization. Nonsmooth differential equations play a role in optimal control, robotic path planning and perturbation analysis. They occur also in electrical circuits with ideal diodes, see Figure 6a. An ideal diode transmits electrical current only in one direction and blocks it in the other. If the electrical current I is reversed, the conductivity of the ideal diode suddenly changes from 0 to ∞. In other words, the electrical current at the diode must be positive at all times, $I(t) \geq 0$. If the current is nonzero, $I(t) > 0$ the voltage must be zero, $V(t) = 0$. If there is a nonzero voltage $V(t) > 0$ at the diode, this means, that the diode is blocking and does not transmit any current, $I(t) = 0$. This translates into the complementarity condition

$$I(t) \geq 0, \qquad V(t) \geq 0, \qquad I(t) \cdot V(t) = 0,$$

at the diode. Which formalism can we use to incorporate inequalities and complementarity conditions into Kirchhoff's circuit laws? Another example are multibody dynamical systems with impacts, where contacts are modeled using a positivity constraint $g(q) \geq 0$ of a signed distance function, see for instance Figure 6b. Without long range and adhesive forces, the contact force f_c between two contacting bodies must be nonnegative. In addition, it must be zero if the contact gap is positive. This again yields a complementarity problem

$$g(q) \geq 0, \qquad f_c \geq 0, \qquad g(q) \cdot f_c = 0.$$

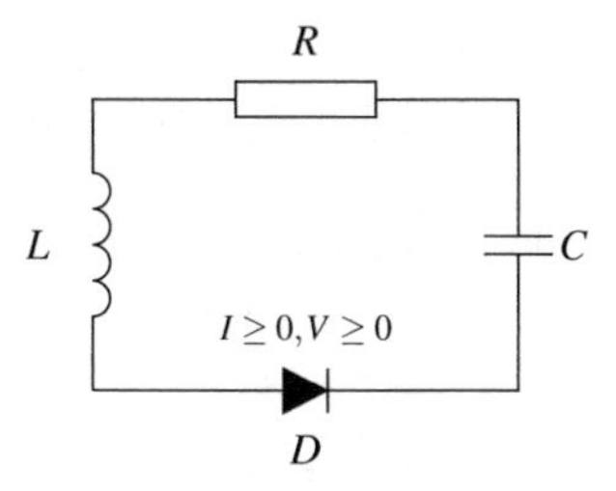

(a) A simple electrical circuit with an ideal diode D

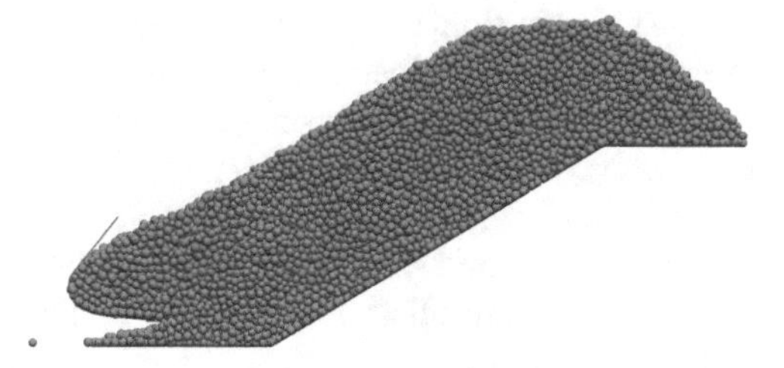

(b) A simulation of granular material with perfect unilateral contacts and friction

Fig. 6: Examples from electrical and mechanical engineering that include nonsmooth phenomena.

It is easy to construct examples, where the velocity of a rigid body must be discontinuous to enforce an inequality constraint. Think of a point mass accelerating under gravity towards the ground. At the moment of impact the velocity of the particle must jump from a negative to a nonnegative value instantaneously. This toy example demonstrates the two main difficulties that must be tackled in nonsmooth dynamical systems.

1. DAEs give us a mathematical structure to describe mechanical systems subject to *algebraic* constraints. How can we incorporate *inequality* constraints $g(q) \geq 0$?
2. Newton's second axiom
$$M\ddot{q} = f(q,\dot{q},t) \tag{49}$$
implies that the trajectory $q(t)$ is at least once continuously differentiable, which is obviously not the case here. Clearly, Equation (49) still has validity almost everywhere, but how can we handle the instances in time with impacts?

In the following, we will take a closer look on the treatment of dynamical systems that are constrained to a feasible set:

$$\begin{aligned} F(q,\dot{q},t) &= \mathbf{0}, \\ q \in K &= \{\, q \mid g(q) \geq \mathbf{0} \,\} \end{aligned} \tag{50}$$

for some appropriate function $g : \mathbb{R}^{n_q} \to \mathbb{R}^{n_\lambda}$.

5.1 A short zoology

This section provides the most important concepts with regard to nonsmooth dynamical systems, including Moreau's sweeping process, differential inclusions, projected dynamical systems, variational inequalities, complementarity dynamic systems, differential variational inequalities and finally measure differential equations and measure differential inclusions.

The list is far from complete and many more authors than the ones cited here have made important contributions. The overview is biased as the subject is approached from the perspective of nonsmooth mechanical problems. Any discussion on this topic is incomplete without a detailed analysis of impact laws and a survey of frictional models. This chapter is not aimed at completeness however, but rather at providing an overview of the mathematical tools to deal with the nonsmooth phenomena that result from such models. For brevity, impact laws and frictional models are excluded all together. The reader is referred to [138] and the references therein for more information.

The tangent space $T_x\mathcal{M}$ and its annihilator $N_x\mathcal{M}$ at a point x of a manifold $\mathcal{M}$, compare Figure 5.

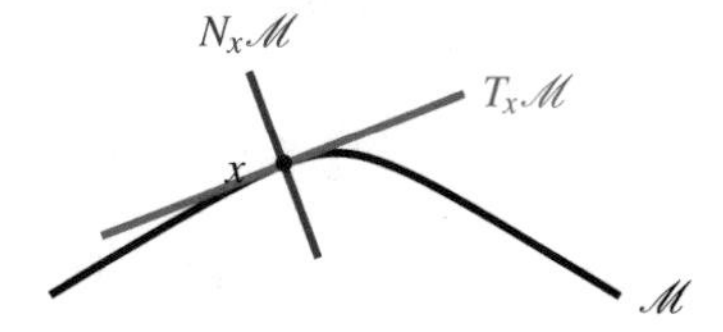

The tangent cones $T_{x_i}K$ and normal cones $N_{x_i}K$ at three points x_i, $i = 1,2,3$ in a set K. Note that $N_{x_3}K = \{0\}$ and $T_{x_3}K$ takes up the whole space.

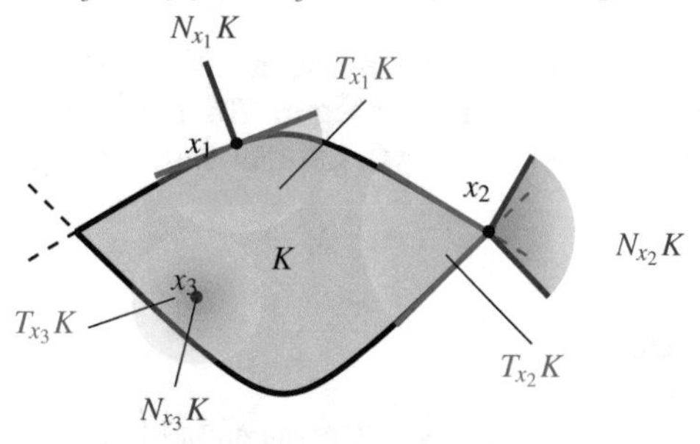

Fig. 7: Comparison of tangent space and annihilator of manifolds to tangent and normal cones of sets.

5.1.1 Unilateral constraints and Moreau's sweeping process

Jean Jacques Moreau studied mechanical systems subject to unilateral constraints since the mid 1960s [114, 115, 117, 118, 119, 120]. A *unilateral constraint* restricts the motion of a dynamical system only in one direction, e.g. by imposing inequality constraints. Imagine a particle's position $x(t)$ is restricted to a moving closed set $K(t)$. Consider a particular point in time t^*, where the particle is located on the boundary of its feasible set, $x(t^*) \in \partial K(t^*)$. Assume for simplicity that the boundary can locally be described as a smooth manifold $\mathcal{M}$. Then we can find a tangent space $T_{x(t^*)}\mathcal{M}$ and a normal direction n to $T_{x(t^*)}\mathcal{M}$. The particle is only allowed to move freely in the half space to one side of the tangent space towards the interior of $K(t^*)$. In other words, the constraint restricts the motion *unilaterally*. In contrast to this, if the particle were to be restricted to the manifold $\mathcal{M}$, the particle would not be allowed to move in either direction along the normal n, i.e. it would be *bilaterally constrained*. The space spanned by the normal vector is often called the annihilator of the tangent space. The assumption that the boundary of the moving set $K(t)$ can locally be described as a manifold is too restrictive in general. It suffices, if we can describe the boundary of the set through so–called tangent cones[2] and normal cones rather than tangent spaces and their associated annihilators, see Figure 7.

A function x is said to be a solution to the *first order sweeping process* for the time–dependent set K, if

[2] See for instance [149] for a definition of tangent cones of sets.

$$x(0) \in K(0), \tag{51a}$$
$$x(t) \in K(t), \tag{51b}$$
$$-\dot{x}(t) \in N_{x(t)}K(t), \tag{51c}$$

where $N_{x(t)}K(t)$ denotes the normal cone to the set $K(t)$ at $x(t)$. It is defined as the polar of the tangent cone $T_{x(t)}K(t)$,

$$N_{x(t)}K(t) = \{ \ \xi \mid \langle \xi, y \rangle \leq 0 \quad \forall y \in T_{x(t)}K(t) \ \}.$$

The *differential inclusion* (DI) in (51c) means that if $x(t)$ is on the boundary of the set, the derivative must point inward, see Figure 7b. In the interior of the set $K(t)$, it holds $N_{x(t)}K(t) = \{0\}$, and thus $\dot{x}(t) = 0$. In other words, the particle is *swept* with the moving boundary.

5.1.2 Projected dynamical systems

In [123], a *projected dynamical system* (PDS) is defined as

$$\dot{x} = \Pi_K(x, -F(x)) \tag{52}$$

where

$$\Pi_K(x, v) = \lim_{\delta \to 0} \frac{\text{prox}_K(x + \delta v) - x}{\delta}$$

and K is a closed convex set defined by constraints on the system. Here, $\text{prox}_K(x)$ denotes the projection operator

$$\text{prox}_K(x) = \arg\min_{z \in K} \|x - z\|. \tag{53}$$

Because it holds

$$\Pi_K(x, v) = \text{prox}_{T_x K}(v),$$

Equation (52) can be rewritten as a projection onto the tangent cone $T_x K$ [28]:

$$\dot{x} = \text{prox}_{T_x K}(-F(x)). \tag{54}$$

Equations (52) and (54) are first order ordinary differential equations with nonsmooth right hand sides. In Moreau's sweeping process, the direction of the derivative is prescribed only by the moving set K, while here, the direction of the derivative is mainly prescribed by the right hand side $-F(x)$. In contrast to Moreau's sweeping process, it holds $\dot{x} = -F(x) \neq 0$ in the interior of the set. The projection operator makes sure, that the system never moves towards the exterior of K if x is on the boundary, see Figure 8. According to [28], any solution to Equation (54) is also a solution of the differential inclusion

$$-\dot{x}(t) \overset{\text{a.e.}}{\in} F(x(t)) + N_{x(t)}K, \tag{55}$$

If $-F(x) \in T_xK$ it holds $\text{prox}_{T_xK}(-F(x)) = -F(x)$ and therefore $\dot{x} = -F(x)$.

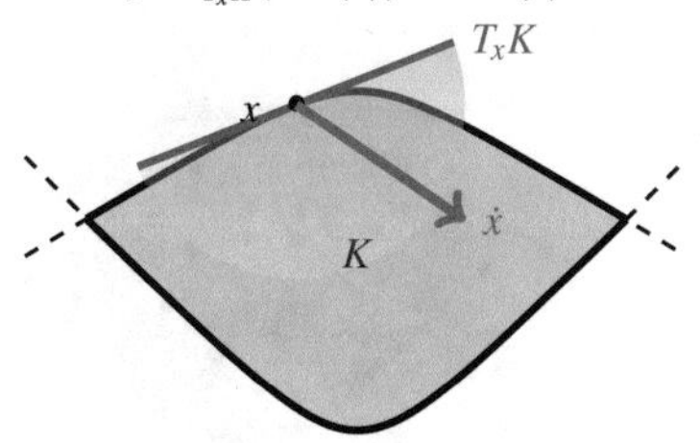

If $-F(x) \notin T_xK$ it holds $\dot{x} = \text{prox}_{T_xK}(-F(x)) \neq -F(x)$ and $-F(x) - \dot{x} \in N_xK$ *(blue dotted line)*.

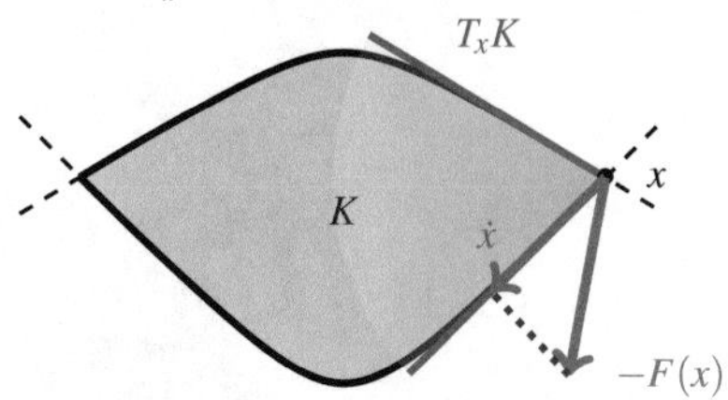

Fig. 8: Two example scenarios for projected dynamical systems.

which is Moreau's first order sweeping process, if $F(x) = 0$.

Standard works on differential equations with discontinuous right hand side and differential inclusions are [19, 48]. Existence and uniqueness to such problems depends on properties of the set–valued map $K(t)$ and its tangent and normal cones. Nagurney and Zhang motivate their definition of projected dynamical systems as a means to unify the theory of dynamical systems and variational inequalities.

5.1.3 Variational inequalities

Stuart Antman recites the emergence of variational inequalities in [13]. In 1959, Antonio Signorini posed the problem of finding the material displacements of a heavy deformable body resting on a rigid frictionless flat ground. This problem, now called the *Signorini contact problem*, is particularly difficult, as the geometry of the contact region between the body and the ground is a priori unknown. Gaetano Fichera, a student of Signorini, studied existence and uniqueness of solutions to this problem using the calculus of variations and published his results in 1963 and 1964. In hindsight, Fichera's solution of Signorini's contact problem relies on the solution to a *variational inequality* (VI). In a general setting, a variational inequality is defined as follows. Let X be a Banach space, $K \subset X$ and $g : K \to X^*$ be a mapping from K to the dual space X^* of X. A variational inequality denotes the problem of finding $u \in K$, such that

$$\langle g(u), u' - u \rangle \geq 0 \quad \forall u' \in K. \tag{56}$$

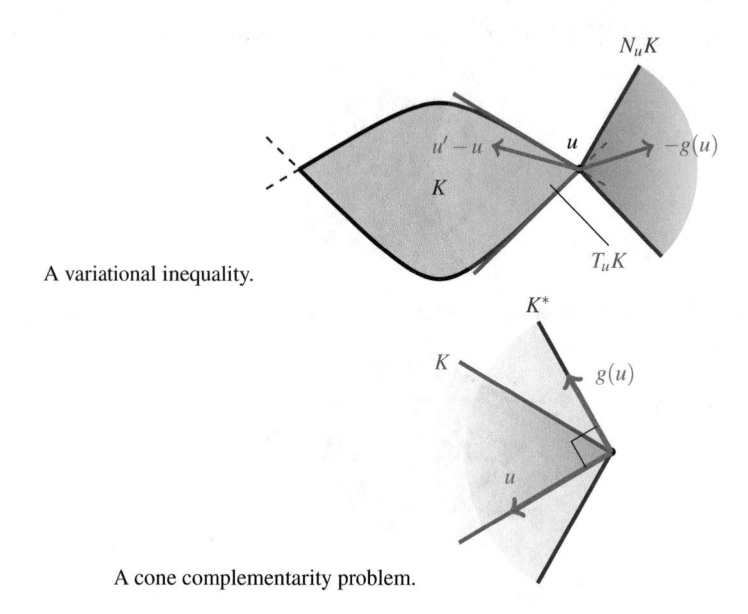

A variational inequality.

A cone complementarity problem.

Fig. 9: Geometric interpretations of a variational inequality and a cone complementarity problem in finite dimensions.

Figure 9a contains a geometric representation of a variational inequality in finite dimensions. Note that (56) translates to $-g(u) \in N_u K$. The term variational inequality was first introduced by Stampacchia in 1965 and a theory around variational inequalities quickly grew during the second half of the 1960s. Most researchers on the subject at that time heavily cited Fichera's work. For more details on how Fichera's solution to Signorini's problem can be solved using variational inequalities and how his work influenced their analysis in the years to come, see [13].

5.1.4 Complementarity dynamic systems

A convex cone K is a set that is closed with respect to positive linear combinations, i.e. if x and y are in the cone K and $\alpha, \beta \geq 0$, it must hold that $\alpha x + \beta y \in K$. The variational inequality (56) is equivalent to a *cone complementarity problem* (CCP)

$$g(u) \in K^*, \qquad u \in K, \qquad \langle g(u), u \rangle = 0, \tag{57}$$

if the set K is a convex cone, which is proved in [79]. Here,

$$K^* = \{ \, y \in X^* \mid \langle y, x \rangle \geq 0 \, \forall x \in K \, \} \subset X^*$$

is the *dual cone* to K. Notice that the dual cone is the negative normal cone to K at 0. Because of the complementarity condition in (57) it is clear, that if either $g(u)$ or

u is in the interior of their respective cones K^* and K, the other of the two must be zero (see Figure 9b), hence the term *complementarity*.

The positive orthant $\mathbb{R}^n_+$ is such a convex cone and it holds $(\mathbb{R}^n_+)^* = \mathbb{R}^n_+$, i.e. it is self–dual. For this specific cone the CCP (57) resembles in part the well–known Karush–Kuhn–Tucker necessary conditions for optimization problems with inequality constraints $g(u) \geq 0$:

$$g(u) \geq 0, \qquad u \geq 0, \qquad u^T g(u) = 0,$$

where the inequality "$\geq$" is to be understood componentwise.

Complementarity dynamic systems are differential equations coupled to a complementarity problem. They were considered by several authors, including J. J. Moreau himself, to tackle mechanical systems with inequality constraints [11, 77, 99, 114, 136, 137]. Their works were heavily influenced by results from nonlinear programming and optimization.

5.1.5 Differential variational inequalities

In [19] a special version of a differential inclusion is mentioned, called a *differential variational inequality*. Given a closed convex subset $K \subset X$ of a vector space X and a set–valued map $F: K \to X$, a differential variational inequality consists of finding a function $x: [0,T] \to X$ satisfying

$$\begin{aligned} & x(t) \in K && \forall\, t \in [0,T], && \text{(58a)}\\ & \dot{x}(t) \in F(x(t)) - N_{x(t)}K && \text{for a.a. } t \in [0,T]. && \text{(58b)} \end{aligned}$$

Notice that the differential inclusion (58b) coincides with (55). The relation to variational inequalities becomes clear, if we write $\dot{x}(t) = f(x(t)) + g(t)$ for $f(x(t)) \in F(x(t))$ and $g(t) \in -N_{x(t)}K$. With this, (58) reads

$$\begin{aligned} & x(t) \in K && \forall\, t \in [0,T], && \text{(59a)}\\ & \langle \dot{x}(t) - f(x(t)), y - x(t) \rangle \geq 0 && \text{for a.a. } t \in [0,T] \text{ and all } y \in K. && \text{(59b)} \end{aligned}$$

A different, slightly more general, definition of a differential variational inequality comes from Pang and Stewart [131]. They define a finite dimensional differential variational inequality as a differential equation coupled to a variational inequality

$$\begin{aligned} & \dot{x} = f(t, x(t), u(t)), && && \text{(60a)}\\ & \langle F(t, x(t), u(t)), u' - u \rangle \geq 0 && \forall u' \in K, && \text{(60b)}\\ & \Gamma(x(0), x(T)) = 0. && && \text{(60c)} \end{aligned}$$

The problem consists of finding functions $x: [0,T] \to \mathbb{R}^{n_x}$ and $u: [0,T] \to K \subset \mathbb{R}^{n_\lambda}$ satisfying Equation (60) given the continuous functions f, F, Γ and a subset $K \subset \mathbb{R}^{n_\lambda}$. In this finite dimensional setting the dual pairing (60b) is the Euclidean scalar product and Equation (60c) is a prescribed initial or boundary condition. If K is a convex cone, the variational inequality (60) can be expressed as a cone complementarity problem. To differentiate their definition (60) from the definition (58) given by Aubin and Cellina, Pang and Stewart call problems of the type (58) *variational inequalities of evolution* (VIE), a convention that is adapted here. Pang and Stewart show in the same article, that the VIE is a special case of a DVI and suggest a unified version of the two, that essentially consists of a differential equation, an algebraic equation and a variational inequality. Finally, it should be noted that if the set K is a convex cone, a DVI and a VIE are conceptually equivalent.

Let $\mathscr{U}$ denote the set of solutions to the variational inequality (60b). Then, the DVI (60) can be rewritten as differential inclusion

$$\dot{x} \in f(t, x(t), \mathscr{U}),$$

The DVI (60) can also be rewritten as a DAE: The function u solves the variational inequality (60b) if and only if

$$0 = g(t,x,u) = u - \operatorname{prox}_K(u - F(t,x,u)),$$

where prox_K is the orthogonal projection onto the set K as defined in (53). Thus, the variational inequality can be replaced by a nonsmooth algebraic constraint. Pang and Stewart plea their case why DVIs deserve a special treatment, even if they can be considered as either DIs or DAEs. They argue, that the DI theory is too general to be of practical use, since many of its assumptions are cumbersome to prove for specific situations. While the theory of DAEs appeals to them, it is problematic as the differentiability of the algebraic constraints plays an important role in it. If the algebraic constraints however are nonsmooth, as is the case if they are constructed from variational inequalities, many results cannot be used.

Still, it is obvious that the authors of [131] were inclined to adopt as much terminology from and analogy to DAEs for their analysis as possible. As for DAEs, the unknowns of a DVI are separated into differential variables x and algebraic variables u. While a DAE can be seen as an ODE coupled to an algebraic equation through the algebraic variables, a DVI is a differential equation coupled to a variational inequality. Even the concept of the index of a DVI is adopted. A DVI of index zero is just an ordinary differential equation. Under certain conditions for the set–valued map F, the algebraic unknown u can be written as a function of x and t using the implicit function theorem. One differentiation then reveals the DVI as system of differential equations in x and u. This is called a DVI of index one. According to the authors, higher index DVIs can be considered as well, but they restrict their analysis to DVIs of index one with absolutely continuous solutions.

5.1.6 Derivatives of functions of bounded variation

Differential variational inequalities and the related concepts help us to describe dynamical systems subject to inequality constraints. We opened the section on nonsmooth dynamical systems with another problem. If the velocity of a mechanical system subject to unilateral constraints is discontinuous, how do we interpret accelerations and forces? The results in the next two sections are taken from [4, 58, 59, 116, 124].

It is physically meaningful to assume that the velocity $v = \dot{x}$ is continuous almost everywhere, except at a countable number of finite jump discontinuities, at which v is not defined per se[3]. Its antiderivative

$$x(t) = x(0) + \int_0^t \dot{x}(\tau)\,\mathrm{d}\tau,$$

is therefore an absolutely continuous function.

How do we define the derivative of a function of bounded variation v? It turns out, that measure theory helps. If the function can have jumps, it changes infinitely fast from its left limit to its right limit at such a discontinuity point. The discontinuity points comprise only a zero–set of the Lebesgue measure $\mathrm{d}t$, but surely the infinitely fast changes in v cannot simply be ignored. We must capture the changes of the function using measures, for which the set of discontinuity points is not a zero set.

We can construct the so–called Lebesgue–Stieltjes measure $\mathrm{d}v$ from a function of bounded variation v, so that it holds

$$v(t) = v(0) + \int_0^t \mathrm{d}v.$$

In this sense, the Lebesgue–Stieltjes measure $\mathrm{d}v$ plays the role of the derivative of v. The measure captures the changes in v regardless of whether they happen smoothly or abruptly. Using the Lebesgue decomposition and Radon–Nikodym theorems, the Lebesgue–Stieltjes measures of v can be decomposed into three parts,

$$\mathrm{d}v = a(t)\mathrm{d}t + \delta_p + \mu_s.$$

Here, $a(t)$ is the so–called Radon–Nikodym density of the continuous part of the measure $\mathrm{d}v$ with respect to the Lebesgue measure $\mathrm{d}t$, δ_p is a sum of Dirac delta measures that captures all jumps of v and μ_s is a singular measure associated with the Cantor part of v. If v has no Cantor part, it is called a function of specially bounded variation and the last term vanishes. This is a valid assumption for a large class of problems. If in addition it is absolutely continuous, the second term vanishes and it holds $\mathrm{d}v = a(t)\mathrm{d}t$ with $a = \dot{v}$.

Note that the Lebesgue–Stieltjes measure $\mathrm{d}v$ corresponds to the distributional derivative of v:

$$\langle \mathrm{D}v, \varphi \rangle = -\int_0^T v\dot{\varphi}\,\mathrm{d}t = -\int_0^T v\,\mathrm{d}\varphi = \int_0^T \varphi\,\mathrm{d}v \quad \forall \varphi \in C_0^\infty(0,T),$$

[3] We could easily define the function $v(t_c)$ at the discontinuity points t_c as either the left limit $v^-(t_c) = \lim\limits_{h\to 0} v(t_c - h)$ or the right limit $v^+(t_c) = \lim\limits_{h\to 0} v(t_c + h)$, with $h > 0$, as convention. While this potentially helps the analysis, it does not change the structure of the problem.

because all functions in $C_0^\infty(0,T)$ are absolutely continuous and thus have a density w.r.t. $\mathrm{d}t$. Here we use integration by parts for Lebesgue–Stieltjes integrals:

$$\int_0^T v\,\mathrm{d}w = [vw]_0^T - \int_0^T w\,\mathrm{d}v \tag{61}$$

for all v, w with bounded variation.

5.1.7 Measure differential equations and measure differential inclusions

A *measure differential equation* (MDE) takes the form

$$\mathrm{d}x = F(t,x)\,\mathrm{d}t + G(t,x)\,\mathrm{d}u, \tag{62}$$

where $\mathrm{d}x$ and $\mathrm{d}u$ denote the Lebesgue–Stieltjes measures of the functions of bounded variation x and u and $\mathrm{d}t$ denotes the Lebesgue measure of time [95, 130]. The corresponding initial value problem consists of finding $x : [0,T] \to \mathbb{R}$ given an initial value $x(0) = x_0$. Measure differential equations were motivated from Optimal Control Theory and the theory of perturbed systems. Here, the function u takes the role of inputs or external perturbations. In control theory, u drives the state x of the system to a certain target. In some applications it might be useful to allow jumps in the input, which in turn result in jumps in the state and vice-versa. In perturbation theory, impulsive perturbations are studied. See [139] for a recent, slightly different approach towards measure differential equations in the context of differential geometry.

One way to read Equation (62) is as a short-form for the variational condition

$$\int_0^T \varphi\,\mathrm{d}x = \int_0^T \varphi\,F(t,x)\,\mathrm{d}t + \int_0^T \varphi\,G(t,x)\,\mathrm{d}u \qquad \forall \varphi \in \mathscr{T}, \tag{63}$$

for some appropriate space of test functions $\mathscr{T}$. If u were absolutely continuous, we could write $\mathrm{d}u = \dot{u}\,\mathrm{d}t$. Then it is obvious from (62), that $\mathrm{d}x$ has a density $F(t,x) + G(t,x)\dot{u}$ with respect to $\mathrm{d}t$ and thus we can write $\mathrm{d}x = \dot{x}\,\mathrm{d}t$. Finally, using the fundamental lemma of calculus of variations, we could derive a differential equation

$$\dot{x} = F(t,x) + G(t,x)\dot{u}$$

from (63).

Moreau was well aware that the sweeping process results in discontinuous velocities. He introduced *measure differential inclusions* (MDI) as an extension to the differential inclusion in (51) [59, 113, 117, 132, 159]. In [113] a *measure differential* inclusion is defined as

$$\mathrm{d}u \in N_u K,$$

for a convex subset $K \subset X$ of a Hilbert space X. It should be noted here, that this definition would have just as much meaning in a Banach space, which will be exploited in the next section. In this abstract setting, the normal cone is a subset of the dual space X^* of X. In the same book, existence and uniqueness results are given for second order measure differential inclusions that result from mechanical systems with unilateral constraints and impacts. See also [5] for higher order MDIs in the context of Moreau's sweeping process.

Up to this point all considerations concerning normal and tangent cones could be directly translated to the finite dimensional spaces in which the functions take their values. In other words, if we consider functions taking values in $\mathbb{R}^n$, the dual pairings $\langle\cdot,\cdot\rangle$ can be interpreted as the Euclidean scalar product in $\mathbb{R}^n$ and the negativity condition in the definition of the normal cone must hold in a pointwise sense. This gives us a geometric interpretation of tangent and normal cones, complementarity problems and variational inequalities. As soon as distributional derivatives play a role, this interpretation becomes more and more difficult and the problems are usually considered in infinite dimensional function spaces.

5.1.8 A very diverse field

In the past 60 years, many researchers devoted themselves to dynamical systems that encounter nonsmooth phenomena. These researchers borrowed results from convex and functional analysis, numerical analysis, nonlinear programming and optimization, measure theory and differential geometry to produce a rich structure and a diverse selection of concepts to describe "*nonsmoothness*" of differential equations. DAEs did not play a major role in the early development of the field. It is only recently that the DAE theory has become an important influence on the field of nonsmooth dynamical systems.

Many of introduced concepts are very similar and under mild assumptions some can be shown to be equivalent [28]. The diversity of the field comes at a small price however, which is the lack of a common language. It can be very hard to judge which of these concepts is most appropriate for the application at hand. It is extremely difficult to find the vocabulary that captures the relevant phenomena of a specific problem, but still leaves enough generality to be useful to a large class of problems. While this can be said about any mathematical theory, it is especially apparent in the field of nonsmooth differential equations.

5.2 Nonsmooth mechanical systems with impacts

We have already established that the trajectory q of a mechanical system with inequality constraints cannot be the solution of Newton's second axiom (49), since we must allow $\dot{q}$ to have jump discontinuities. To understand the physics behind the system, we have to fall back to more general mechanical principles.

5.2.1 Hamilton's principle as a differential inclusion

Classical mechanics offers a few physical principles as governing equations, that are formulated in a variational setting rather than as a second order ordinary differential equation. Hamilton's principle of least action (16) is an example. While there have been several attempts to extend the principle for discontinuous solutions since its development, most of them were undertaken quite recently with the help of the new tools provided by the growing field of nonsmooth dynamical systems, see [45, 47, 96, 129] and the references therein. It is unclear whether William Hamilton had nonsmooth mechanical systems on his mind in the 1830s. But his principle is formulated with just enough generality to be extended to the nonsmooth case.

To begin, let us consider the problem in a function space setting. While the (generalized) coordinates q are not continuously differentiable, they must still be continuous: A material point of a mechanical system cannot just disappear and reappear elsewhere. The velocity $\dot{q}$ is continuous almost everywhere, except for a countable number of time points, where the velocity undergoes finite jumps. In mathematical terms, $q : \mathbb{R} \supset [0,T] \to \mathbb{R}^{n_q}$ is an absolutely continuous function and $\dot{q} : [0,T] \to \mathbb{R}^{n_q}$ is a function of specially bounded variation,

$$q \in \mathrm{AC}([0,T]) \quad \text{and} \quad \dot{q} \in \mathrm{SBV}([0,T]) .$$

Introducing inequality constraints $g(q) \geq 0$ into the system translates to finding a feasible physical solution q in the set

$$\mathscr{S} = \{ \, q \in \mathrm{AC}([0,T]) \mid g(q(t)) \geq 0 \quad \forall t \in [0,T] \, \} .$$

The classical principle of virtual work and the closely related principle of D'Alembert can be used to examine systems, where the feasible set is a manifold

$$\mathscr{M} = \{ \, q \in \mathrm{AC}([0,T]) \mid g(q(t)) = 0 \quad \forall t \in [0,T] \, \}$$

with a sufficiently smooth function g. The principle requires the constraint forces $f_c(q)$ to be orthogonal to the tangent space $T_q\mathscr{M}$ of the manifold at q at all times, i.e.

$$-f_c(q) \in N_q\mathscr{M} ,$$

where $N_q\mathscr{M}$ is the annihilator space of $T_q\mathscr{M}$, see Figure 7a. This concept can be extended for the inequality case, $q \in \mathscr{S}$, simply by replacing the tangent space with the tangent cone and the annihilator with the normal cone to $\mathscr{S}$ at q, see Figure 7b [96]:

$$-f_c(q) \in N_q\mathscr{S} . \tag{64}$$

Notice, that the normal cone

$$N_q\mathscr{S} = \{ \, \xi \in \mathrm{AC}([0,T])^* \mid \langle \xi, d \rangle \leq 0 \quad \forall d \in T_q\mathscr{S} \, \}$$

is a subset of the *dual space* of the space of absolutely continuous functions.

Leine, Aeberhard and Glocker use the nonsmooth version (64) of the principle of virtual work and some further mild assumptions[4] to derive a version of Hamilton's principle of least action that is valid in the presence of inequality constraints and discontinuous velocities [96]. The classical version of this principle states that the Fréchet–derivative of the action vanishes at q,

$$\delta S = \delta \int_0^T L(q,\dot{q})\mathrm{d}t = 0 \tag{65}$$

where $L(q,\dot{q}) = T(q,\dot{q}) - U(q)$ denotes the difference between kinetic and potential energy. In the nonsmooth case, the principle reads

$$\delta S \in N_q \mathscr{S}. \tag{66}$$

If the feasible set takes up the whole space, $\mathscr{S} = \mathrm{AC}([0,T])$, the normal cone to $\mathscr{S}$ at any $q \in \mathscr{S}$ contains only the zero, $N_q\mathscr{S} = \{0\}$. In this case, (66) reappears as its classical pendant (65).

5.2.2 Forces and Accelerations are Measures

It would be convenient to have a form similar to the Lagrange equations of the first kind (14) for inequality constraints that makes the constraint forces explicit. To reach this goal at the end of Section 5.2.3, we must first accept that the constraint forces of inequality constraints are not necessarily classical functions anymore. If the velocity has a jump, its derivative at that point in time does not exist classically. We can find a weak derivative at the discontinuity point, that is a Dirac delta distribution. This Dirac delta distribution is a measure rather than a function.

Recall that the constraint forces for inequality constraints are from the dual space of absolutely continuous functions

$$-f_c(q) \in N_q\mathscr{S} \subset (\mathrm{AC}([0,T]))^*.$$

The space of absolutely continuous functions equipped with the weak norm

$$\|q\| = \max\left\{ \sup_{t\in[0,T]} \|q(t)\|, \ \sup_{t\in[0,T]} \|\dot{q}(t)\| \right\}$$

is a Banach space. Its dual space is given by the signed Radon measures [58, 124], in the sense that any functional $f \in (\mathrm{AC}([0,T]))^*$ can be written as

$$\langle f,x\rangle = \int_0^T x\,\mathrm{d}p,$$

[4] Particularly, the feasible set $\mathscr{S}$ does not have to be convex, but it is required to be tangentially regular.

where $\mathrm{d}p$ is the Lebesgue–Stieltjes measure of a function of bounded variation p. It is remarkable how impulsive forces are a direct consequence only of the principle of virtual work applied to the correct function space.

5.2.3 Existence of Lagrangian multipliers

Up to this point, $\mathscr{S}$ could be any tangentially regular subset of $\mathrm{AC}([0,T])$. We haven't exploited yet that the set $\mathscr{S}$ is defined by the inequality constraints $g(q) \geq 0$. In doing so we can characterize the problem using Lagrangian multipliers.

Before the Polish mathematician Stanislav Kurcyusz died in a tragic accident at a young age in 1978, he made important contributions to optimization in Banach spaces subject to operator inequality constraints. His theory translates very nicely to the variational formulation of nonsmooth mechanics. Using the results published in [89, 170] the following theorem can be formulated, that rewrites (66) in terms of Lagrangian multipliers [81] and a complementarity condition:

Theorem 5.1 *Let $g : \mathbb{R}^{n_q} \to \mathbb{R}^{n_\lambda}$ be continuously differentiable and let*

$$\mathscr{S} = \{\, q \in AC([0,T]) \mid g(q(t)) \geq 0 \quad \forall t \in [0,T] \,\}$$

denote the set of admissible trajectories. Let $q \in \mathscr{S}$, let $\mathscr{S}$ be tangentially regular and assume that the Robinson regularity condition

$$\mathrm{im}(G(q)) + \mathrm{span}(g(q)) - \mathbb{R}^{n_\lambda}_{+} = \mathbb{R}^{n_\lambda}$$

holds, where $\mathrm{im}(G(q))$ denotes the image of the Jacobian of g at q, $\mathrm{span}(g(q))$ the space spanned by $g(q)$ and $\mathbb{R}^{n_\lambda}_{+}$ the positive orthant in $\mathbb{R}^{n_\lambda}$. Then there exists a non–negative measure $\mathrm{d}\lambda$, such that

$$0 = \delta S(\delta q) + \int_0^T \delta q^T G(q)^T \mathrm{d}\lambda \tag{67a}$$

$$0 = \int_0^T g(q)^T \mathrm{d}\lambda. \tag{67b}$$

for all variations $\delta q \in AC([0,T])$ with $\delta q(0) = \delta q(T) = 0$.

We can now bring Equation (67a) in a more recognizable form that resembles the Lagrange equations of first kind (14). Using $\delta \dot{q} \mathrm{d}t = \mathrm{d}(\delta q)$, integration by parts for Lebesgue–Stieltjes measures (61) and $\delta q(0) = \delta q(T) = 0$, we can rewrite the Fréchet derivative $\delta S(\delta q)$ of the action S as

$$
\begin{aligned}
\delta S(\delta q) &= \int_0^T \delta q^T \frac{\partial L}{\partial q} \mathrm{d}t + \int_0^T \delta \dot{q}^T \frac{\partial L}{\partial \dot{q}} \mathrm{d}t \\
&= \int_0^T \delta q^T \frac{\partial L}{\partial q} \mathrm{d}t + \int_0^T \frac{\partial L}{\partial \dot{q}}^T \mathrm{d}(\delta q) \\
&= \int_0^T \delta q^T \frac{\partial L}{\partial q} \mathrm{d}t + \underbrace{\left[\delta q^T \frac{\partial L}{\partial \dot{q}}\right]_0^T}_{=0} - \int_0^T \delta q^T \mathrm{d}\left(\frac{\partial L}{\partial \dot{q}}\right) \\
&= \int_0^T \delta q^T \frac{\partial L}{\partial q} \mathrm{d}t - \int_0^T \delta q^T \mathrm{d}\left(\frac{\partial L}{\partial \dot{q}}\right).
\end{aligned} \tag{68}
$$

Plugging (68) into (67a) yields

$$
0 = \int_0^T \delta q^T \frac{\partial L}{\partial q} \mathrm{d}t - \int_0^T \delta q^T \mathrm{d}\left(\frac{\partial L}{\partial \dot{q}}\right) + \int_0^T \delta q^T G(q)^T \mathrm{d}\lambda \tag{69}
$$

or, equivalently expressed as an equality of measures,

$$
\mathrm{d}\left(\frac{\partial L}{\partial \dot{q}}\right) = \frac{\partial L}{\partial q} \mathrm{d}t + G(q)^T \mathrm{d}\lambda. \tag{70}
$$

This is the nonsmooth version of the Euler–Lagrange equations. As long as the momentum $\partial L / \partial \dot{q}$ has discontinuous jumps, i.e. is not absolutely continuous, it holds

$$
\mathrm{d}\left(\frac{\partial L}{\partial \dot{q}}\right) \neq \frac{\mathrm{d}}{\mathrm{d}t} \frac{\partial L}{\partial \dot{q}} \mathrm{d}t.
$$

Therefore we cannot derive an ordinary differential equation from the variational problem (69) using the fundamental lemma of calculus, as is usually done to arrive at the Euler–Lagrange equations. The physics can only be described variationally, in this case in the form of a measure differential equation (70).

Finally, by plugging in the Lagrange function of a multibody system

$$
L(q, \dot{q}) = \frac{1}{2} \dot{q}^T M(q) \dot{q} - U(q)
$$

and using

$$\begin{aligned}\frac{\partial L}{\partial q} &= \frac{1}{2}\dot{q}^T \frac{\partial M(q)}{\partial q}\dot{q} - \frac{\partial U(q)}{\partial q},\\ \mathrm{d}\left(\frac{\partial L}{\partial \dot{q}}\right) &= M(q)\mathrm{d}\dot{q} + \dot{q}^T \mathrm{d}M(q)\\ &= M(q)\mathrm{d}\dot{q} + \dot{q}^T \dot{M}(q)\mathrm{d}t\\ &= M(q)\mathrm{d}\dot{q} + \dot{q}^T \frac{\partial M(q)}{\partial q}\dot{q}\mathrm{d}t,\\ f(q,\dot{q},t) &:= -\frac{1}{2}\dot{q}^T \frac{\partial M(q)}{\partial q}\dot{q} - \frac{\partial U(q)}{\partial q},\end{aligned}$$

Equation (70) becomes

$$M(q)\ \mathrm{d}\dot{q} = f(q,\dot{q},t)\ \mathrm{d}t + G(q)^T\ \mathrm{d}\lambda. \tag{71}$$

Replacing Equation (67a) in Theorem 5.1 with Equation (71) yields the Lagrange equations of first kind for the nonsmooth case with impacts

$$M(q)\ \mathrm{d}\dot{q} = f(q,\dot{q},t)\ \mathrm{d}t + G(q)^T\ \mathrm{d}\lambda, \tag{72a}$$
$$0 \le g(q), \tag{72b}$$
$$0 \le \mathrm{d}\lambda, \tag{72c}$$
$$0 = \int_0^T g(q)^T\ \mathrm{d}\lambda. \tag{72d}$$

The differential equation (14a) in the smooth case is replaced by the MDE (72a) in the nonsmooth version and the equality constraint (14b) turns into the complementarity conditions (72b)–(72d). Due to the positivity (72b) and (72c), (72d) means that the measure $\mathrm{d}\lambda$ is nonzero only for subsets of $[0,T]$ on which g is zero.

5.3 Numerical solution strategies

Nonsmooth dynamical systems have been researched rigorously since the late 1950s and many theoretical results were developed during the 1970s and 1980s. However, based on the number of publications on the subject, the development of numerical methods only slowly took up pace in the 1990s and led to a boom in the field just recently during the 2000s and 2010s.

In the 1990s, DAEs were a hot topic and many multibody simulation tools were developed. Nonsmooth rigid contact models arose as an alternative to the penalty models used thus far. The later penalize interpenetrations of rigid bodies with a stiff spring–damper element at the contact point. While this is a very useful model for many applications, it can be somewhat restrictive for mechanical systems with many contacts.

As more computational power became available, industrial scale simulations of granular material became viable. The classical Discrete Element Method of Cundall and Strack [40] is based on a penalized contact model. Because of the frequent changes in the contact states of the granular particles and because higher order stiff integrators are computationally too expensive for such large systems, very small time steps must be used to maintain a stable simulation. Using the new techniques from nonsmooth dynamical systems, where contacts are modeled via inequality constraints, became an attractive alternative to deal with contacts and collisions in multibody and granular simulations. With the new approach, it is not necessary to capture the change in contact states exactly and the "infinitely stiff" character of the contact laws do not yield stiff differential systems. Much larger time steps can be used.

5.3.1 Event–driven and event–capturing methods

There is a very intuitive method for dealing with nonsmooth events when solving the equations of motion of a dynamical system. Remember, that the velocity is a function of bounded variation, and as such is smooth almost everywhere except at a countable number of discontinuity points. *Event–driven* integrators calculate the smooth trajectory of the system using available ODE or DAE solvers, until such an event is detected. The time integration stops, the event can be handled, e.g. by evaluating a contact model, and the integrator is restarted with new initial conditions until the next nonsmooth event is detected. Today, many ODE and DAE solvers integrate event–detection features. These techniques are useful if the overall number of nonsmooth events is not too large and the time points of the events can be predicted effectively.

But already simple mechanical problems, such as a rigid bouncing ball can be problematic for this technique. If the ball dissipates energy with each impact, the maximum height reached in each bounce decreases and impacts with the ground become more frequent. Eventually there will be an accumulation point in time with infinitely many impacts in between which the ball bounces to ever decreasing heights.

A rigorous event–driven time integrator would not be able to pass this accumulation point without any additional trickery. In addition to this, there are numerous applications where the frequency of nonsmooth events in time is high and the numerical overhead of event–detection becomes the bottleneck of the entire simulation. This is especially true for dynamic simulations of granular matter, where each rigid particle is subject to frequent changes in the contact state. For this reason, this section focuses solely on time integration schemes that can step over nonsmooth events and capture the net movement during time steps, regardless of whether this movement is due to smooth motion or a nonsmooth phenomenon. Integration schemes of this type are often denoted as *event–capturing* or *time–stepping* methods.

Time–Stepping methods are commonly split into two categories, namely the Paoli–Schatzman [133, 134] and the Moreau–Jean time–stepping schemes [77, 78]. The main difference between the two is that the first aims at finding a solution on position level while the latter tackle the problem on a velocity level. In both methods projections onto certain sets must be performed. In the first, projections onto arbitrary convex sets are applied, while in the latter only projection onto the normal and tangent cones to these sets play a role, which are easier to compute in general [4]. In the following, we will concentrate the discussion on Moreau–Jean time–stepping.

5.3.2 Nonsmooth time–stepping

Consider a discrete version of Equation (72a), that can be obtained by integrating over the time interval $[t_n, t_{n+1}]$:

$$M(q_n)(v_{n+1} - v_n) = k_n + G(q_n)p_{n+1} \tag{73}$$

where $q_n \approx q(t_n)$ and $v_n \approx v(t_n) := \dot{q}(t_n)$,

$$p_{n+1} = \int_{t_n}^{t_{n+1}} \mathrm{d}\lambda$$

is an impulse that appears as a new unknown and

$$k_n \approx \int_{t_n}^{t_{n+1}} f(q(s), v(s), s)\, \mathrm{d}s$$

is an approximation of the force integral that can be obtained using a suitable quadrature rule. The generalized position of the mechanical system can be calculated for instance using a Θ–method

$$q_{n+1} = q_n + \tau(\Theta v_{n+1} + (1 - \Theta)v_n), \tag{74}$$

where $\tau = t_{n+1} - t_n$. With $\Theta = 0$, Equation (74) corresponds to the explicit Euler method, $\Theta = \frac{1}{2}$ yields the trapezoidal rule and $\Theta = 1$ the implicit Euler method.

For simplicity, it is assumed here, that the mass matrix M and constraint Jacobian G remain almost constant during a time step and that $f(q(s), v(s), s)$ can be approximated or extrapolated for the time interval $[t_n, t_{n+1}]$. For applications in linear elasticity, f is often linear in q and v, e.g.

$$f = f_{\text{ext}} - Kq - Dv,$$

with stiffness and damping matrices K and D. If in this case the Θ–method is used for the approximation of the force integral k_n, no further extrapolation is needed, see for instance [4]. Convergence results for this time–stepping scheme are provided for instance in [4, 113, 160].

In [83, 151, 152] a more rigorous derivation of the discrete time–stepping equations is provided. Similarly to the way finite element methods are constructed from a weak form of a partial differential equation, the authors consider equation (72) as a variational problem and apply a discontinuous Galerkin approximation in time to obtain the above mentioned time stepping equations.

The Lagrangian multiplier appears as an impulse p_{n+1}, that has the unit of momentum. No approximations for the accelerations and forces exist. The dynamics are solved directly on velocity level. The new unknown p_{n+1} does not distinguish between smooth or nonsmooth parts of the acceleration. This can potentially be a problem in applications, where particularly the loads due to contact forces are of interest [82].

In order to advance the time–stepping scheme, we need to solve for p_{n+1}. The complementarity problem (72b) – (72c) for $\mathrm{d}\lambda$ translates directly to a complementarity problem for the new unknown p_{n+1}.

$$g(q_{n+1}) \geq 0, \qquad p_{n+1} \geq 0, \qquad p_{n+1}^T g(q_{n+1}) = 0. \tag{75}$$

In general, this is a *nonlinear complementarity problem* (NCP). For $\Theta = 1$, Equations (73),(74) and the NCP (75) amount to a nonsmooth version of the SHAKE integrator (44) [66, 67]. The position q_{n+1} is updated using the velocity v_{n+1}, which is calculated in such a way, that the constraint is satisfied at the end of the time step from t_n to t_{n+1}.

Figure 10 shows a simulation of two–dimensional rotationless circular particles. In this simulation, the SHAKE integrator was used. The velocity is piecewise constant and the position is continuous and piecewise linear.

5.3.3 Dealing with collisions

How can we make sure, that the constraints are satisfied at the end of the time step, if we do not use any event detection in our numerical method? There are two common strategies to do this at least approximately.

In the first strategy, we allow the constraints to be violated and assume that the time step size is small enough, so that these violations remain small. Let $\mathscr{I} = \{i_1, ..., i_m \in \mathbb{N}\}$ denote the index set of the active and the violated inequality constraints at time t_n,

$$g_s(q_n) \leq 0, \qquad s \in \mathscr{I}.$$

For this *active set* of constraints it it clear, that the contact velocity must be non–negative by the next time step, so that the constraint is not violated any further,

$$\dot{g}_s(q_{n+1}) = \frac{\partial g_s(q_{n+1})}{\partial q}^T \cdot v_{n+1} \geq 0 \qquad s \in \mathscr{I}. \tag{76}$$

Thus, with the additional assumption that the constraint Jacobian

$$G_{n,s} = \frac{\partial g_s(q_n)}{\partial q} \approx \frac{\partial g_s(q_{n+1})}{\partial q}$$

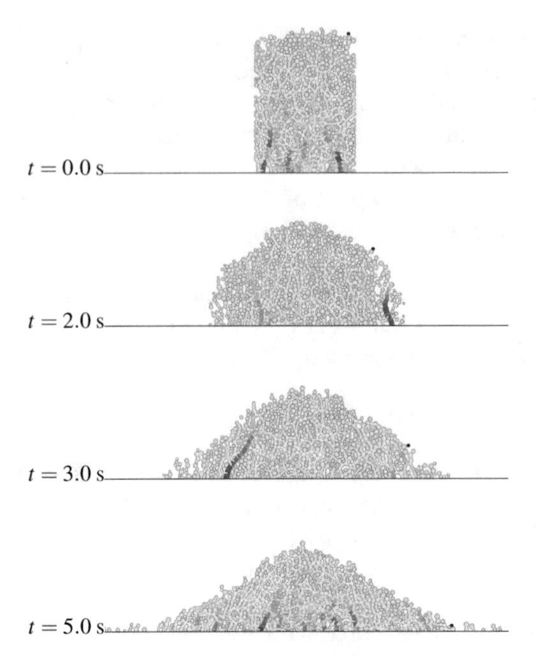

The continuous, piecewise linear position and discontinuous, piecewise constant velocity of a single particle in vertical direction.

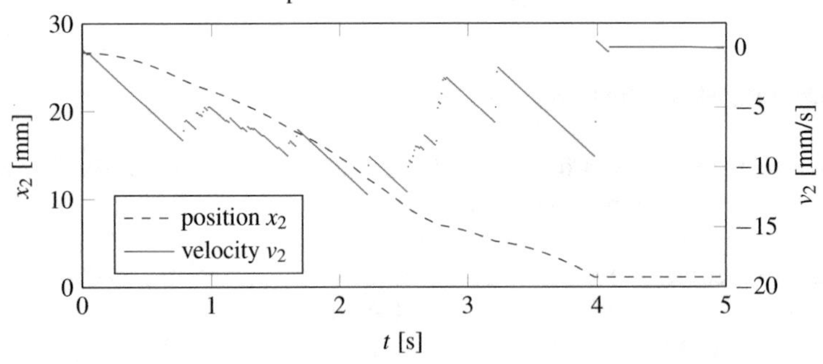

Fig. 10: A simulation of a collapsing pile of 1000 rotationless disks in two dimensions. Figures 10a–10d show snapshots at different time points in the simulation. The images contain visualizations of the so–called *force chains* in the granular material. The vertical position and velocity of the highlighted particle is shown in Figure 10e.

remains almost constant during one time step we can replace the NCP (75) with the following complementarity problem on velocity level, that is solved only for the active set of inequality constraints

$$G_{n,s} \cdot v_{n+1} \geq 0, \qquad p_{n+1,s} \geq 0, \qquad p_{n+1,s} G_{n,s} \cdot v_{n+1} = 0 \;\; \forall\, s \in \mathscr{I}.$$

Here $p_{n+1,s}$ denotes the component of p_{n+1} associated to the inequality constraint with index s. Note that initial violations remain, as the reaction impulse $p_{n+1,s}$ is just large enough to not decrease the value of g_s any further.

The other strategy consists of linearizing the inequality constraint around q_n,

$$\begin{aligned} g(q_{n+1}) &= g\left(q_n + \tau\left(\Theta v_{n+1} + (1-\Theta) v_n\right)\right) \\ &\approx g(q_n) + \tau \frac{\partial g(q_n)}{\partial q}^T \left(\Theta v_{n+1} + (1-\Theta) v_n\right) \geq 0. \end{aligned} \tag{77}$$

In this strategy we are looking ahead: We can calculate the velocity v_{n+1} in such a way, that the inequality constraint is satisfied at the end of the next time step. We do not need to separate the constraints into an active and inactive set. If the value of a linearized inequality constraint is positive and far away from zero, the NCP (75) implies that the corresponding contact impulse must be zero. Dividing the linearized inequality constraint (77) by the time step size τ reveals it as an inequality constraint on velocity level with an additional stabilization term [9]

$$u_{n+1} = \frac{g(q_n)}{\tau} + \frac{\partial g(q_n)}{\partial q}^T \left(\Theta v_{n+1} + (1-\Theta) v_n\right) \geq 0, \tag{78}$$

very similar to Baumgarte stabilization of velocity based bilateral constraints [21].

The NCP (75) is replaced by

$$u_{n+1} \geq 0, \qquad p_{n+1} \geq 0, \qquad p_{n+1}^T u_{n+1} = 0.$$

As the constraints are considered on velocity level, these two strategies are very similar to index reduction methods for DAEs.

Both approaches amount to perfectly inelastic collisions. The unilateral constraint can be modified so that partially or perfectly elastic collisions are modeled, e.g. by incorporating Newton's impact law. In general however, we must separate the collision into a compression and decompression phase to accomplish this without introducing additional errors. This requires the solution of two complementarity problems per time step [11]. And even then, this only models elastic collisions between two bodies. modeling inelastic shocks through a network of contacting rigid bodies, such as for example Newton's cradle, is still a challenge in event–capturing strategies [127].

5.3.4 Dealing with complementarity

Since the early days of numerical methods for nonsmooth dynamical systems and regardless of which concept from Section 5.1 is used for the description of the problem, most authors formulate the discrete numerical method in terms of a complementarity problem similar to the one from the previous section [99, 100, 114]. Posing the inequality constraints (76) or (78) on velocity level has a big advantage: they are linear in the velocity v_{n+1} and therefore also in the unknown impulse p_{n+1}. Solving (73) for v_{n+1} and inserting the result into (78) reveals this,

$$
\begin{aligned}
u_{n+1} =& \frac{g(q_n)}{\tau} + G_n^T\left(\Theta v_{n+1} + (1-\Theta)v_n\right) \\
=& \frac{g(q_n)}{\tau} + G_n^T\left(v_n + \Theta M_n^{-1}k_n\right) \\
& + \Theta G_n^T M_n^{-1} G_n p_{n+1} \\
=& b_n + A_n p_{n+1},
\end{aligned}
$$

where $G_n = G(q_n)$ is the constraint Jacobian, $M_n = M(q_n)$ is the mass matrix,

$$
b_n = \frac{g(q_n)}{\tau} + G_n^T\left(v_n + \Theta M_n^{-1}k_n\right)
$$

and

$$
A_n = \Theta G_n^T M^{-1} G_n
$$

is the *Delassus* matrix. The NCP (75) is transformed to a *linear complementarity problem* (LCP)

$$
u_{n+1} = A_n p_{n+1} + b_n \geq 0, \qquad p_{n+1} \geq 0, \qquad p_{n+1}^T u_{n+1} = 0, \tag{79}
$$

which is equivalent to the *quadratic program* (QP)

$$
\begin{aligned}
\min_{p_{n+1}}\ & p_{n+1}^T A_n p_{n+1} + p_{n+1}^T b_n \\
\text{s.t. } & 0 \leq A_n p_{n+1} + b_n, \\
& 0 \leq p_{n+1}.
\end{aligned} \tag{80}
$$

These LCPs or QPs are often solved using Lemke's algorithm [121], see for example [10, 11, 20, 100, 161]. Lemke's algorithm a pivoting strategy for LCPs similar in spirit to the Simplex algorithm [39]. Lemke's algorithm has a computational complexity of $\mathcal{O}(n^3)$, which makes it impractical as soon as the complementarity problem becomes quite large. This is the case when many bodies are connected through a network of contacts, as for instance in granular assemblies.

5.3.5 Augmented Lagrangian and projected Gauß–Seidel

It has already been established that the complexity of Lemke's algorithm is too restrictive for large problems. The publication [93] includes an alternative approach to deal with frictionless contact problems in the context of finite element methods based on an augmented Lagrangian method. This idea was later adapted by Alart and Curnier in [6] to frictional contact problems. The authors formulate a variational minimization problem at each time step. An energy functional $\varphi(q)$ must be minimized by a trajectory q without violating inequality constraints $g_i(q) \geq 0$, $i = 1, ..., n_\lambda$. This variational inequality–constrained minimization problem is solved using the augmented Lagrangian method. The basic idea is to recast this as an unconstrained optimization problem. For the frictionless case, this leads to

$$\min \Phi_r(q,\gamma) = \varphi(q) - \frac{1}{2r}\sum_{i=1}^{n_\lambda}\left(\|\gamma_i\|^2 - \mathrm{dist}^2(\gamma_i + r g_i(q), \mathbb{R}_+)\right), \tag{81}$$

where

$$\mathrm{dist}(x,C) = x - \mathrm{prox}_C(x)$$

denotes the Euclidean distance of a point x from a set C and $r > 0$ is a numerical factor controlling the steepness of Φ_r. The complementarity conditions and inequality constraints are eliminated from (81). The unknown γ is an approximation of the Lagrangian multiplier associated with the inequality–constrained variational problem. A solution to the minimization problem (81) is sought by solving the saddle point problem $\nabla\Phi_r(q,\gamma) = 0$. In the context of time–stepping methods, the saddle point problem is solved together with (73), where γ takes on the role of the reaction impulse p_{n+1}. Alart and Curnier suggest a generalized Newton method for the solution, that boils down to the following iterative projection method [6, 97, 137, 163]. Let $i = 0$. Until convergence is achieved in $p_{n+1}^{(i)}$, repeat the following steps:

1. $v_{n+1} = v_n + M_n^{-1}(k_n + G_n p_{n+1}^{(i)})$, where $p_{n+1}^{(i)}$ is given from a previous iteration or from an initial guess.
2. Calculate u_{n+1} from v_{n+1} using (78).
3. $p_{n+1}^{(i+1)} = \mathrm{prox}_{\mathbb{R}_+^{n_\lambda}}(p_{n+1}^{(i)} - r u_{n+1})$.
4. $i \leftarrow i+1$.

Note, that for any $r > 0$ the following equivalence holds

$$p = \mathrm{prox}_{\mathbb{R}_+^{n_\lambda}}(p - ru) \quad \Leftrightarrow \quad p \geq 0, \quad u \geq 0, \quad p^T u = 0. \tag{82}$$

In other words, if the iteration converges, it converges to a solution of the complementarity problem (79). For brevity, friction has not been considered here, details can be found in the before mentioned publications. Equation (82) is written as a vector equation. We can rewrite the projection step as

$$p_{n+1}^{(i+1)} = \mathrm{prox}_{\mathbb{R}_+^{n_\lambda}}\left(p_{n+1}^{(i)} - r\left(A_n p_{n+1}^{(i)} + b_n\right)\right),$$

which reveals it as the Gauß–Jacobi method for the solution of linear systems of equations with an additional projection, called the *projected Gauß–Jacobi method method* (PGJ). Of course, given the previous iterate $p_{n+1}^{(i)}$, the equation can be evaluated component–wise and the newly calculated components $j = 1, ..., l < n_\lambda$ can directly be used in the right–hand side of the equation for the calculation for the component with index $l+1$. This method is called the *projected Gauß–Seidel method* (PGS). It has slightly better convergence properties than PGJ, but PGJ lends itself for a parallel implementation. The convergence of both PGJ and PGS can be controlled with a *successive overrelaxation* parameter α,

$$\begin{aligned} \widetilde{p}_{n+1}^{(i+1)} &= \operatorname{prox}_{\mathbb{R}_+^{n_\lambda}} \left(p_{n+1}^{(i)} - r\left(A_n p_{n+1}^{(i)} + b_n \right) \right) \\ p_{n+1}^{(i+1)} &= \alpha \cdot \widetilde{p}_{n+1}^{(i+1)} + (1-\alpha) \cdot p_{n+1}^{(i)}, \end{aligned} \tag{83}$$

yielding the class of *projected successive overrelaxation* (PSOR) methods.

The augmented Lagrangian method is applied to granular simulations in [49]. In [12, 165, 166] a numerical method based on the PGJ, PGS and PSOR schemes is proposed for general cone complementarity problems. The authors use this method to solve for the normal and tangential reaction impulses involved in Coulomb friction simultaneously by directly projecting the three-dimensional contact force onto the Coulomb friction cone. They demonstrate a large number of numerical examples including the simulation of large granular assemblies and provide implementation details in an HPC context.

5.3.6 Recent developments

Over the course of the last two decades several simulation codes were developed around nonsmooth mechanical systems. It is not our intention to compile a complete list, but a few of these codes deserve to be mentioned.

Jean and Dubois initiated the open source software LMGC90 [42, 78, 98] which is currently being developed at the University of Montpellier. SICONOS [157] is an open source software developed at INRIA in Grenoble by the TRITOP team led by Vincent Acary, following the previous works of Bernard Brogliato's BIPOP team. DynamY [162] is another example for a C++ library for nonsmooth mechanical systems that was developed during the course of the PhD thesis [163] at the ETH Zürich. Algoryx [7] is a spin–off software company from Umeå University, that develops commercial software for the simulation of nonsmooth mechanical systems. This software builds on the results on discrete variational methods and the SPOOK stepper developed by Claude Lacoursière and his colleagues [90, 156]. Another free simulation code is the PE Rigid Body Physics Engine [44, 140] developed at the University of Erlangen. Finally, the open source software Chrono [37] is currently being developed by a large team at the Università di Parma and the University of Wisconsin by Tasora, Negrut, Serban and a team of their students.

Recent research in the field of nonsmooth mechanical systems concentrates either on solving large complementarity problems in a short time, or on increasing the order of the integration methods. The first is useful for applications with many rigid bodies in close contact, as is the case for the simulation of granular material. The latter is interesting for general flexible multibody simulation scenarios with a small number of contacts. Here, the presence of unilateral contact and friction decreases the overall order of the integration methods compared to standard DAE solvers, if the Moreau–Jean or Paoli–Schatzman integrators must be used to resolve the collisions consistently.

Fast solvers for complementarity problems

So far, only two strategies have been discussed to deal with the complementarity conditions. The first was the direct solution of an LCP using a pivoting strategy such as Lemke's method. The other was an augmented Lagrangian approach that uses a fix point iteration for the Lagrangian multipliers, which involves a projection onto a feasible set.

Other numerical tools from the field of continuous optimization than Lemke's method can be borrowed to solve the complementarity problems in multibody simulations. For this purpose, it makes sense to introduce two categories of iterative solvers. Consider for this the QP (80), where the objective function takes on the form

$$f(p) = p^T Ap + p^T b.$$

The first category consists of those methods, that only require the gradient $\nabla f(p) = Ap + b$ in every iteration. The second category consists of those methods, that need to solve linear systems involving the Hessian A in every iteration. The intuitive expectation is, that methods from the first category have numerically cheap iterations, but superlinear convergence at best. The second category on the other hand, that includes all Newton–type solution strategies, requires the solution of several large linear systems but has the potential for quadratic convergence.

The PGJ, PGS and PSOR solvers can be seen as methods from the first category. Other methods have been suggested as well, such as projected gradient methods and spectral methods [74, 145]. The numerical results obtained with Nesterov's accelerated projected gradient descent method are very promising [110].

Methods from the second category typically use so–called complementarity functions [46, 105, 167]. A complementarity function is a function that is zero, if and only if the associated NCP is solved. This way, the complementarity problem is recast into the form of a root finding problem. Complementarity functions are typically nonsmooth and the gradient is not guaranteed to exist in general. Therefore the root finding problem must be solved using a semismooth Newton method, where in place of the gradients an arbitrary element from the subgradient of the complementarity function is taken.

In [41] a hybrid strategy is proposed. The authors use a per–contact version of the Fischer–Burmeister complementarity function to model exact Coulomb friction. A semismooth Newton method per contact is used in an inner loop, while the global NCP is solved using a Gauß–Seidel–like outer loop.

An alternative strategy to overcome the nonsmoothness of the complementarity function is to approximate it using a smooth function. This is done by introducing a smoothing parameter α in such a way, that as $\alpha \to 0$, the smoothed function converges to the original complementarity function. Then a series of smooth root finding problems can be solved with a classical Newton method. Each subproblem is solved approximately, possibly with just one Newton iteration, before the smoothing parameter is decreased and the previous iterate serves as an initial guess for the next iteration. This strategy, though motivated completely differently, is very similar in spirit to interior point methods and path–following algorithms [35, 164].

The use of Interior Point Methods (IPM) is proposed in [81, 84, 86, 106]. The strategy consists of using a logarithmic version of the objective function in the QP (80) and adding a logarithmic penalty term, also called potential. The penalty term is infinitely large at the boundary (except at the exact solution) and drives the current iterate towards the interior of the feasible set. The zero–set of the potential is a smooth curve, called the central path. It can be parametrized using a parameter α in such a way, that the smooth curve passes through the solution of the complementarity problem at $\alpha = 0$. The strategy of interior point methods is to use a Newton–iteration to step towards the central path at a given α and then decrease the parameter iteratively. This way, the solution is approached from within a close neighborhood of the central path. A big shortcoming of IPMs is, that the condition of the linear systems involved in the Newton–steps is known to diverge as $\alpha \to 0$. A remedy for this is preconditioning of the linear systems. Typically, a block–diagonal preconditioner is already very effective. IPMs are numerically costly, as a series of linear systems must be solved. But, especially in early iterations, the linear systems must not be solved to a high accuracy, so that inexact search directions can be calculated using Krylov subspace methods with generous tolerances. In [38] a compression based direct linear solver is proposed for the linear systems arising in IPMs in the context of nonsmooth rigid body dynamics.

Comparisons of some of the recent numerical methods for complementarity problems in the context of multibody simulations are provided in [3, 73, 111, 112].

Towards higher order integration

The Moreau–Jean method discussed so far is attractive because it integrates nonsmooth motion at a fixed time step size for an arbitrary number of contacts, while maintaining robustness. Its main shortcoming is the global accuracy of order one, even in smooth phases of motion such as free flight without impacts[5]. If no unilateral contacts are present in a multibody simulation, standard DAE solvers can be

[5] We should be careful when using the notion of the order of truncation error in the context of nonsmooth problems, as this concept only makes sense for differential equations with sufficiently

used with favorable properties such as second order accuracy or more, unconditional stability and controlled numerical damping.

Recently, some effort has been put into finding a compromise between both worlds. This is a mixed strategy that consists of a consistent time–stepping scheme to deal with unilateral contacts and a higher order integration method for the smooth phases of a simulation [2, 36, 152, 153, 163].

The general idea is to separate the time integration into two parts for smooth and nonsmooth motion. The smooth motion is integrated using a higher order DAE–solver on velocity level, i.e. using an index-2 formulation. If a unilateral constraint $g_i(q) \geq 0$ switches from inactive, $g_i(t_{n-1}) > 0$, to active, $g_i(t_n) \leq 0$, during the time step $[t_{n+1}, t_n]$, the impact equations for this unilateral constraint are solved and the smooth motion is updated by the impulse contribution of the impact.

During all time steps that contain a discontinuous velocity jump, the order cannot be expected to be higher than one. Therefore the global order of accuracy of the mixed time–stepping approach cannot exceed one. Locally, during time intervals without impacts however, the accuracy order is that of the DAE solver used for the smooth motion.

The above mentioned literature contains numerical experiments with the Newmark and Hilber–Hughes–Taylor integrators, the generalized α–method as well as several variants of the RADAU and Lobatto Runge–Kutta methods. In the context of higher order integration methods, constraint stabilization using the Gear–Gupta–Leimkuhler algorithm [55], see (22), is discussed in [29, 30, 154].

Summary

From the simulation of electrical circuits to the simulation of large granular assemblies, the advances in the fields of DAEs and nonsmooth dynamical systems in the last few decades were always influenced by applications in industry. New numerical methods such as the multibody formalisms tremendously reduced the time to market of newly developed products, especially in the automotive industry.

Vice versa, the use of these methods and simulation tools in industry has been a driver for the constant improvement in accuracy and efficiency. Their application to real world problems opened up new theoretical questions, many of which could not have been answered without diving deeper into seemingly unrelated mathematical subjects, such as differential geometry, measure theory and convex optimization.

---

smooth right hand sides. The important point here is that when only considering isolated intervals without any nonsmooth events, the Moreau–Jean method is of low order.

Acknowledgments

We would like to thank Alessandro Tasora, Dan Negrut, and Klaus Dreßler for fruitful discussions and valuable hints on constrained mechanical systems and nonsmooth extensions.

References

[1] Abraham, R., Marsden, J.E., Ratiu, T.: Manifolds, Tensor Analysis, and Applications. Springer (1988)
[2] Acary, V.: Higher order event capturing time-stepping schemes for nonsmooth multibody systems with unilateral constraints and impacts. Applied Numerical Mathematics **62**(10), 1259 – 1275 (2012). DOI 10.1016/j.apnum.2012.06.026. URL https://hal.inria.fr/inria-00476398
[3] Acary, V., Brémond, M., Huber, O.: On solving contact problems with coulomb friction: formulations and numerical comparisons. In: Advanced Topics in Nonsmooth Dynamics, pp. 375–457. Springer (2018)
[4] Acary, V., Brogliato, B.: Numerical Methods for Nonsmooth Dynamical Systems: Applications in Mechanics and Electronics. Springer Verlag (2008)
[5] Acary, V., Brogliato, B., Goeleven, D.: Higher order moreau's sweeping process: mathematical formulation and numerical simulation. Mathematical Programming **113**(1), 133–217 (2008). DOI 10.1007/s10107-006-0041-0. URL https://doi.org/10.1007/s10107-006-0041-0
[6] Alart, P., Curnier, A.: A mixed formulation for frictional contact problems prone to Newton like solution methods. Computer Methods in Applied Mechanics and Engineering **92**(3), 353 – 375 (1991)
[7] Algoryx: Homepage. https://www.algoryx.se/ (2018). Accessed: 2018-11-01
[8] Ali, G., Bartel, A., Günther, M., Tischendorf, C.: Elliptic partial differential-algebraic multiphysics models in electrical network design. Mathematical Models and Methods in Applied Sciences **13**(09), 1261–1278 (2003)
[9] Anitescu, M., Hart, G.D.: A constraint-stabilized time-stepping approach for rigid multibody dynamics with joints, contact and friction. International Journal for Numerical Methods in Engineering **60**(14), 2335 – 2371 (2004)
[10] Anitescu, M., Hart, G.D.: A fixed-point iteration approach for multibody dynamics with contact and small friction. Mathematical Programming **101**(1), 3 – 32 (2004). DOI 10.1007/s10107-004-0535-6. URL https://doi.org/10.1007/s10107-004-0535-6
[11] Anitescu, M., Potra, F.A.: Formulating dynamic multi-rigid-body contact problems with friction as solvable linear complementarity problems. Nonlinear Dynamics **14**, 231 – 247 (1997)

[12] Anitescu, M., Tasora, A.: An iterative approach for cone complementarity problems for nonsmooth dynamics. Computational Optimization and Applications **47**(2), 207 – 235 (2010)
[13] Antman, S.S.: The influence of elasticity on analysis: Modern developments. Bull. Amer. Math. Soc. (N.S.) **9**(3), 267 – 291 (1983). URL `https://projecteuclid.org:443/euclid.bams/1183551288`
[14] Arnold, M., Murua, A.: Non-stiff integrators for differential–algebraic systems of index 2. Numerical Algorithms **19**(1-4), 25–41 (1998)
[15] Arnold, V.I.: Ordinary Differential Equations. MIT Press, Cambridge (1981)
[16] Ascher, U., Chin, H., Petzold, L., Reich, S.: Stabilization of constrained mechanical systems with DAEs and invariant manifolds. J. Mech. Struct. Machines **23**, 135–158 (1995)
[17] Ascher, U., Lin, P.: Sequential regularization methods for nonlinear higher index DAEs. SIAM J. Scient. Comput. **18**, 160–181 (1997)
[18] Ascher, U.M., Petzold, L.R.: Projected implicit Runge-Kutta methods for differential-algebraic equations. SIAM J. Numer. Anal. **28**, 1097–1120 (1991)
[19] Aubin, J.P., Cellina, A.: Differential Inclusions: Set-Valued Maps and Viability Theory. Springer-Verlag, Berlin, Heidelberg (1984)
[20] Baraff, D.: Dynamic simulation of non-penetrating rigid bodies. Ph.D. thesis (1992)
[21] Baumgarte, J.: Stabilization of constraints and integrals of motion in dynamical systems. Comp. Meth. in Appl. Mechanics **1**, 1 – 16 (1972)
[22] Benner, P., Losse, P., Mehrmann, V., Voigt, M.: Numerical linear algebra methods for linear differential-algebraic equations. In: Surveys in Differential-Algebraic Equations III, A. Ilchmann and T. Reis (Eds.), Springer DAE-Forum, pp. 117–175 (2015)
[23] Brasey, V., Hairer, E.: Half-explicit Runge–Kutta methods for differential-algebraic systems of index 2. SIAM J. Numer. Anal. **30**, 538–552 (1993)
[24] Brenan, K.E., Campbell, S.L., Petzold, L.R.: The Numerical Solution of Initial Value Problems in Ordinary Differential-Algebraic Equations. SIAM, Philadelphia (1996)
[25] Brezzi, F., Fortin, M.: Mixed and Hybrid Finite Element Methods. Springer, New York (1991)
[26] Brizard, A.: An Introduction to Lagrangian Mechanics. World Scientific Publishing (2008)
[27] Brogliato, B.: Nonsmooth Mechanics – Models, Dynamics and Control. Springer Verlag (2016). DOI 10.1007/978-3-319-28664-8
[28] Brogliato, B., Daniilidis, A., Lemarechal, C., Acary, V.: On the equivalence between complementarity systems, projected systems and unilateral differential inclusions. Systems and Control Letters **55**(1), 45 – 51 (2006)
[29] Brüls, O., Acary, V., Cardona, A.: Simultaneous enforcement of constraints at position and velocity levels in the nonsmooth generalized-α scheme. Computer Methods in Applied Mechanics and Engineering **281**, 131 – 161 (2014). DOI 10.1016/j.cma.2014.07.025. URL `https://hal.inria.fr/hal-01059823`

[30] Brüls, O., Acary, V., Cardona, A.: On the Constraints Formulation in the Nonsmooth Generalized-α Method. In: S.I. Publishing (ed.) Advanced Topics in Nonsmooth Dynamics. Transactions of the European Network for Nonsmooth Dynamics, pp. 335 – 374 (2018). DOI 10.1007/978-3-319-75972-2_9. URL https://hal.inria.fr/hal-01878550
[31] Callies, R., Rentrop, P.: Optimal control of rigid-link manipulators by indirect methods. GAMM-Mitteilungen **31**(1), 27–58 (2008)
[32] Campbell, S., Gear, C.: The index of general nonlinear DAEs. Numer. Math. **72**, 173–196 (1995)
[33] Campbell, S.L.: Singular Systems of Differential Equations II. Research Notes in Mathematics 61. Pitman (1982)
[34] Campbell, S.L.: Least squares completions for nonlinear differential-algebraic equations. Numer. Math. **65**, 77–94 (1993)
[35] Chen, B., Chen, X., Kanzow, C.: A penalized Fischer-Burmeister NCP-function. Mathematical Programming **88**(1), 211 – 216 (2000)
[36] Chen, Q.Z., Acary, V., Virlez, G., Brüls, O.: A nonsmooth generalized- α scheme for flexible multibody systems with unilateral constraints. International Journal for Numerical Methods in Engineering **96**(8), 487 – 511 (2013). DOI 10.1002/nme.4563. URL https://onlinelibrary.wiley.com/doi/abs/10.1002/nme.4563
[37] Chrono, P.: C++ library for multi-physics simulation. https://github.com/projectchrono/chrono (2018). Accessed: 2018-11-01
[38] Corona, E., Rahimian, A., Zorin, D.: A tensor-train accelerated solver for integral equations in complex geometries. Journal of Computational Physics **334**, 145 – 169 (2017). DOI https://doi.org/10.1016/j.jcp.2016.12.051. URL http://www.sciencedirect.com/science/article/pii/S0021999116307185
[39] Cottle, R., Pang, J., Stone, R.: The Linear Complementarity Problem. Classics in Applied Mathematics. Society for Industrial and Applied Mathematics (SIAM, 3600 Market Street, Floor 6, Philadelphia, PA 19104) (1992). URL https://books.google.de/books?id=bGM80_pSzNIC
[40] Cundall, P.A., Strack, O.D.L.: A discrete numerical model for granular assemblies. Géotechnique **29**(1), 47 – 65 (1979)
[41] Daviet, G., Bertails-Descoubes, F., Boissieux, L.: A hybrid iterative solver for robustly capturing Coulomb friction in hair dynamics. ACM Transactions on Graphics (TOG) - Proceedings of ACM **30**(6) (2011)
[42] Dubois, F., Jean, M., Renouf, M., Mozul, R., Martin, A., Bagnéris, M.: LMGC90. In: 10e colloque national en calcul des structures, p. Clé USB. Giens, France (2011). URL https://hal.archives-ouvertes.fr/hal-00596875
[43] Eich, E.: Convergence results for a coordinate projection method applied to constrained mechanical systems. SIAM J. Numer. Anal. **30**(5), 1467–1482 (1993)
[44] Engine, P.P.: Homepage. https://www.cs10.tf.fau.de/research/software/pe/ (2018). Accessed: 2018-11-01

[45] Erdmann, G.: Ueber unstetige lösungen in der variationsrechnung. Journal für die reine und angewandte Mathematik **82**, 21–30 (1877)
[46] Ferris, M.C., Kanzow, C.: Complementarity and related problems: A survey. Mathematical Programming Technical Report **17** (1998)
[47] Fetecau, R.C., Marsden, J.E., Ortiz, M., West, M.: Nonsmooth Lagrangian mechanics and variational collision integrators. SIAM Journal on Applied Dynamical Systems **2**(3), 381 – 416 (2003)
[48] Filippov, A.F.: Differential equations with discontinuous righthand sides: control systems, *Mathematics and its Applications*, vol. 18. Springer Science & Business Media (1988)
[49] Fortin, J., Hjiaj, M., de Saxcé, G.: An improved discrete element method based on a variational formulation of the frictional contact law. Computers and Geotechnics **29**(8), 609 – 640 (2002). DOI https://doi.org/10.1016/S0266-352X(02)00016-2. URL `http://www.sciencedirect.com/science/article/pii/S0266352X02000162`
[50] Führer, C., Leimkuhler, B.: Numerical solution of differential-algebraic equations for constrained mechanical motion. Numer. Math.. **59**, 55–69 (1991)
[51] Gantmacher, F.: Matrizenrechnung, Teil 2. VEB Deutscher Verlag der Wissenschaften, Berlin (1959)
[52] Garcia de Jalón, J., Bayo, E.: Kinematic and Dynamic Simulation of Multibody Systems. Springer (1994)
[53] Gear, C.: Numerical Initial Value Problems in Ordinary Differential Equations. Prentice-Hall (1971)
[54] Gear, C.: Simultaneous numerical solution of differential-algebraic equations. IEEE Trans. Circuit Theory **CT-18**(1), 89 – 95 (1971)
[55] Gear, C., Gupta, G., Leimkuhler, B.: Automatic integration of the Euler-Lagrange equations with constraints. J. Comp. Appl. Math. **12 & 13**, 77–90 (1985)
[56] Gear, C.W.: Differential-algebraic equation index transformation. SIAM J. Sci. & Statist. Comp. **9**, 39–47 (1988)
[57] Gear, C.W.: Differential-algebraic equations, indices, and integral algebraic equations. SIAM J. Numer. Anal. **27**, 1527–1534 (1990)
[58] Gerdts, M.: Optimal Control of ODEs and DAEs. De Gruyter (2012)
[59] Glocker, C.: Set-valued force laws in rigid body dynamics. Habilitation thesis, Lehrstuhl für Mechanik, Technische Universität München (2000)
[60] Griepentrog, E., März, R.: Differential-Algebraic Equations and Their Numerical Treatment. Teubner-Texte zur Mathematik No. 88. Teubner Verlagsgesellschaft, Leipzig (1986)
[61] Günther, M.: Partielle differential-algebraische Systeme in der numerischen Zeitbereichsanalyse elektrischer Schaltungen. VDI-Verlag, Reihe 20, Düsseldorf (2001)
[62] Günther, M., Feldmann, U.: CAD based electric circuit modeling in industry I: mathematical structure and index of network equations. Surv. Math. Ind. **8**, 97–129 (1999)

[63] Günther, M., Hoschek, M., Rentrop, P.: Differential-algebraic equations in electric circuit simulation. Int. J. Electron. Commun. **54**, 101–107 (2000)
[64] Hairer, E., Lubich, C., Roche, M.: Error of Runge-Kutta methods for stiff problems studied via differential algebraic equations. BIT Numerical Mathematics **28**(3), 678–700 (1988)
[65] Hairer, E., Lubich, C., Roche, M.: The Numerical Solution of Differential-Algebraic Equations by Runge-Kutta Methods. Lecture Notes in Mathematics Vol. 1409. Springer, Heidelberg (1989)
[66] Hairer, E., Lubich, C., Wanner, G.: Geometric Numerical Integration. Springer-Verlag, Berlin (2002)
[67] Hairer, E., Lubich, C., Wanner, G.: Geometric numerical integration illustrated by the Störmer–Verlet method. Acta Numerica **12**, 399 – 450 (2003)
[68] Hairer, E., Nørsett, S., Wanner, G.: Solving Ordinary Differential Equations I: Nonstiff Problems. Springer-Verlag, Berlin (1993)
[69] Hairer, E., Wanner, G.: Solving Ordinary Differential Equations II: Stiff and Differential-Algebraic Problems. Springer-Verlag, Berlin (1996)
[70] Hairer, E., Wanner, G.: Stiff differential equations solved by Radau methods. J. Comp. Appl. Math. **111**, 93–111 (1999)
[71] Hanke, M.: On the regularization of index 2 differential-algebraic equations. J. Math. Anal. Appl. **151**(1), 236–253 (1990)
[72] Haug, E.: Computer-Aided Kinematics and Dynamics of Mechanical Systems. Allyn and Bacon, Boston (1989)
[73] Heyn, T.: On the modeling, simulation, and visualization of many–body dynamics problems with friction and contact. Ph.D. thesis, Uiversity of Wisconsin–Madison (2013)
[74] Heyn, T., Anitescu, M., Tasora, A., Negrut, D.: Using Krylov subspace and spectral methods for solving complementarity problems in many-body contact dynamics simulation. International Journal for Numerical Methods in Engineering **95**(7), 541 – 561 (2013)
[75] Higham, D.J., Mao, X., Stuart, A.M.: Strong convergence of Euler-type methods for nonlinear stochastic differential equations. SIAM Journal on Numerical Analysis **40**(3), 1041–1063 (2002)
[76] Hindmarsh, A.C., Brown, P.N., Grant, K.E., Lee, S.L., Serban, R., Shumaker, D.E., Woodward, C.S.: Sundials: Suite of nonlinear and differential/algebraic equation solvers. ACM Transactions on Mathematical Software **31(3)**, 363–396 (2005)
[77] Jean, M.: The non-smooth contact dynamics method. Computer Methods in Applied Mechanics and Engineering **177**(3 – 4), 235 – 257 (1999)
[78] Jourdan, F., Alart, P., Jean, M.: A gauss-seidel like algorithm to solve frictional contact problems. Computer Methods in Applied Mechanics and Engineering **155**(1), 31 – 47 (1998). DOI https://doi.org/10.1016/S0045-7825(97)00137-0. URL `http://www.sciencedirect.com/science/article/pii/S0045782597001370`
[79] Karamardian, S.: Generalized complementarity problem. Journal of Optimization Theory and Applications **8**, 161 – 168 (1971)

[80] Kirchhoff, G.: Ueber die Auflösung der Gleichungen, auf welche man bei der Untersuchung der linearen Vertheilung galvanischer Ströme geführt wird. Annalen der Physik **148**(12), 497–508 (1847)
[81] Kleinert, J.: Simulating granular material using nonsmooth time–stepping and a matrix–free interior point method. Ph.D. thesis, Technische Universität Kaiserslautern (2015)
[82] Kleinert, J., Obermayr, M., Balzer, M.: Modeling of large scale granular systems using the discrete element method and the non-smooth contact dynamics method: A comparison. In: Proceedings of the ECCOMAS Multibody Dynamics Conference, Zagreb (2013)
[83] Kleinert, J., Simeon, B., Dreßler, K.: Nonsmooth contact dynamics for the large-scale simulation of granular material. Journal of Computational and Applied Mathematics **316**, 345 – 357 (2017). DOI https://doi.org/10.1016/j.cam.2016.09.037. URL `http://www.sciencedirect.com/science/article/pii/S0377042716304575`. Selected Papers from NUMDIFF-14
[84] Kleinert, J., Simeon, B., Obermayr, M.: An inexact interior point method for the large-scale simulation of granular material. Computer Methods in Applied Mechanics and Engineering **278**(0), 567 – 598 (2014)
[85] Körkel, S., Kostina, E., Bock, H.G., Schlöder, J.P.: Numerical methods for optimal control problems in design of robust optimal experiments for nonlinear dynamic processes. Optimization Methods and Software **19**(3-4), 327–338 (2004)
[86] Krabbenhoft, K., Lyamin, A.V., Huang, J., da Silva, M.V.: Granular contact dynamics using mathematical programming methods. Computers and Geotechnics **43**(0), 165 – 176 (2012)
[87] Kronecker, L.: Algebraische Reduktion der Schaaren bilinearer Formen. Akademie der Wissenschaften Berlin **III**, 141–155 (1890)
[88] Kunkel, P., Mehrmann, V.: Differential-Algebraic Equations – Analysis and Numerical Solution. EMS Publishing House (2006)
[89] Kurcyusz, S.: On the existence and nonexistence of Lagrange multipliers in banach spaces. Journal of Optimization Theory and Applications **20**(1), 81 – 110 (1976)
[90] Lacoursière, C.: Ghosts and machines : regularized variational methods for interactive simulations of multibodies with dry frictional contacts. Ph.D. thesis, Umeå University, Computing Science (2007)
[91] Lagrange, J.L.: Méchanique analytique. Libraire chez la Veuve Desaint, Paris (1788)
[92] Lamour, R., März, R., Tischendorf, C.: Differential-Algebraic Equations: A Projector Based Analysis (Differential-Algebraic Equations Forum). Springer (2013)
[93] Landers, J.A., Taylor, R.L.: An augmented lagrangian formulation for the finite element solution of contact problems. Tech. rep., CALIFORNIA UNIV BERKELEY DEPT OF CIVIL ENGINEERING (1986)

[94] Lang, J.: Adaptive multilevel solution of nonlinear parabolic pde systems - theory, algorithm, and applications. In: Lecture Notes in Computational Science and Engineering (2001)
[95] Leela, S.: Stability of measure differential equations. Pacific J. Math. **55**(2), 489 – 498 (1974). URL https://projecteuclid.org:443/euclid.pjm/1102910983
[96] Leine, R.I., Aeberhard, U., Glocker, C.: Hamilton's principle as variational inequality for mechanical systems with impact. Journal of Nonlinear Science **19**(6), 633 – 664 (2009)
[97] Leine, R.I., Nijmeijer, H.: Bifurcations of equilibria in non-smooth continuous systems. In: Dynamics and Bifurcations of Non-Smooth Mechanical Systems, pp. 125–176. Springer (2004)
[98] LMGC90: Simulation platform. https://git-xen.lmgc.univ-montp2.fr/lmgc90/lmgc90_user (2018). Accessed: 2018-11-01
[99] Lötstedt, P.: Mechanical systems of rigid bodies subject to unilateral constraints. SIAM J. Appl. Math. **42**, 281–296 (1982)
[100] Lötstedt, P.: Numerical simulation of time-dependent contact and friction problems in rigid body mechanics. SIAM journal on scientific and statistical computing **5**(2), 370 – 393 (1984)
[101] Lötstedt, P., Petzold, L.: Numerical solution of nonlinear differential equations with algebraic constraints i: Convergence results for BDF. Math.Comp. **46**, 491–516 (1986)
[102] Lubich, C.: h^2 extrapolation methods for differential-algebraic equations of index-2. Impact Comp. Sci. Eng. **1**, 260–268 (1989)
[103] Lubich, C.: Integration of stiff mechanical systems by Runge-Kutta methods. ZAMP **44**, 1022–1053 (1993)
[104] Lubich, C., Engstler, C., Nowak, U., Pöhle, U.: Numerical integration of constrained mechanical systems using MEXX. Mech. Struct. Mach. **23**, 473–495 (1995)
[105] Mangasarian, O.L.: Equivalence of the complementarity problem to a system of nonlinear equations. SIAM Journal on Applied Mathematics **31**(1), 89 – 92 (1976)
[106] Mangoni, D., Tasora, A., Garziera, R.: A primal–dual predictor–corrector interior point method for non-smooth contact dynamics. Computer Methods in Applied Mechanics and Engineering **330**, 351 – 367 (2018). DOI https://doi.org/10.1016/j.cma.2017.10.030. URL http://www.sciencedirect.com/science/article/pii/S004578251730703X
[107] März, R., Tischendorf, C.: Recent results in solving index-2 differential-algebraic equations in circuit simulation. SIAM J. Sci. Comp. **18**, 139–159 (1997)
[108] Mattson, S., Söderlind, G.: Index reduction in differential-algebraic equations using dummy derivatives. SIAM Sc.Stat.Comp. (3), 677–692 (1993)
[109] Mattsson, S.E., Elmqvist, H., Otter, M.: Physical system modeling with modelica. Control Engineering Practice **6**(4), 501–510 (1998)

[110] Mazhar, H., Heyn, T., Negrut, D., Tasora, A.: Using nesterov's method to accelerate multibody dynamics with friction and contact. ACM Trans. Graph. **34**(3), 1 – 14 (2015). DOI 10.1145/2735627. URL http://doi.acm.org/10.1145/2735627
[111] Melanz, D., Fang, L., Jayakumar, P., Negrut, D.: A comparison of numerical methods for solving multibody dynamics problems with frictional contact modeled via differential variational inequalities. Computer Methods in Applied Mechanics and Engineering **320**, 668 – 693 (2017). DOI https://doi.org/10.1016/j.cma.2017.03.010. URL http://www.sciencedirect.com/science/article/pii/S0045782516317005
[112] Melanz, D.J.: Physics-based contact using the complementarity approach for discrete element applications in vehicle mobility and terramechanics. Ph.D. thesis, The University of Wisconsin-Madison (2016)
[113] Monteiro Marques, M.: Differential Inclusions in Nonsmooth Mechanical Problems: Shocks and Dry Friction. Progress in Nonlinear Differential Equations and Their Applications. Birkhäuser Basel (1993)
[114] Moreau, J.J.: Quadratic Programming in Mechanics: Dynamics of One-Sided Constraints. SIAM Journal on Control and Optimization **4**(1), 153 – 158 (1966). DOI 10.1137/0304014. URL https://hal.archives-ouvertes.fr/hal-01379713
[115] Moreau, J.J.: Evolution problem associated with a moving convex set in a Hilbert space. Journal of Differential Equations **26**(3), 347 – 374 (1977)
[116] Moreau, J.J.: Bounded variation in time. In: J.J. Moreau, P.D. Panagiotopoulos, G. Strang (eds.) Topics in nonsmooth mechanics, pp. 1 – 74. Birkhäuser Verlag (1988)
[117] Moreau, J.J.: Unilateral Contact and Dry Friction in Finite Freedom Dynamics, pp. 1 – 82. Springer Vienna, Vienna (1988). DOI 10.1007/978-3-7091-2624-0_1. URL https://doi.org/10.1007/978-3-7091-2624-0_1
[118] Moreau, J.J.: Some numerical methods in multibody dynamics: application to granular materials. European Journal of Mechanics-A/Solids **13**(4-suppl), 93 – 114 (1993)
[119] Moreau, J.J.: Numerical aspects of the sweeping process. Computer methods in applied mechanics and engineering **177**, 329 – 349 (1999)
[120] Moreau, J.J., Panagiotopoulos, P.D.: Nonsmooth mechanics and applications. ZAMM - Journal of Applied Mathematics and Mechanics / Zeitschrift für Angewandte Mathematik und Mechanik **70**(10), 472 – 472 (1990)
[121] Murty, K.G.: Linear Complementarity, Linear and Nonlinear Programming. Sigma series in applied mathematics. Heldermann Verlag (1988)
[122] Nagel, L.W., Pederson, D.: Spice (simulation program with integrated circuit emphasis). Tech. Rep. UCB/ERL M382, EECS Department, University of California, Berkeley (1973). URL http://www.eecs.berkeley.edu/Pubs/TechRpts/1973/22871.html
[123] Nagurney, A., Zhang, D.: Projected Dynamical Systems and Variational Inequalities with Applications. Innovations in Financial Markets and Institutions.

Springer US (1995). URL https://books.google.de/books?id=EaG3Wvc_MbgC

[124] Natanson, I., Boron, L.: Theory of Functions of a Real Variable. No. Bd. 1 in Teoria functsiy veshchestvennoy peremennoy. Frederick Ungar (1964). URL https://books.google.de/books?id=l4dCmgEACAAJ

[125] Nedialkov, N.S., Pryce, J.D.: Solving differential-algebraic equations by taylor series (i): Computing taylor coefficients. BIT Numerical Mathematics **45**(3), 561–591 (2005)

[126] Nedialkov, N.S., Pryce, J.D.: Solving differential-algebraic equations by taylor series (ii): Computing the system jacobian. BIT Numerical Mathematics **47**(1), 121–135 (2007)

[127] Nguyen, N.S., Brogliato, B.: Comparisons of Multiple-Impact Laws For Multibody Systems: Moreau's Law, Binary Impacts, and the LZB Approach, pp. 1 – 45. Springer International Publishing, Cham (2018). DOI 10.1007/978-3-319-75972-2_1. URL https://doi.org/10.1007/978-3-319-75972-2_1

[128] O'Malley, R.E.: Introduction to Singular Perturbations. Academic Press, New York (1974)

[129] Panagiotopoulos, P.D., Glocker, C.: Inequality constraints with elastic impacts in deformable bodies. the convex case. Archive of Applied Mechanics **70**(5), 349 – 365 (2000)

[130] Pandit, S., Deo, S.: Differential Systems Involving Impulses. Lecture Notes in Mathematics. Springer Berlin Heidelberg (1982). URL https://books.google.de/books?id=n8gZAQAAIAAJ

[131] Pang, J.S., Stewart, D.E.: Differential variational inequalities. Mathematical Programming **113**, 345 – 424 (2008)

[132] Paoli, L.: An existence result for non-smooth vibro-impact problems. Journal of Differential Equations **211**(2), 247 – 281 (2005). DOI https://doi.org/10.1016/j.jde.2004.11.008. URL http://www.sciencedirect.com/science/article/pii/S0022039604005133

[133] Paoli, L., Schatzman, M.: A numerical scheme for impact problems i: The one-dimensional case. SIAM Journal on Numerical Analysis **40**(2), 702–733 (2002)

[134] Paoli, L., Schatzman, M.: A numerical scheme for impact problems ii: The multidimensional case. SIAM journal on numerical analysis **40**(2), 734–768 (2002)

[135] Petzold, L.: A description of DASSL: A differential/algebraic system solver. In: Proc. 10th IMACS World Congress, August 8-13 Montreal 1982 (1982)

[136] Pfeiffer, F.: The idea of complementarity in multibody dynamics. Archive of Applied Mechanics **72**(11), 807 – 816 (2003). DOI 10.1007/s00419-002-0256-3. URL https://doi.org/10.1007/s00419-002-0256-3

[137] Pfeiffer, F., Foerg, M., Ulbrich, H.: Numerical aspects of non-smooth multibody dynamics. Computer Methods in Applied Mechanics and Engineering **195**(50 – 51), 6891 – 6908 (2006)

[138] Pfeiffer, F., Glocker, C.: Multibody Dynamics with Unilateral Contacts. Wiley Series in Nonlinear Science. Wiley (1996). URL https://books.google.de/books?id=XrZN_TBfiGcC
[139] Piccoli, B.: Measure differential equations. arXiv preprint arXiv:1708.09738 (2017)
[140] Preclik, T.: Models and algorithms for ultrascale simulations of non-smooth granular dynamics. doctoralthesis, Friedrich-Alexander-Universität Erlangen-Nürnberg (FAU) (2014)
[141] Pryce, J.D.: A simple structural analysis method for daes. BIT Numerical Mathematics **41**(2), 364–394 (2001)
[142] Rabier, P., Rheinboldt, W.: Theoretical and numerical analysis of differential-algebraic equations. In: P. Ciarlet, J. Lions (eds.) Handbook of Numerical Analysis, Volume VIII. Elsevier, Amsterdam (2002)
[143] Reich, S.: On a geometric interpretation of DAEs. Circ. Syst., Sig. Processing **9**, 367–382 (1990)
[144] Reis, T., Stykel, T.: Stability analysis and model order reduction of coupled systems. Mathematical and Computer Modelling of Dynamical Systems **13**(5), 413–436 (2007)
[145] Renouf, M., Bonamy, D., Dubois, F., Alart, P.: Numerical simulation of two-dimensional steady granular flows in rotating drum: On surface flow rheology. Physics of Fluids **17**(10) (2005)
[146] Rentrop, P., Roche, M., Steinebach, G.: The application of Rosenbrock-Wanner type methods with stepsize control in differential-algebraic equations. Num. Math. **55**, 545–563 (1989)
[147] Rheinboldt, W.: Differential - algebraic systems as differential equations on manifolds. Math. Comp. **43**(168), 2473–482 (1984)
[148] Rheinboldt, W.: Manpak: A set of algorithms for computations on implicitly defined manifolds. Computers and Math. Applic. **32**, 15–28 (1996)
[149] Rockafellar, R.T., Wets, R.J.B.: Variational Analysis. Grundlehren der mathematischen Wissenschaften. Springer Berlin Heidelberg (1998)
[150] Schiehlen, W. (ed.): Multibody System Handbook. Springer, Heidelberg (1990)
[151] Schindler, T., Acary, V.: Application of timestepping schemes based on time discontinuous galerkin methods to multi-dimensional examples. In: Euromech 514-New trends in Contact Mechanics (2012)
[152] Schindler, T., Acary, V.: Timestepping schemes for nonsmooth dynamics based on discontinuous Galerkin methods: Definition and outlook. Mathematics and Computers in Simulation **95**(0), 180 – 199 (2014)
[153] Schindler, T., Rezaei, S., Kursawe, J., Acary, V.: Half-explicit timestepping schemes on velocity level based on time-discontinuous galerkin methods. Computer Methods in Applied Mechanics and Engineering **290**, 250 – 276 (2015)
[154] Schoeder, S., Ulbrich, H., Schindler, T.: Discussion of the Gear–Gupta–Leimkuhler method for impacting mechanical systems. Multibody System Dynamics **31**(4), 477 – 495 (2014)
[155] Schwerin, R.: Multibody System Simulation. Springer, Berlin (1999)

[156] Servin, M., Wang, D., Lacoursiere, C., Bodin, K.: Examining the smooth and nonsmooth discrete element approaches to granular matter. International Journal for Numerical Methods in Engineering **97**(12), 878–902 (2014)

[157] Siconos: Simulation framework for nonsmooth dynamical systems. `https://github.com/siconos/siconos` (2018). Accessed: 2018-11-01

[158] Simeon, B.: Computational Flexible Multibody Dynamics: A Differential-Algebraic Approach. Springer (2013)

[159] Stewart, D.: Reformulations of measure differential inclusions and their closed graph property. Journal of Differential Equations **175**(1), 108 – 129 (2001). DOI https://doi.org/10.1006/jdeq.2000.3968. URL `http://www.sciencedirect.com/science/article/pii/S0022039600939688`

[160] Stewart, D.E.: Convergence of a time-stepping scheme for rigid-body dynamics and resolution of Painlevé's problem. Archive for Rational Mechanics and Analysis **145**(3), 215 – 260 (1998)

[161] Stewart, D.E., Trinkle, J.C.: An implicit time-stepping scheme for rigid body dynamics with inelastic collisions and Coulomb friction. International Journal for Numerical Methods in Engineering **39**(15), 2673 – 2691 (1996)

[162] Studer, C.: C++ class library for the simulation of mechanical systems with unilateral contacts and friction. `http://www.zfm.ethz.ch/dynamY/` (2008). Accessed: 2018-11-01

[163] Studer, C.W.: Augmented time-stepping integration of non-smooth dynamical systems. Ph.D. thesis, ETH Zürich (2008)

[164] Sun, D., Qi, L.: On NCP-functions. Computational Optimization and Applications **13**(1 – 3), 201 – 220 (1999). DOI 10.1023/A:1008669226453

[165] Tasora, A., Anitescu, M.: A matrix-free cone complementarity approach for solving large-scale, nonsmooth, rigid body dynamics. Computer Methods in Applied Mechanics and Engineering **200**, 439 – 453 (2011)

[166] Tasora, A., Negrut, D., Anitescu, M.: Large-scale parallel multi-body dynamics with frictional contact on the graphical processing unit. Proceedings of the Institution of Mechanical Engineers, Part K: Journal of Multi-body Dynamics **222**(4), 315 – 326 (2008)

[167] Wanzke, C.: Partikelsimulation mit halbglatten Newton-Verfahren. Master's thesis, Technische Universität Kaiserslautern (2013)

[168] Weierstrass, K.: Zur Theorie der bilinearen und quadratischen Formen. Monatsber. Akad. Wiss. Berlin pp. 310–338 (1868)

[169] Winkler, R.: Stochastic differential algebraic equations of index 1 and applications in circuit simulation. Journal of computational and applied mathematics **157**(2), 477–505 (2003)

[170] Zowe, J., Kurcyusz, S.: Regularity and stability for the mathematical programming problem in Banach spaces. Applied Mathematics and Optimization **5**(1), 49 – 62 (1979)

Fast Numerical Methods to Compute Periodic Solutions of Electromagnetic Models

Alfredo Bermúdez, Dolores Gómez, Marta Piñeiro and Pilar Salgado

Abstract The numerical simulation of electrical machines is of crucial interest in electrical engineering, where design optimization, time-to-market and cost effectiveness have become major concerns in order to face competition in the global market. In the case of squirrel-cage induction machines, one of the challenges is to reduce the CPU time needed to reach the steady-state during the computations, which can take several days whereas engineers are only interested in the last machine revolution needing just a few minutes. Thus, being able to reduce this simulation time is crucial to get a competitive tool at the design stage. This chapter summarizes the history of how a real industrial problem led to the development of a novel numerical algorithm for the solution of the above problem, since the mathematical methods existing at that moment were not completely satisfactory, in terms of computational time, to approximate the steady-state behaviour of squirrel-cage induction machines. Starting from a brief historical background, the authors explain the main problem to solve, the limitations of the existing techniques and the novel methodology, illustrated with very promising numerical results. The method still has different aspects to be exploited that would allow not only to expand the range of applications in the field of electrical engineering, but also in other areas such as acoustics or hydrodynamics.

Alfredo Bermúdez, Dolores Gómez and Pilar Salgado
Dpto. de Matemática Aplicada & Instituto de Matemáticas (IMAT) & Instituto Tecnológico de Matemática Industrial (ITMATI), Universidade de Santiago de Compostela, ES-15782 Santiago de Compostela, Spain. e-mail: alfredo.bermudez@usc.es, mdolores.gomez@usc.es, mpilar.salgado@usc.es

Marta Piñeiro
Dpto. de Matemática Aplicada, Universidade de Santiago de Compostela, ES-15782 Santiago de Compostela, Spain. e-mail: marta.pineiro@usc.es

M. Günther, W. Schilders (eds.), *Novel Mathematics Inspired by Industrial Challenges*, Mathematics in Industry 38, https://doi.org/10.1007/978-3-030-96173-2_5

1 Starting Point in Electrical Engineering

During the last two decades, the authors of this chapter, as members of the Research Group in Mathematical Engineering (mat+i; `www.usc.es/ingmat`) of the Universidade de Santiago de Compostela (Spain), have been working in the mathematical analysis and numerical simulation of different industrial problems. In particular, one of their main research lines is related to the study of low-frequency electromagnetic problems. During year 2012, they aimed their attention at the numerical solution of periodic problems exhibiting long-lasting transient states, eventually converging to the sought stationary regime. For instance, this phenomenon arises when solving nonlinear transient magnetic problems coupled with electrical circuit equations, with sources given in terms of periodic voltage drops in conductors. Indeed, this system has an underdamped-forced response, so that, when the simple step-by-step procedure is chosen for the solution, reaching the steady-state may need integrating the system along many periods, leading to very high computational costs. For instance, Fig. 1 shows the typical behaviour over time of the current through a conductor in which a sinusoidal voltage drop source is defined. In this particular case, the model needed about 60 periods to reach the steady-state. To overcome this drawback, the

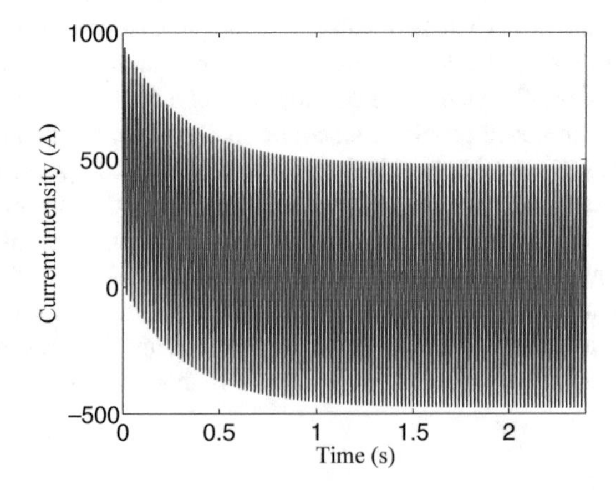

Fig. 1: Current (A) vs. time (s) through a conductor in which a voltage drop source is given.

authors in [1] introduced a method allowing for the fast computation of the stationary solution in the case of a 2D transient magnetic model neglecting eddy currents, written in terms of the magnetic vector potential. The proposed methodology was based on computing suitable initial conditions in conductors in which voltage drops are given, by solving a nonlinear system of equations with a small number of unknowns. The results were presented in an international conference and attracted the attention of some researchers from Robert Bosch GmbH. These researchers were

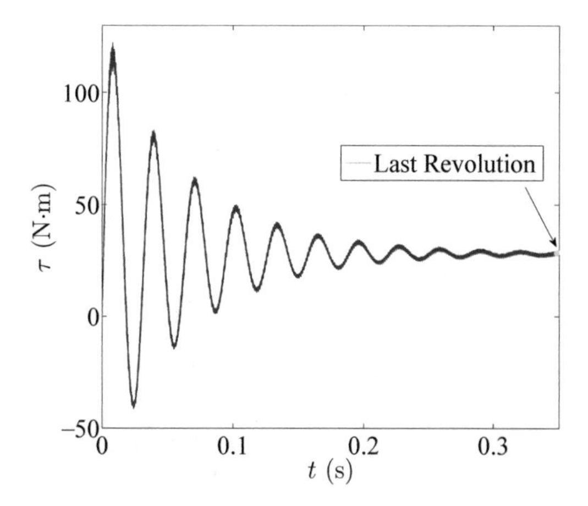

Fig. 2: Torque (τ) vs. time in an induction machine starting with null initial conditions.

dealing with the simulation of an induction machine for which the computation of the stationary solution took several days, although they were only interested in the last machine revolution needing just a few minutes (see Fig. 2). Thus, being able to reduce this simulation time could be crucial to get a competitive tool for the numerical simulation of this kind of machines. As a consequence, the company contacted the research group hoping it was possible to adapt the technique to the case of an induction motor. This was the beginning of a collaborative project lasting several months granted by Robert Bosch GmbH, and carried out through the Technological Institute of Industrial Mathematics (ITMATI; `www.itmati.com`).

1.1 Methodology Motivation from a Toy Model

In the subsequent sections we will describe how the proposed problem was addressed and why the existing mathematics were not able to solve it. However, we first present a very simple problem that illustrates both the problematic and the fundamental idea behind the developed methodology. More precisely, in this section we will analyze the response of a simple *RL* linear circuit in terms of the current initial condition. Let us consider the circuit shown in Fig. 3, which is composed by an inductor of inductance L, a resistor of resistance R and an electric generator with source voltage $e(t)$, all of them connected in series. The current $i(t)$ that flows through the circuit satisfies the initial value problem given in (1)–(2).

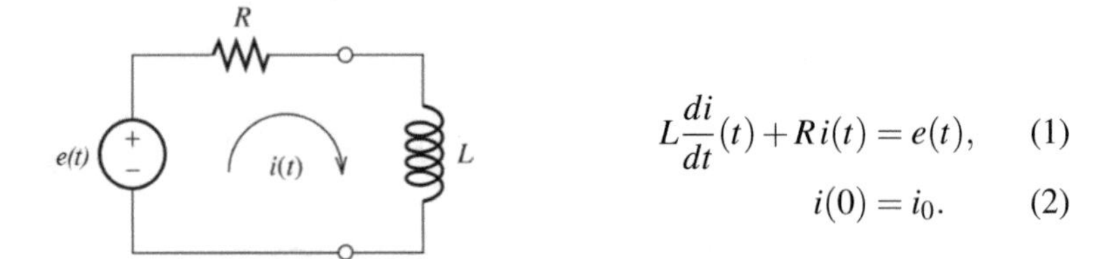

$$L\frac{di}{dt}(t)+Ri(t)=e(t), \quad (1)$$

$$i(0)=i_0. \quad (2)$$

Fig. 3: RL circuit.

When the source is harmonic, that is, if

$$e(t)=E\cos(\omega t+\alpha)=\mathrm{Re}\left(E\mathrm{e}^{\mathrm{i}\alpha}\mathrm{e}^{\mathrm{i}\omega t}\right),$$

the general solution to (1) is

$$i(t)=A\mathrm{e}^{-\frac{R}{L}t}+\frac{E}{\sqrt{R^2+\omega^2L^2}}\cos(\omega t+\alpha-\varphi), \quad (3)$$

with $\varphi=\arctan(\omega L/R)$ and A a constant that depends on the initial condition i_0. Thus, depending on the initial current, we will have a different amplitude A for the exponential term in (3), which is the transient term to be extinguished in order to compute the stationary solution (the second term in (3)). For instance, let us take $\alpha=-\pi/2$ (that is, $e(t)=E\sin\omega t$). Then,

- if $i_0=0$, $\quad i(t)=\dfrac{E}{\sqrt{R^2+\omega^2L^2}}\mathrm{e}^{-\frac{R}{L}t}-\dfrac{E}{\sqrt{R^2+\omega^2L^2}}\cos\omega t,$
- if $i_0=-\dfrac{E}{\sqrt{R^2+\omega^2L^2}}$, $\quad i(t)=-\dfrac{E}{\sqrt{R^2+\omega^2L^2}}\cos\omega t.$

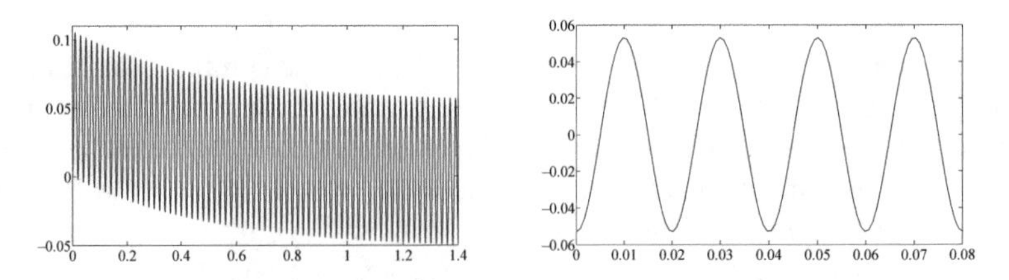

Fig. 4: $i(t)$ with $i_0=0$ (left) and $i_0=-\frac{E}{\sqrt{R^2+\omega^2L^2}}$ (right).

We notice that the time to attain the stationary solution depends on the ratio RT/L, where $T=2\pi/\omega$ is the period of the source $e(t)$. Two main cases can be considered: if $RT/L>>1$, the exponential vanishes very quickly independently of the initial condition i_0. On the contrary, if $RT/L<<1$, the transient part of the solution strongly depends on the initial condition and its extinction can last a long time. This behaviour can be seen in Fig. 4 (left), while in Fig. 4 (right) we show that just by taking the right initial condition we are able to completely remove the transient part of the solution, attaining the steady-state from the beginning. This idea is the one exploited by the authors in the methodology.

2 Statement of the Problem. Mathematical Modelling

When the research project started, the authors got precise information from the Bosch researchers on the problem of accelerating the computation of the stationary torque in a squirrel-cage induction motor. As illustrated in Fig. 5, this kind of machine has a stator winding in which the source is imposed, and a rotor consisting of a cylinder of steel laminations, with embedded aluminum or copper bars connected on both ends to an end-ring to make a closed path; see, for instance [4].

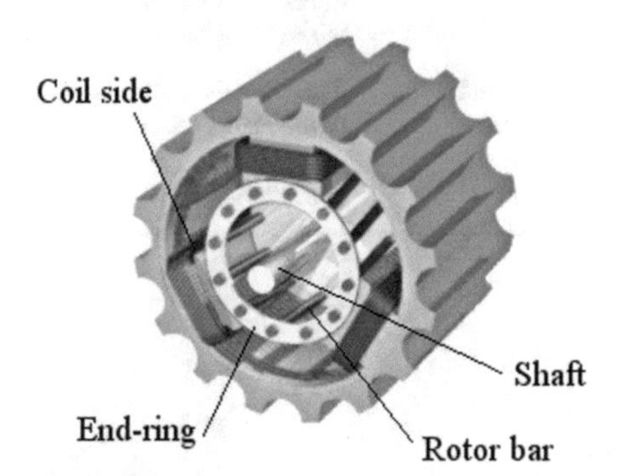

Fig. 5: Main parts integrating an squirrel-cage induction rotor. From Wikimedia Commons by Mtodorov 69 under license CC-BY-SA-3.0.

The electromagnetic model of these devices is often based on describing the active zone of the motor as a 2D distributed nonlinear eddy current or transient magnetic problem (see, for instance, [7, 13]). Indeed, in order to reduce electromagnetic losses, the magnetic cores of electric machines are laminated media orthogonal to the direction of the currents traversing the stator coil sides. Then, the usual simulation model consists in an electromagnetic problem defined on a cross-section of the machine, while the end regions of the stator windings and the end-rings are eventually modeled by circuit elements; in this way, the distributed 2D model would be coupled with a lumped one. On the other hand, the interplay between the magnetic fields in stator and rotor gives rise to a force that causes the latter to rotate around the machine axis. Therefore, the result is a transient eddy current problem defined in a moving geometry with prescribed speed.

According to the previous considerations, let Ω be a cross-section of an induction motor perpendicular to the direction of its shaft (see Fig. 5). Hence, domain Ω consists of n_c connected conductors (stator coil sides and rotor bars), the ferromagnetic core (in rotor and stator), the air between rotor and stator (air-gap), and the rotor shaft. In the sequel, we will use the notation summarized in Tables 1 and 2. Moreover, we will assume that the magnetic reluctivity ν is equal to ν_0 in Ω_0, a positive constant ν_c in conductors $\Omega_c := \cup_{i=1}^{n_c} \Omega_i$ and a nonlinear function of the magnetic flux, $\nu_{mc}(|\boldsymbol{B}|)$, in the magnetic core Ω_{mc}.

Table 1: Subdomain notation.

Symbol	Color in Fig. 6	Description
Ω		cross-section of an induction motor perpendicular to the direction of its shaft
$\Omega_0 = \Omega_0^{sta} \cup \Omega_0^{rot}$	White	domain occupied by air
$\Omega_i,\ i = 1,\ldots,n_b$	Grey	linear conductors representing the cross-sections of the rotor bars
$\Omega_i,\ i = n_b+1,\ldots,n_c$	Blue, yellow & red	linear conductors representing the cross-sections of the stator coil sides
$\Omega_{mc} = \Omega_{mc}^{sta} \cup \Omega_{mc}^{rot}$	Brown	non-conducting nonlinear magnetic cores

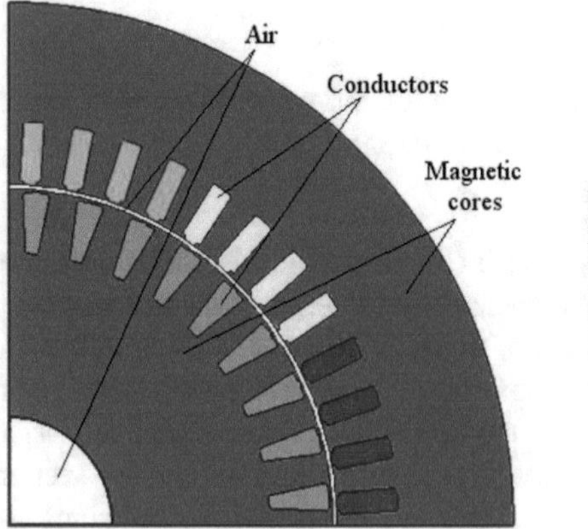

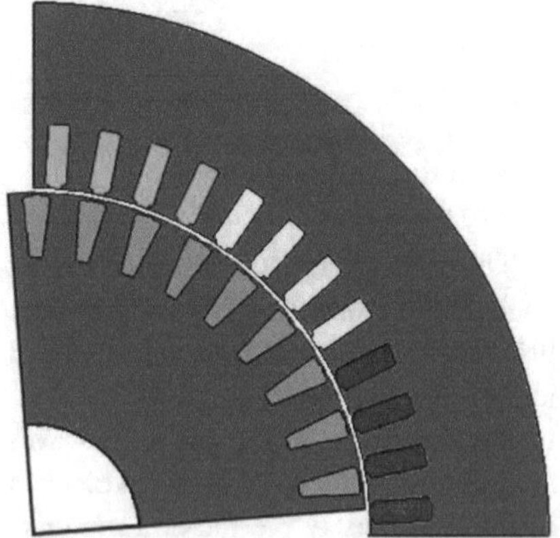

Fig. 6: A quarter of domain Ω at time $t = 0$ (left) and $t > 0$ (right). Modification of a picture provided by Robert Bosch GmbH.

Finally, we will choose a moving reference frame moving with the rotor, so that the subdomains corresponding to the stator undergo a rotational motion, r_t, with an angular velocity opposite to that of the rotor. Moreover, we neglect the end regions of the stator windings and these windings are assumed to be stranded conductors, which means that the source currents are supposed to be uniformly distributed on the cross-section of the corresponding conductors.

Thus, the formulation used to describe the electromagnetic behaviour of the induction motor is based on the coupling of a bidimensional transient eddy current model, written in terms of the axial component A_z of the magnetic vector potential, with circuit equations that model the squirrel-cage. On the one hand, if we call $y_i(t)$, $i = 1, \ldots, n_b$, and $I_i(t)$, $i = n_b + 1, \ldots, n_c$, the current intensities along the rotor bars and stator coil sides, respectively, and $C_i(t)$ the potential drop per unit length along direction z in conductor Ω_i, $i = 1, \ldots, n_b$, the transient magnetic model reads:

$$\begin{aligned}
-\operatorname{div}(\nu_0 \,\mathbf{grad}\, A_z) &= 0 && \text{in } r_t\left(\Omega_0^{\mathrm{sta}}\right) \cup \Omega_0^{\mathrm{rot}}, \\
\sigma \frac{\partial A_z}{\partial t} - \operatorname{div}(\nu_c \,\mathbf{grad}\, A_z) &= -\sigma C_i(t) && \text{in } \Omega_i, i = 1, \ldots, n_b, \\
-\operatorname{div}(\nu_c \,\mathbf{grad}\, A_z) &= \frac{I_i(t)}{\operatorname{meas}(\Omega_i)} && \text{in } r_t\left(\Omega_i\right), i = n_b + 1, \ldots, n_c, \\
-\operatorname{div}(\nu_{mc}(|\mathbf{grad}\, A_z|)\, \mathbf{grad}\, A_z) &= 0 && \text{in } r_t\left(\Omega_{mc}^{\mathrm{sta}}\right) \cup \Omega_{mc}^{\mathrm{rot}}, \\
A_z &= 0 && \text{on } \partial\Omega, \\
\frac{d}{dt}\int_{\Omega_i} \sigma A_z(x,y,t)\, dxdy + y_i(t) &= -C_i(t)\sigma \operatorname{meas}(\Omega_i), && i = 1, \ldots, n_b,
\end{aligned}$$

where σ denotes the electric conductivity in conductors.

As we have mentioned before, the magnetic field in the squirrel-cage is induced by the one in the stator and, therefore, neither currents in the bars, $y_i(t)$, nor potential drops per unit length, $C_i(t)$, $i = 1, \ldots, n_b$, are known in advance. To be able to compute them, we have to take into account that all bars are connected to each other through the end-rings. Since we cannot include these end-rings in the 2D model of the cross-section of the motor, we will write a lumped model for the squirrel-cage electrical circuit, and couple it with the distributed one. The topology of this circuit is modelled as a directed graph. If we denote by $\mathcal{A}$ the incidence matrix of this graph, the first Kirchhoff's law, which states that the charge is conserved at the nodes of the circuit, can be written as follows:

$$\mathcal{A}\mathbf{y}(t) = \mathbf{0}, \tag{4}$$

where $\mathbf{y}(t) \in \mathbb{R}^{n_{\mathrm{edg}}}$ denotes the vector of currents along the edges of the graph. Notice that

$$\mathbf{y} = \begin{pmatrix} \mathbf{y}^b \\ \mathbf{y}^r \end{pmatrix},$$

$\mathbf{y}^b$ being the currents along the rotor bars and $\mathbf{y}^r$ those corresponding to the squirrel-cage end-rings. In a similar way, two blocks can be distinguished in the incidence matrix $\mathcal{A}$ which will be denoted by $\mathcal{A}^b$ and $\mathcal{A}^r$.

Let us introduce the vector of nodal electric potentials at time t, denoted by $\mathbf{v}(t) \in \mathbb{R}^{n_{\text{nod}}}$, and the resistance of the i-th bar per unit length in the z space direction, denoted by

$$\alpha_i := \frac{1}{\sigma \operatorname{meas}(\Omega_i)},$$

$i = 1, \ldots, n_{\text{b}}$. Then, the constitutive equations for the circuit elements can be written as (see, for instance, [3])

$$\mathcal{D}\mathbf{y}(t) + \mathcal{A}^\top \mathbf{v}(t) = \mathbf{0}, \tag{5}$$

where $\mathcal{D}$ denotes the diagonal operator given by $(\mathcal{D}\mathbf{y}(t))_i = \mathcal{D}_i(y_i(t))$, with

$$\mathcal{D}_i(y_i(t)) = \begin{cases} R_i \dfrac{d}{dt} \displaystyle\int_{\Omega_i} \sigma A_z(t) + R_i y_i(t) & i = 1, \ldots, n_{\text{b}}, \\ R_i y_i(t) & i = n_{\text{b}} + 1, \ldots, n_{\text{edg}}, \end{cases}$$

being R_i, $i = n_{\text{b}} + 1, \ldots, n_{\text{edg}}$, the resistance of the i-th edge of the graph, $R_i = \ell_i \alpha_i$, $i = 1, \ldots, n_{\text{b}}$, the resistance of the i-th rotor bar and ℓ_i its length.

Thus, if the currents are known in the stator coils, the problem to be solved is the following:

Problem 1 *Given the currents along the coil sides $I_i(t)$, $i = n_{\text{b}} + 1, \ldots, n_{\text{c}}$, the vector of initial currents along the bars $\mathbf{y}^{b,0} = (y_1^0, \ldots, y_{n_{\text{b}}}^0)$ and an initial magnetic vector potential $A^0(x,y)$ in the bars, find, for every $t \in [0,T]$, a field $A_z(x,y,t)$, currents $y_i(t)$, $i = 1, \ldots, n_{\text{edg}}$, along the edges of the graph and voltages $v_i(t)$, $i = 1, \ldots, n_{\text{nod}}$, at the nodes of the graph such that*

$$-\operatorname{div}(\nu_0 \operatorname{\mathbf{grad}} A_z) = 0 \quad \text{in } \Omega_0^{\text{rot}} \cup r_t\left(\Omega_0^{\text{sta}}\right), \tag{6}$$

$$\sigma \frac{\partial A_z}{\partial t} - \operatorname{div}(\nu_{\text{c}} \operatorname{\mathbf{grad}} A_z) = -\frac{\sigma}{\ell_i} \left((\mathcal{A}^b)^\top \mathbf{v}(t) \right)_i \quad \text{in } \Omega_i,\ i = 1, \ldots, n_{\text{b}}, \tag{7}$$

$$-\operatorname{div}(\nu_{\text{c}} \operatorname{\mathbf{grad}} A_z) = \frac{I_i(t)}{\operatorname{meas}(\Omega_i)} \quad \text{in } r_t(\Omega_i),\ i = n_{\text{b}} + 1, \ldots, n_{\text{c}}, \tag{8}$$

$$-\operatorname{div}(\nu_{\text{mc}}(|\operatorname{\mathbf{grad}} A_z|) \operatorname{\mathbf{grad}} A_z) = 0 \quad \text{in } \Omega_{\text{mc}}^{\text{rot}} \cup r_t\left(\Omega_{\text{mc}}^{\text{sta}}\right), \tag{9}$$

$$A_z = 0 \quad \text{on } \partial\Omega, \tag{10}$$

$$\mathcal{D}\mathbf{y}(t) + \mathcal{A}^\top \mathbf{v}(t) = \mathbf{0}, \tag{11}$$

$$\mathcal{A}\mathbf{y}(t) = \mathbf{0}, \tag{12}$$

$$\mathbf{y}(0) = \mathbf{y}^{b,0}, \tag{13}$$

$$A_z(x,y,0) = A^0(x,y) \quad \text{in } \Omega_i,\ i = 1, \ldots, n_{\text{b}}. \tag{14}$$

We notice that this transient eddy current model needs two different quantities as initial conditions in the rotor bars: a distributed field, the magnetic vector potential A^0, and a vector of currents $\mathbf{y}^{b,0}$.

We refer the reader to [11, 13] for more information on this model.

Table 2: Notation summary.

Symbol	Description
n_b	number of bars
n_c	total number of conductors (bars + coil sides)
n_edg	edges of the squirrel-cage associated graph
n_nod	nodes of the squirrel-cage associated graph
$y_i(t)$	currents through the rotor bars
$I_i(t)$	currents through the stator coil sides
$C_i(t)$	potential drops per unit-length along the rotor bars
$v_i(t)$	potential drops along the rotor bars
ℓ_i	characteristic length of the motor in the z-direction
R_i	resistance of the squirrel-cage associated graph edges
α_i	resistance per unit-length of the squirrel-cage bars
$\mathcal{A}$	incidence matrix of the squirrel-cage associated graph
ν_0, ν_c, ν_mc	magnetic reluctivities in air, conductors and cores, respectively
σ	electric conductivity
r_t	rotation with angular velocity opposite to that of the rotor

3 Existing Mathematics

The need to shorten the transient part of the solution in numerical simulation arises for a wide range of applications in electromagnetism. Indeed, when a time-dependent problem having a periodic solution in the stationary regime is not provided with suitable initial conditions for its numerical solution, a huge amount of CPU time may have to be spent in order to reach the steady-state. Usually, these suitable initial conditions are not known in advance what makes this drawback not easily avoidable. As a consequence, different methods have been proposed in order to reduce the computational cost derived from the need of overcoming this transient regime.

Most approaches try to take advantage, either directly or indirectly, of the fact that, in the steady regime and under certain conditions, the solution is periodic. For instance, one of the first introduced strategies is the *Time Periodic Finite Element Methods* (TPFEM) which has been subjected to thorough study and improvement during the past two decades (see [5, 10, 15, 16]). One of the key points of this methodology is writing the problem in a time-interval in which its solution is periodic.

In Problem 1, the currents along the stator coil sides are generally harmonic functions of the same frequency f_c (in Hz), and the rotor moves at a constant angular velocity n_r (in rpm). Depending on the values of f_c and n_r, the magnetic fields in rotor and stator may oscillate at different frequencies. This happens when the rotation velocity n_r is different from the synchronous speed, $n_s = (60 f_c)/p$ (in rpm), p being the number of pole-pairs of the machine. In this case, we can express the difference in terms of the so-called slip, $s = (n_s - n_r)/n_s$. On the one hand, the slip being different from zero is a requirement for the current to be induced in the rotor bars because, otherwise, these would not see the oscillation of the stator currents. However, on the other hand, the slip affects the problem time-periodicity. Indeed, it can be shown that the main frequency of the currents induced in the rotor bars is

$$f_b = s f_c$$

(see, for instance, [4]). Therefore, if we try to exploit the above methodologies for the induction motor problem, we have two frequencies to take into account, f_c and f_b. Still, if f_c/f_r is a rational number, with $f_r := n_r/60$ (in Hz), it would be possible to speak about a common period T_e. This value is known as the effective period, and can be very large. This fact makes the application of these techniques quite impractical.

Another approach is accelerating the convergence by using error correction techniques which rely on the previous knowledge of the error sources, in order to effectively correct them as the transient simulation advances. For instance, the *Time Periodic - Explicit Error Correction Methods* (TP-EEC), (see [6]), and the *Time Differential Correction* (TDC), (see [9]), are based on this principle. However, these techniques have been shown to be only partially effective, and therefore have to be combined with other approaches. Moreover, these methods usually share some properties with TPFEM, exhibiting therefore the above inconvenience with respect to the solution period.

A completely different approach is trying to approximate suitable and physically-compatible initial conditions, in the sense that solving the original problem leads very quickly to a periodic solution. The idea behind this methodology is that the transient eddy current model can be approximated with the corresponding harmonic one by changing the magnetic constitutive law and the conductivity of the rotor bars (see [14]). However, these harmonic approximations are not able to take into account many physical aspects of the problem (for instance, the rotor motion), and therefore the transient eddy current problem initialized with the obtained values still presents an important transient part.

Taking into account the absence of a completely satisfactory solution in the literature, a novel methodology was developed inspired in the techniques introduced in [1], and having the advantage of making use of the periodicity condition only in the rotor bars. As a consequence, the limitations concerning the presence of several frequencies or the size of the effective period do not apply. Moreover, the computational cost of the novel approach does not depend on the size of this period, and the number of unknowns is very small in comparison with the previous methods.

4 A Novel and Efficient Methodology to Solve the Problem

As mentioned before, the method proposed to give an answer to the company is inspired in the techniques introduced in [1] for a 2D transient magnetic model without either eddy currents or motion. A key point in that methodology consisted in re-writing the problem in such a way that the only unknowns are the currents in the conductors with voltage drop sources. To get a similar procedure in the case of the induction machine, it was necessary to assume that the rotor bars are stranded conductors, that is, that the induced currents in these conductors are uniformly distributed. Under this assumption, the parabolic equation (7) is transformed into an elliptic one, and the problem to be solved is the following (see [2] for further details):

Problem 2 *Given the currents along the coil sides $I_i(t)$, $i = n_{\mathrm{b}}+1,\dots,n_{\mathrm{c}}$, the vector of initial currents along the bars $\mathbf{y}^{b,0} = (y_1^0,\dots,y_{n_{\mathrm{b}}}^0)$ find, for every $t \in [0,T]$, a field $A_z(x,y,t)$, currents $y_i(t)$, $i = 1,\dots,n_{\mathrm{edg}}$, along the edges of the graph and voltages $v_i(t)$, $i = 1,\dots,n_{\mathrm{nod}}$, at the nodes of the graph such that*

$$-\operatorname{div}(\nu_0 \,\mathbf{grad} A_z) = 0 \quad \text{in } \Omega_0^{\mathrm{rot}} \cup r_t\left(\Omega_0^{\mathrm{sta}}\right), \tag{15}$$

$$-\operatorname{div}(\nu_{\mathrm{c}} \,\mathbf{grad} A_z) = \frac{y_i(t)}{\mathrm{meas}(\Omega_i)} \quad \text{in } \Omega_i,\ i = 1,\dots,n_{\mathrm{b}}, \tag{16}$$

$$-\operatorname{div}(\nu_{\mathrm{c}} \,\mathbf{grad} A_z) = \frac{I_i(t)}{\mathrm{meas}(\Omega_i)} \quad \text{in } r_t(\Omega_i),\ i = n_{\mathrm{b}}+1,\dots,n_{\mathrm{c}}, \tag{17}$$

$$-\operatorname{div}(\nu_{\mathrm{mc}}(|\mathbf{grad} A_z|)\,\mathbf{grad} A_z) = 0 \quad \text{in } \Omega_{\mathrm{mc}}^{\mathrm{rot}} \cup r_t\left(\Omega_{\mathrm{mc}}^{\mathrm{sta}}\right), \tag{18}$$

$$A_z = 0 \quad \text{on } \partial\Omega, \tag{19}$$

$$\mathcal{D}\mathbf{y}(t) + \mathcal{A}^\top \mathbf{v}(t) = \mathbf{0}, \tag{20}$$

$$\mathcal{A}\mathbf{y}(t) = \mathbf{0}, \tag{21}$$

$$\mathbf{y}(0) = \mathbf{y}^{b,0}. \tag{22}$$

4.1 Reduced Problem

The next step is to obtain an equivalent formulation to Problem 2 by eliminating the vector of voltages, $\mathbf{v}(t)$, and the vector of currents along the end-rings, $\mathbf{y}^r(t)$. For this purpose, let us first introduce some notations that will allow us to write Problem 2 in a more compact form. Let $\mathcal{F}:[0,T]\times\mathbb{R}^{n_\mathrm{b}}\longrightarrow\mathbb{R}^{n_\mathrm{b}}$ be the nonlinear operator given by

$$\mathcal{F}(t,\mathbf{w}):=\left(\int_{\Omega_1}\sigma A_z(x,y,t)\,dx\,dy,\ldots,\int_{\Omega_{n_\mathrm{b}}}\sigma A_z(x,y,t)\,dx\,dy\right)^\top\in\mathbb{R}^{n_\mathrm{b}},$$

with $A_z(x,y,t)$ the solution to the following nonlinear magnetostatic problem:

Problem 3 *Given a fixed $t\in[0,T]$, currents along the coil sides $I_i(t)$, $i=n_\mathrm{b}+1,\ldots,n_\mathrm{c}$, and $\mathbf{w}\in\mathbb{R}^{n_\mathrm{b}}$, find a field $A_z(x,y,t)$ such that*

$$\begin{aligned}
-\operatorname{div}(\nu_0\,\mathbf{grad}A_z)&=0 && \text{in } \Omega_0^{\mathrm{rot}}\cup r_t\left(\Omega_0^{\mathrm{sta}}\right),\\
-\operatorname{div}(\nu_\mathrm{c}\,\mathbf{grad}A_z)&=\frac{w_i}{\operatorname{meas}(\Omega_i)} && \text{in } \Omega_i,\ i=1,\ldots,n_\mathrm{b},\\
-\operatorname{div}(\nu_\mathrm{c}\,\mathbf{grad}A_z)&=\frac{I_i(t)}{\operatorname{meas}(\Omega_i)} && \text{in } r_t(\Omega_i),\ i=n_\mathrm{b}+1,\ldots,n_\mathrm{c},\\
-\operatorname{div}(\nu_\mathrm{mc}(|\mathbf{grad}A_z|)\,\mathbf{grad}A_z)&=0 && \text{in } \Omega_\mathrm{mc}^{\mathrm{rot}}\cup r_t\left(\Omega_\mathrm{mc}^{\mathrm{sta}}\right),\\
A_z&=0 && \text{on } \partial\Omega.
\end{aligned}$$

By distinguishing the blocks corresponding to the rotor bars and end-rings in equations (20) and (21), equations (15)–(21) can be rewritten in the more compact manner

$$\mathcal{R}^b\frac{d}{dt}\mathcal{F}\left(t,\mathbf{y}^b(t)\right)+\mathcal{R}^b\mathbf{y}^b(t)+\left(\mathcal{A}^b\right)^\top\mathbf{v}(t)=\mathbf{0}, \tag{23}$$

$$\mathcal{R}^r\mathbf{y}^r(t)+(\mathcal{A}^r)^\top\mathbf{v}(t)=\mathbf{0}, \tag{24}$$

$$\mathcal{A}^b\mathbf{y}^b(t)+\mathcal{A}^r\mathbf{y}^r(t)=\mathbf{0}. \tag{25}$$

Moreover, it can be shown that there exists a matrix $\mathcal{B}^{-1}$ and a scalar function $\lambda(t)$ such that

$$\mathbf{v}(t)=\mathcal{B}^{-1}\mathcal{A}^b\mathbf{y}^b(t)+\lambda(t)\begin{pmatrix}\mathbf{0}\\ \mathbf{e}\end{pmatrix},$$

(see [2]), and then $\mathbf{y}^r(t)$ and $\mathbf{v}(t)$ can be eliminated from the system, obtaining the following system of differential-algebraic equations (DAE).

Problem 4 *Given the currents along the coil sides $I_i(t)$, $i=n_\mathrm{b}+1,\ldots,n_\mathrm{c}$, and the initial currents along the bars $\mathbf{y}^{b,0}=(y_1^0,\ldots,y_{n_\mathrm{b}}^0)$, find, for every $t\in[0,T]$, currents $y_i(t)$, $i=1,\ldots,n_\mathrm{b}$, along the bars and $\lambda(t)\in\mathbb{R}$ such that*

$$\mathcal{R}^b \frac{d}{dt}\mathcal{F}\left(t,\mathbf{y}^b(t)\right) + \left(\mathcal{R}^b + \left(\mathcal{A}^b\right)^\top \mathcal{B}^{-1}\left(\mathcal{A}^b\right)\right)\mathbf{y}^b(t)$$

$$+\lambda(t)\left(\mathcal{A}^b\right)^\top \begin{pmatrix}\mathbf{0}\\ \mathbf{e}\end{pmatrix} = \mathbf{0}, \qquad (26)$$

$$\mathcal{A}^b\mathbf{y}^b(t)\cdot\begin{pmatrix}\mathbf{0}\\ \mathbf{e}\end{pmatrix} = 0, \qquad (27)$$

$$\mathbf{y}^b(0) = \mathbf{y}^{b,0}, \quad (28)$$

where operator $\mathcal{F}$ is defined in terms of $I_i(t)$, $i = n_b+1,\ldots,n_c$, through the solution to Problem 3.

4.2 Approximating the Initial Currents in Rotor Bars

Let us assume that the currents along the stator coil sides are periodic functions of the same frequency f_c. Following the methodology introduced in [1], the next step consists in integrating in time the first equation in the reduced problem, first in $[0,t]$, and then in $[0,T_b]$, with T_b the period in the rotor bars. Changing the order of integration in the last two terms, we obtain

$$\mathcal{R}^b\left(\int_0^{T_b}\mathcal{F}\left(t,\mathbf{y}^b(t)\right)dt - T_b\mathcal{F}\left(0,\mathbf{y}^{b,0}\right)\right)$$
$$+\left(\mathcal{R}^b + (\mathcal{A}^b)^\top\mathcal{B}^{-1}(\mathcal{A}^b)\right)\int_0^{T_b}(T_b - t)\mathbf{y}^b(t)\,dt$$
$$+\left(\int_0^{T_b}(T_b-t)\lambda(t)\,dt\right)(\mathcal{A}^b)^\top\begin{pmatrix}\mathbf{0}\\ \mathbf{e}\end{pmatrix} = \mathbf{0}. \quad (29)$$

In [1], this double-integration technique, along with some simplifications, led to a nonlinear system of equations with a small number of unknowns that approximated the sought initial currents. In the present case, we will obtain a similar result by making the two following assumptions that, together with considering the rotor bars as stranded conductors, represent the key points for the development of the proposed technique:

(I) The first term on the left-hand side in (29), namely,

$$\mathcal{R}^b\int_0^{T_b}\mathcal{F}\left(t,\mathbf{y}^b(t)\right)dt$$

can be neglected because, in real situations, it is much smaller than the other terms. From a physical point of view, this assumption means that the flux linkages of the rotor bars have approximately zero mean over one period of the fundamental frequency of currents in the rotor bars.

(II) The currents in rotor bars are approximated by their respective fundamental harmonics assuming that, for symmetry reasons, their amplitudes are the same (denoted by Y),

$$y_i^b(t) \simeq \left[(\mathcal{A}^b)^\top \begin{pmatrix} \mathbf{0} \\ \mathbf{e} \end{pmatrix} \right]_i Y \cos(2\pi f_b t + \beta_i), \tag{30}$$

and with phase-shifts β_i between the current source in the coil sides and the current of the bars given by

$$\beta_i = \beta_1 + (i-1)\gamma, \quad i = 1, \ldots, n_\mathrm{b}, \tag{31}$$

with $\gamma := (2\pi p)/n_\mathrm{b}$. In particular, the approximations of the initial currents are

$$y_i^{b,0} = Y \left[(\mathcal{A}^b)^\top \begin{pmatrix} \mathbf{0} \\ \mathbf{e} \end{pmatrix} \right]_i \cos\beta_i. \tag{32}$$

We notice assumption ((II)) allows us to avoid calculating $\mathbf{y}^b(t)$ along the interval $(0, T_b]$, because now the corresponding integral in (29) can be analytically computed. In this way we approximate the original problem by means of a time-independent one. This is essential because, otherwise, we would have to solve the full model!

Since our goal was computing the initial currents in the rotor bars, only two unknowns remain: Y and β_1. In order to compute them, we solve the following system of n_b nonlinear equations in the least-square sense:

Problem 5 *Given periodic currents along the coil sides $I_i(t)$, $i = n_\mathrm{b}+1, \ldots, n_\mathrm{c}$, find $Y \in \mathbb{R}$ and $\beta_1 \in [0, 2\pi)$ such that,*

$$T_b \left[\frac{1}{a} (\mathcal{A}^b)^\top \begin{pmatrix} \mathbf{0} \\ \mathbf{e} \end{pmatrix} \otimes \left(\left(\mathcal{R}^b + (\mathcal{A}^b)^\top \mathcal{B}^{-1} \mathcal{A}^b \right)^{-\top} (\mathcal{A}^b)^\top \begin{pmatrix} \mathbf{0} \\ \mathbf{e} \end{pmatrix} \right) - \mathcal{I} \right] \mathcal{R}^b \mathcal{F}(0, Y\mathbf{u})$$
$$- Y \frac{T_b^2}{2\pi} (\mathcal{R}^b + (\mathcal{A}^b)^\top \mathcal{B}^{-1} \mathcal{A}^b) \mathbf{w} = \mathbf{0}. \tag{33}$$

In the above problem, $\mathbf{u}$ and $\mathbf{w}$ are the column vectors whose respective i-th components are

$$u_i := \left[(\mathcal{A}^b)^\top \begin{pmatrix} \mathbf{0} \\ \mathbf{e} \end{pmatrix} \right]_i \cos\beta_i \quad \text{and} \quad w_i := \left[(\mathcal{A}^b)^\top \begin{pmatrix} \mathbf{0} \\ \mathbf{e} \end{pmatrix} \right]_i \sin\beta_i,$$

$i = 1, \ldots, n_\mathrm{b}$, $\mathcal{I}$ denotes the identity matrix and

$$a := \left(\mathcal{R}^b + (\mathcal{A}^b)^\top \mathcal{B}^{-1} \mathcal{A}^b \right)^{-1} (\mathcal{A}^b)^\top \begin{pmatrix} \mathbf{0} \\ \mathbf{e} \end{pmatrix} \cdot (\mathcal{A}^b)^\top \begin{pmatrix} \mathbf{0} \\ \mathbf{e} \end{pmatrix}.$$

We notice that β_1 appears in the previous system through $\mathbf{u}$ and $\mathbf{w}$. Similarly, $\mathcal{F}(0, Y\mathbf{u})$ is defined in terms of $I_i(0)$, $i = n_\mathrm{b}+1, \ldots, n_\mathrm{c}$, through the solution to Problem 3. For further technical details, we refer the reader to [2].

We notice that the steps given above can be reproduced for the toy model: if we perform the double integration and assume that the current is harmonic, we obtain an equation that is exactly solved by the initial condition which actually eliminates the transient state, i.e., by $i_0 = -E/\sqrt{R^2+\omega^2 L^2}$ (see Section 1.1).

4.3 Numerical Results

In this section we present the numerical results obtained for a particular induction machine with squirrel-cage rotor. We first use Problem 5 to estimate suitable initial currents in the bars of the induction motor and then solve a transient eddy current problem using the obtained currents as initial condition. We notice that, in the second step, we solve a formulation similar to that appearing in Problem 1. In particular, we model the rotor bars as solid (instead of stranded) solid conductors, which means that the induced currents are not uniformly distributed. The comparison between the numerical results obtained with the initial currents provided by our methodology and the ones starting with null initial conditions allow us to illustrate the reduction of the time needed to achieve the steady-state.

The cross-section of the induction machine used in the numerical results is sketched in Fig. 7. A three-phase current is supplied having the following stator winding distribution: red, yellow and blue slots correspond to phases A, B and C, respectively. These source currents are defined as

$$\begin{aligned}
I_A(t) &= \sqrt{2} I_c \cos(2\pi f_c t),\\
I_B(t) &= \sqrt{2} I_c \cos(2\pi f_c t + 2\pi/3),\\
I_C(t) &= \sqrt{2} I_c \cos(2\pi f_c t - 2\pi/3).
\end{aligned}$$

We notice that we have interpreted angle β_1 appearing in (31) as the phase-shift between the current through the first bar and the current corresponding to phase A.

We have considered two operating points corresponding to different electrical sources in the stator and rotational speed; see Table 3. In particular, we notice that the period of the current in the rotor , T_b, is one order of magnitude smaller in the second operating point with respect to the first one. However, the methodology to compute the initial currents is time-independent, and therefore its computational cost does not depend on these sizes.

Table 3: Characteristics of the different operating points.

	f_c (Hz)	n_r (rpm)	I_c (A_{RMS})	T_b (s)
Op. Point 1	171.2	5000	314	0.221
Op. Point 2	632.0	18000	531	0.031

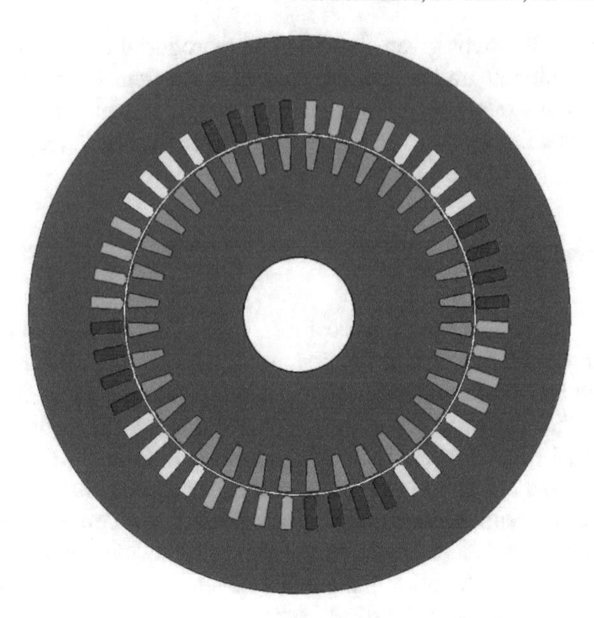

Fig. 7: Computational domain. Courtesy of Robert Bosch GmbH.

For each operating point, we have found the minimum value of $\|\mathbf{f}(Y,\beta_1)\|^2$ by using the Matlab function `lsqnonlin`, with $\mathbf{f}$ defined as the left-hand side in (33), that is,

$$\begin{aligned}\mathbf{f}(Y,\beta_1) := &-Y\frac{T_b^2}{2\pi}(\mathcal{R}^b+(\mathcal{A}^b)^\top\mathcal{B}^{-1}\mathcal{A}^b)\mathbf{w}\\ &+T_b\frac{1}{a}(\mathcal{A}^b)^\top\begin{pmatrix}\mathbf{0}\\ \mathbf{e}\end{pmatrix}\otimes\left(\left(\mathcal{R}^b+(\mathcal{A}^b)^\top\mathcal{B}^{-1}\mathcal{A}^b\right)^{-\top}(\mathcal{A}^b)^\top\begin{pmatrix}\mathbf{0}\\ \mathbf{e}\end{pmatrix}\right)\mathcal{R}^b\mathcal{F}(0,Y\mathbf{u})\\ &-T_b\mathcal{F}(0,Y\mathbf{u}).\end{aligned} \quad (34)$$

Table 4 shows the optimal values obtained for Y and β_1 and the residual at each operating point.

Table 4: Optimal values for the different operating points.

	Op. Point 1	Op. Point 2
Y (A)	477.78	942.20
β_1 (rad)	3.34	2.87
$\|\mathbf{f}(Y,\beta_1)\|^2$	1.12e−08	3.80e−10

In order to show that the time needed to reach the steady-state in a transient simulation strongly depends on the choice of the initial currents in the rotor bars of the machine, we performed transient simulations starting with the initial currents defined from the computed values of Y and β_1 and we compared the results with those corresponding to null initial currents.

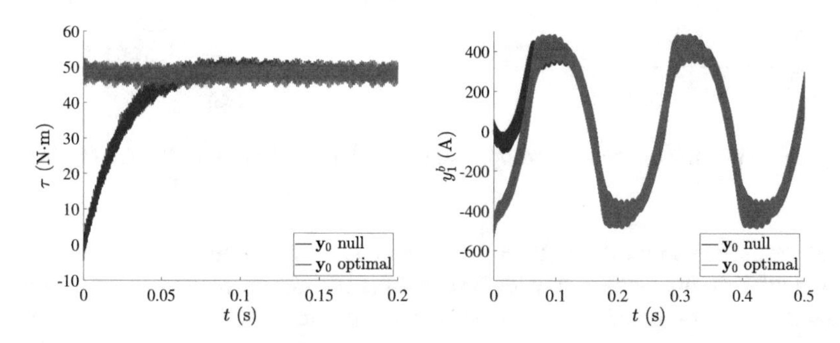

Fig. 8: Op. Point 1. Torque vs. time (left). Current in bar 1 vs. time (right).

To analyze if the solution of the transient eddy current model has reached the steady-state, it is usual to study the torque in the rotor and the currents in the bars of the squirrel-cage. Thus, Figs. 8 and 9 show the electromagnetic torque in rotor and the current along the first bar over time for the two operating points. In these figures, $\tau : \mathbb{R}^+ \longrightarrow \mathbb{R}$ denotes the scalar function that expresses the electromagnetic torque as a function of time; moreover, the red curve corresponds to null initial values for the current along the bars, while the blue ones have been obtained by using the values provided by the proposed methodology.

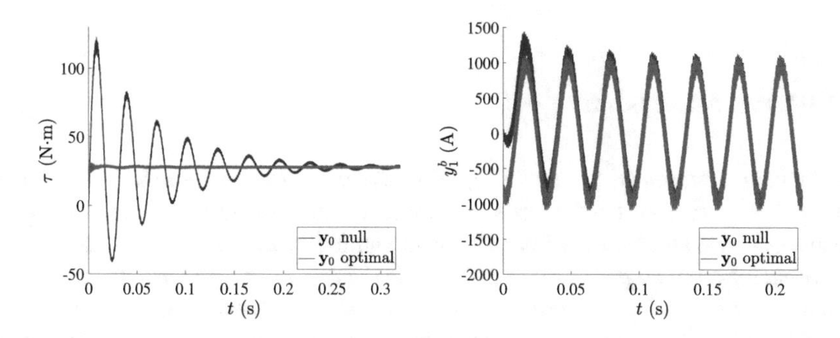

Fig. 9: Op. Point 2. Torque vs. time (left). Current in bar 1 vs. time (right).

Fig. 10 shows the torque over time for the considered operating points. In both cases, the time needed to reach the steady-state has been indicated with vertical lines:

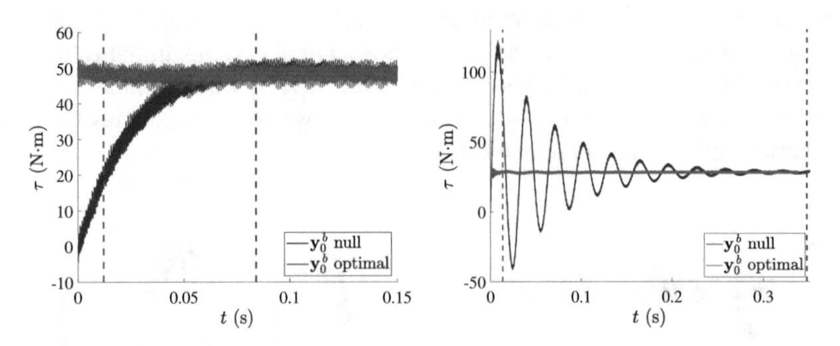

Fig. 10: Time to steady-state comparison. Op. Point 1 (left). Op. Point 2 (right).

in red for initial conditions $\mathbf{y}^b(0) = \mathbf{0}$ and in blue for initial conditions $\mathbf{y}^b(0) = Y\mathbf{u}$. We remark that, in comparison with the case of null initial currents, starting with the initial currents computed with the proposed methodology leads to a very important computational saving, specially for the second operating point (see also Table 5).

Table 5: Time to reach the steady-state for different operating points.

	Initial condition	T_{steady} (s)	Number of revolutions m	Saving (%)
Op. Point 1	$\mathbf{y}^b(0) = \mathbf{0}$	0.0840	7	86
	$\mathbf{y}^b(0) = Y\mathbf{u}$	0.0120	1	
Op. Point 2	$\mathbf{y}^b(0) = \mathbf{0}$	0.3467	104	96
	$\mathbf{y}^b(0) = Y\mathbf{u}$	0.0133	4	

5 Current State of Art

The technique presented in previous section was developed some time before this book publication, and it has already given rise to some publications. In particular, the authors have applied for a Spanish patent entitled *Procedimiento y producto de programa informático para acelerar el cálculo del estado estacionario de un motor de inducción de jaula de ardilla* (Procedure and software product to accelerate the steady state calculation of a squirrel-cage induction motor). This patent was granted mid-2019 (ES2687868 B2). Moreover, a paper was accepted for publication in an international journal [2] and the research is part of the PhD thesis of the third author. However, from our point of view, the procedure still offers relevant challenges and opportunities, some of which are explained below.

Firstly, we recall that one of the key points of the methodology proposed is approximating the currents along the rotor bars by their fundamental harmonics (see equation (30)). However, if we analyze the steady-state currents in the rotor bars, it is possible to appreciate the presence of other frequency components. For instance, in Figure 11–right we show an approximation of the so-called energy spectral density corresponding to the steady-state current in the first bar for operating point 2, shown in Figure 11–left. This graph describes the distribution of energy into frequency components f of y_1^b (see, for instance, [12]).

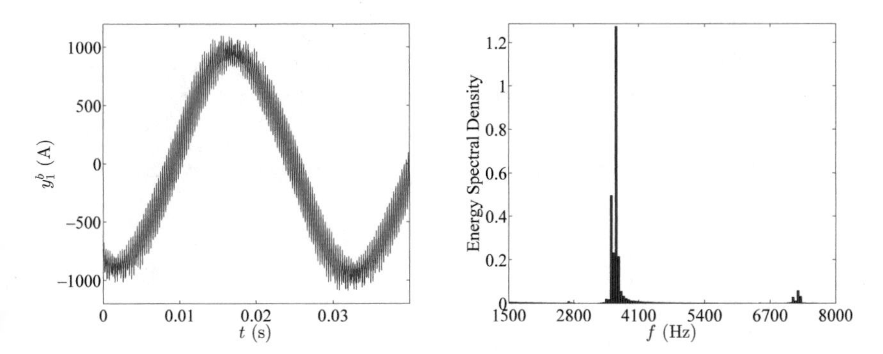

Fig. 11: Current $\mathbf{y}^b(t)$ (left) and energy spectral density (right).

Taking this analysis into account, we are working on the generalization of the method with a multiharmonic approximation of the currents in the rotor bars. For this purpose, the approximation (30) would be replaced by:

$$y_n^b(t) \simeq \left[(\mathcal{A}^b)^{\mathrm{T}} \begin{pmatrix} \mathbf{0} \\ \mathbf{e} \end{pmatrix} \right]_n \sum_{m=1}^{M} Y_m \cos(2\pi f_m t + \beta_{m,n}), \tag{35}$$

where f_m are the frequencies to take into account and, similarly to the previous case, $\beta_{m,n}$ are the phase-shifts between the currents in the stator and the m-th harmonic of the current along the n-th bar. We hope that this would improve the approximation of these currents, and thereby the associated initial conditions that accelerate the steady-state.

On the other hand, we plan to test the methodology performance for slightly different induction motor cases. For instance, one possibility would be taking into account the squirrel-cage end-ring inductances, what would involve adding a term in the end-ring circuit equations. Another example of interest would be wounded induction motors, which have windings in the rotor instead of a squirrel-cage (see, for instance, [8]). For this purpose, the same partial differential equation (PDE) model as for the squirrel-cage case would be valid, but some modifications may be needed depending on the winding connection-type.

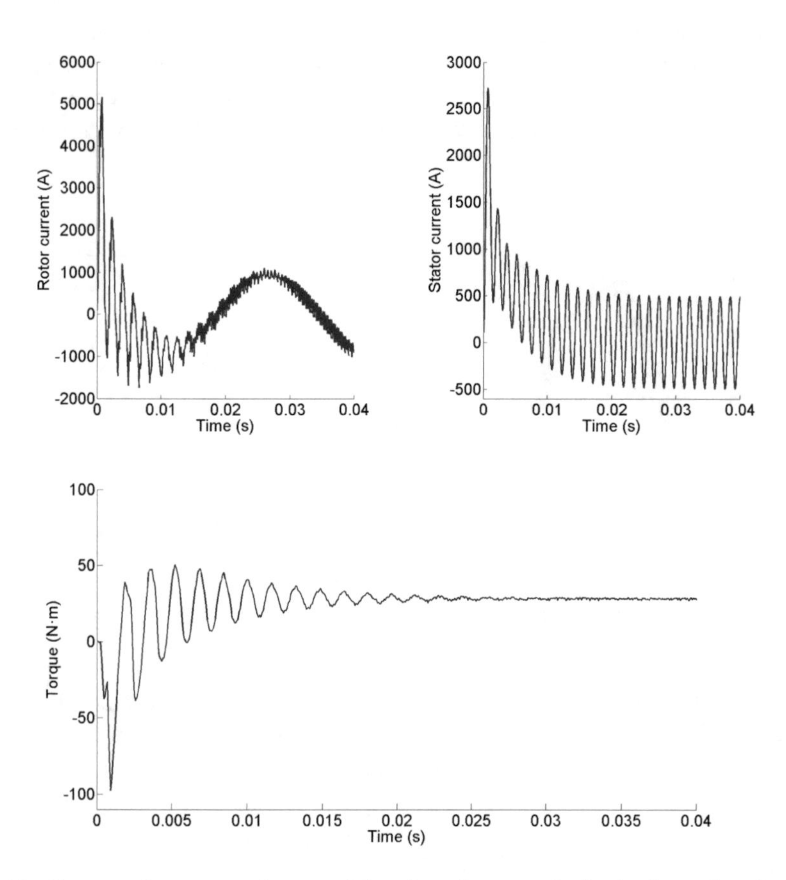

Fig. 12: Currents in rotor and stator (above) and torque (below) along time in an induction machine fed by voltage drop excitations. Null initial currents in both rotor and stator.

Furthermore, the described methodology may be enhanced to cover the case in which the induction motor is fed by voltage drop sources instead of current ones, as many papers cited in Section 3 also address it. In this case, the time to reach the steady-state may also be important, as illustrated in Fig. 12. Moreover, this time can be specially long if the voltage source is given by a PWM signal, which requires using very small time steps.

We highlight that the case with voltage drops given as sources in the stator coils, the problem becomes more involved because we also need suitable initial currents for these conductors. Namely, even if the stator coils are assumed to be stranded conductors, it is necessary to add an equation relating currents and voltages in these subdomains, similar to that of the rotor bars, i.e.,

$$\frac{d}{dt}\int_{\Omega_i} \sigma A_z(x,y,t)\,dxdy + I_i(t) = -C_i(t)\sigma\,\mathrm{meas}(\Omega_i), \quad i = n_\mathrm{b}+1,\ldots,n_\mathrm{c},$$

with $C_i(t)$, $i = n_\mathrm{b}+1,\ldots,n_\mathrm{c}$, being the source voltage drops per unit length. As a consequence, in this case the reduced problem has to be written in terms of the currents in the rotor bars and in the stator coils. Therefore, its solution will present two different periods, corresponding to frequencies f_c (stator) and f_b (rotor); see again Fig. 12–above. Thus, it is not clear wether the methodology could be adapted to these conditions.

Finally, we would like to consider the extension of the methodology developed in this chapter to fit under more general ordinary differential equation (ODE) and PDE problems with periodic stationary solution; in particular, to second order in time equations. Applications could take place in nonlinear models for acoustics or water waves.

6 Conclusions

This chapter presents a novel mathematical technique for solving a practical real problem appearing in electrical engineering. The challenge was to develop a new mathematical method to reduce the long CPU time needed to reach the steady-state in the numerical simulation of squirrel-cage induction machines, when the source in the stator coil sides is given in terms of periodic currents. The particular characteristics of this kind of devices makes the existing methods not well-adapted to deal with this problem, so a novel strategy was required. The new method proposed here is aimed to compute suitable initial currents for the transient magnetic problem that describes the behavior of the machine and which is written as a PDE nonlinear model coupled with circuit equations. As a consequence of this research, a Spanish patent was granted (ES2687868 B2) and a paper in an international journal published [2].

After the success of this collaboration, the authors plan to extend the method to other types of induction motor as wounded induction motors, which have windings in the rotor instead of a squirrel-cage squirrel-cage. Furthermore, the method may be enhanced to cover the case in which the induction motor is fed by voltage sources instead of current ones. In this case, the problem is more involved since suitable initial currents are needed not only in the rotor bar, but also in the stator coils. An extension of the method to fit under more general ODE and PDE problems with periodic stationary solutions would also be interesting in order to deal with applications in nonlinear models for acoustic or water waves.

References

[1] Bermúdez, A., Domínguez, O., Gómez, D., Salgado, P.: Finite element approximation of nonlinear transient magnetic problems involving periodic potential drop excitations. Comput. Math. Appl. **65**, 1200–1219 (2013)
[2] Bermúdez, A., Gómez, D., Piñeiro, M., Salgado, P.: A novel numerical method for accelerating the computation of the steady-state in induction machines. Comput. Math. Appl. **79**, 274–292 (2020)
[3] Bermúdez, A., Gómez, D., Salgado, P.: Mathematical Models and Numerical Simulation in Electromagnetism. New York, Springer (2014)
[4] Boldea I., Nasar, S. A.: The Induction Machine Handbook, CRC Press (2002)
[5] Hara, T., Naito, T., Umoto, J.: Time-periodic finite element method for nonlinear diffusion equations. IEEE Trans. Magn. **21**(6), 2261–2264 (1985)
[6] Katagiri, H., Kawase, Y., Yamaguchi, T., Tsuji, T., Shibayama, Y.: Improvement of convergence characteristics for steady-state analysis of motors with simplified singularity decomposition-explicit error correction method. IEEE Trans. Magn. **47**, 1786–1789 (2011)
[7] Meunier, G.: The Finite Element Method for Electromagnetic Modeling. ISTE - Wiley (2008)
[8] Mezani, S., Hamiti, T., Belguerras, L., Lubin, T., Gerada, C.: Computation of wound rotor induction machines based on coupled finite elements and circuit equation under a first space harmonic approximation. IEEE Trans. Magn. **52**(3), 1–4 (2016)
[9] Miyata, K.: Fast analysis method of time-periodic nonlinear fields. J. Math-for-Ind. **3**, 131–140 (2011)
[10] Nakata, T., Takahashi, N., Fujiwara, K., Muramatsu, K., Ohashi, H., Zhu, H. L.: Practical analysis of 3-D dynamic nonlinear magnetic field using time-periodic finite element method. IEEE Trans. Magn., **31**, 1416–1419 (1995)
[11] Rachek, M., Merzouki, T., Finite Element Method Applied to the Modelling and Analysis of Induction Motors. In: Dr. Peep Miidla (Ed.), Numerical Modelling, InTech (2012)
[12] Randall, R. B.: Frequency Analysis. Brüel & Kjaer (1987)
[13] Savov, V. N., Georgiev, Zh. D., Bogdanov, E. S.: Analysis of cage induction motor by means of the finite element method and coupled of field, circuit and motion equations. Electr. Eng., **80**, 21–28 (1997)
[14] Stermecki, A., Bíró, O., Preis, K., Rainer, S., Ofner, G.: Numerical analysis of steady-state operation of three-phase induction machines by an approximate frequency domain technique. Elektrotech. Inftech. **128**, 81–85 (2011)
[15] Takahashi, Y., Iwashita, T., Nakashima, H., Tokumasu, T., Fujita, M., Wakao, S., Fujiwara, K., Ishihara, Y.: Parallel time-periodic finite-element method for steady-state analysis of rotating machines. IEEE Trans. Magn. **48**(2), 1019–1022 (2012)
[16] Takahashi, Y., Tokumasu, T., Fujita, M., Iwashita, T., Nakashima, H., Wakao, S., Fujiwara, K.: Time-domain parallel finite-element method for fast magnetic field analysis of induction motors. IEEE Trans. Magn., **49**, 2413–2416 (2013)

Challenges in the Simulation of Radio Frequency Circuits

Kai Bittner, Hans Georg Brachtendorf, and Roland Pulch

Abstract Radio frequency (RF) circuits are ubiquitous in daily live. However, they are extremely hard to design. Even experienced RF designers typically need several re-designs before the circuit fulfills the specifications. This repetitive procedure leads to increased expenses and, even worse, a time-to-market delay. There are several reasons, which cause re-designs. At ever increasing frequencies, the modelling of the constitutive circuit devices is extremely cumbersome and error prone. Even the lumped model assumptions are often not valid anymore, delays and crosstalks of the interconnects are modelled imprecisely and the device tolerances and mismatches are not sufficiently taken into account. Moreover, for the reason of cost reduction, the analog front-end and the digital signal processing is realized on the same die nowadays, leading to clock feedthrough via substrate coupling. Through silicon vias are a modelling challenge of its own, as 3D stacking in general. Hence, the rapid increase of center frequencies and the miniaturization cause a gap between design methodologies and tools on the one hand and challenges as well as requirements of RF circuit designs on the other hand. RF or bandpass signals are characterized by the fact that the bandwidth B of the signal's spectrum is orders of magnitudes smaller than the center frequency f_c, i.e., $B \ll f_c$. From the Whittaker-Kotelnikow-Shannon sampling theorem it is known that the sampling rate must be at least twice as high as the highest relevant frequency, i.e., $R > 2\,f_c$ for its discrete representation. Numerical initial value solvers are therefore prohibitively slow, since they represent signals locally by algebraic polynomials and hence a sampling rate even $R \gg 2\,f_c$ is required in practice, leading to a very small time-step for keeping the truncation error small. This chapter deals with mathematical techniques to overcome these problems in circuit simulation specifically with emphasis on RF circuits, operating at increasingly

Kai Bittner, Hans Georg Brachtendorf
University of Applied Sciences of Upper Austria, Softwarepark 11, Hagenberg im Mühlkreis, 4232, Austria, e-mail: brachtd@fh-hagenberg.at, Kai.Bittner@fh-hagenberg.at

Roland Pulch
Institute of Mathematics and Computer Science, University of Greifswald, Walther-Rathenau-Str. 47, Greifswald, 17489, Germany, e-mail: roland.pulch@uni-greifswald.de

M. Günther, W. Schilders (eds.), *Novel Mathematics Inspired by Industrial Challenges*, Mathematics in Industry 38, https://doi.org/10.1007/978-3-030-96173-2_6

higher frequencies. Basically, different time-scales together with artifical so-called multiscale signals are introduced, leading to the multirate partial differential equation (MPDE) technique, whereas the underlying problem is the numerical solution of an ordinary differential-algebraic equation (DAE). These multiscale signals do not have an immediate physical meaning, unlike the solution of the DAE. However, the MPDEs can be solved efficiently to obtain the multiscale signals, since they overcome the bottleneck of the sampling theorem. The MPDE is of hyperbolic type, which includes characteristic curves. Since only a specific solution along one charactristic curve of the MPDE has a physical meaning, this let to novel numerical techniques for solving this type of PDEs.
Multirate simulation problems also occur in the field of power electronics. Pulse-width modulated (PWM) signals exhibit similar numerical problems as RF circuits. Electric power drives employ PWM and similar waveforms for motor control. In recent time, these simulation problems were also tackled by the MPDE approach, representing these specific multiscale signals in a suitable compact basis for an efficient numerical treatment.

1 Introduction

Radio frequency (RF) integrated circuits (ICs) are ubiquitous in our daily life. In 2020 there are expected to be 7 trillion wireless devices. The communication traffic increases therefore exponentially, reaching the limits of the channel capacity of existing mobile communication infrastructure. Moreover business models in the field of Internet of Things (IoT) and Industry 4.0 require real time responses, which cannot be guaranteed by state-of-the-art communication standards.

These challenges are addressed by several means, including

a) opening the spectrum towards higher frequencies,
b) improving/optimizing the modulation waveforms and coding techniques towards higher bandwidth and power efficiency, Peak-to-Average Ratio (PAR), etc.,
c) increasing the channel capacities by multiple-input-multiple-output (MIMO) systems, and
d) network planning, e.g., ad-hoc networks.

The demands can only be tackled by mathematical tools from various fields, such as Harmonic Analysis for optimizing the modulation waveforms and MIMO channels, Algebra as the mathematical basis of coding theory, Information Theory combined with Harmonic Analysis for reaching the theoretical limits of the channel capacity, e.g., for MIMO techniques. Moreover, Vector Analysis is needed for electromagnetic field theory and wave propagation as the suitable mathematical means for Maxwell's theory. Nowadays, transceivers (transmitter and receiver) shift as much as possible of the signal generation at the transmitter and the channel equalization and decoding at the receiver side to the digital domain, since modern Digital Signal Processing (DSP) is easy to implement in digital hardware, easy to reconfigure, and practically not subject to mismatches or device tolerances, aging, etc. This shift is referred to as Software Defined Radio (SDR). However, the foundations of DSP lie in Harmonic Analysis, Calculus of Probability and Applied Numerical Algebra. Methods like the singular value decomposition (SVD), the QR algorithm, and the Levinson-Durban algorithm for solving Toeplitz matrices represent other prominent examples.

This chapter deals with the simulation of analog RF circuits and electronic devices. Therefore, we immerse ourselves into the numerical solution of a special sort of differential-algebraic equations and partial differential-algebraic equations, which result from the mathematical modelling of electronics. Electronic circuits are described by their physical voltages and currents, resulting typically in a system of nonlinear (ordinary) differential-algebraic equations. Since the emergence of circuit simulators like SPICE2 [1] and ASTAP [2] in the early 70s, the stationary DC operating point, the AC or phasor analysis of (linearized) circuits, as well as the numerical solution of initial value problems, i.e., transient analysis in engineer's terms became state-of-the-art. The 80s and 90s came up with the Harmonic Balance technique, e.g. [3, 4, 5, 6], which is a generalization of the phasor analysis for the periodic steady state analysis of nonlinear circuits. Another direction of research was the development of faster simulators specifically for digital circuitry such as the waveform relaxation method [7], timing analysis [8], the employment of the fast Krylov technique based accelerated iterative methods [9, 10], stabilizing the convergence of Newton's method by homotopy techniques or continuation methods [11]. We refer to the excellent tutorial work of Günther and Feldmann [12], which reproduces the various branches of circuit simulation comprehensively.

The challenges of RF circuit simulation at very high center frequencies is considered in detail now. Such circuits are very difficult to simulate, since the time step of numerical integration formulas or, vice versa, the sampling rate is limited by the sampling theorem [13], attributed to Whittacker, Kotelnikov and Shannon. The sampling theorem states that the sampling rate must be at least twice as high as the maximum relevant frequency component of the spectrum. In practice, however, the sampling rate must be at least a magnitude higher, since conventional numerical integration methods employ approximations of the waveforms by algebraic polynomials, which are not a good fit for oscillating signals encountered in RF circuits. Hence traditional initial value solvers are prohibitively slow. This lower bound on the sampling rate represented a bottleneck in the numerical simulation of oscillating circuits for a long time.

The problem can be overcome by a multirate technique, which was developed since the end of the 90s, see [14, 15, 16]. Therein, a multivariate description separates the high-frequency carrier oscillation from the low-frequency oscillation containing the signal information. A new mathematical object was derived: the multirate partial differential-algebraic equation (MPDE), which replaces the (ordinary) differential-algebraic equation (DAE) modelling the circuit. This multirate technique is the topic of the subsequent sections. The presentation below adopts the nomenclature from [17].

In recent time, the MPDE technique has been employed also in power electronic applications, namely DC-DC converters, power drives employing pulse-width modulated (PWM) signals, etc. [18, 19].

2 Network equations

In this section, the system of DAEs resulting from electronic circuits is derived. It is assumed that the circuit consists purely of lumped devices, i.e., there are only interactions between the devices through their terminal currents and voltages. According to Kirchhoff, electric networks can be modelled by graphs with z branches and n nodes. Furthermore, we assume that the network graph is connected, i.e., there is always a path between all distinct nodes. Without loss of generality, one node is the ground node with index "0". There are two matrices: the node incidence matrix $A \in \mathbb{R}^{(n-1)\times z}$ and the branch incidence matrix $B^T \in \mathbb{R}^{(z-n+1)\times z}$. The coefficients of these two matrices are either 0 or ± 1. A system of linear independent loops can be obtained by choosing a tree in the graph. A tree is a subgraph, which contains all nodes but no loops of the underlying graph. A connected graph has a tree with $n-1$ branches, the so-called tree branches. The remaining branches are referred to as interconnection branches. One can construct a system of $z-n+1$ linear independent loops by selecting the loops containing only one interconnection branch. The remaining branches are tree branches.

Definition 2.1 [20, 21] A pair of matrices (A,B), $A \in \mathbb{R}^{n\times m}$, $B \in \mathbb{R}^{m\times \ell}$ is called *exact, iff the following conditions hold*

$$AB = 0$$
$$rank(A) + rank(B) = m.$$

Theorem 2.1 *The node incidence matrix A and the branch incidence matrix B form an exact pair* (A,B).

By introducing the branch currents $i \in \mathbb{R}^z$ and branch voltages $u \in \mathbb{R}^z$, Kirchhoff's equations can be written as

$$\begin{pmatrix} A & 0 \\ 0 & B^T \end{pmatrix} \begin{pmatrix} i \\ u \end{pmatrix} = 0. \tag{1}$$

Remark 2.1 The kernel of the system is referred to as the Kirchhoff space

$$\mathcal{K} = \left\{ \begin{pmatrix} i \\ u \end{pmatrix} \mid \begin{pmatrix} A & 0 \\ 0 & B^T \end{pmatrix} \begin{pmatrix} i \\ u \end{pmatrix} = 0 \right\}.$$

Below the loop currents j and node voltages v are introduced, which span the Kirchhoff space characterized by the following lemma.

Lemma 2.1 *The pair of matrices* (A,B), $A \in \mathbb{R}^{n\times m}$, $B \in \mathbb{R}^{m\times(m-n)}$, $n < m$ *is assumed to be exact. The solution space of the homogeneous equations* $A\,i = 0$ *and* $B^T u = 0$ *can be written in the form*

$$\begin{pmatrix} i \\ u \end{pmatrix} = \begin{pmatrix} B & 0 \\ 0 & A^T \end{pmatrix} \begin{pmatrix} j \\ v \end{pmatrix}, \quad \textit{with } j \in \mathbb{R}^{m-n} \textit{ and } v \in \mathbb{R}^n.$$

Proof The exactness of the matrix pair (A, B) implies $AB = 0$ as well as $B^T A^T = 0$. It follows that $A\,i = A\,B\,j \equiv 0$, $B^T\,u = B^T\,A^T\,v \equiv 0$. Since B, A^T have full column ranks, the Kirchhoff space is spanned completely. □

The solution space of the homogeneous Kirchhoff's equation is equivalent with the image $\mathscr{S}$ of the above matrix defined by

$$\mathscr{S} = \left\{ \begin{pmatrix} i \\ u \end{pmatrix} \mid \begin{pmatrix} i \\ u \end{pmatrix} \in \text{im} \begin{pmatrix} B & 0 \\ 0 & A^T \end{pmatrix} \right\}$$

and therefore $\mathscr{K} = \mathscr{S}$. Consequently, the solution space of Kirchhoff's equations is given by

$$\begin{pmatrix} i \\ u \end{pmatrix} = \begin{pmatrix} B & 0 \\ 0 & A^T \end{pmatrix} \begin{pmatrix} j \\ v \end{pmatrix}, \; j \in \mathbb{R}^{z-n+1}, \; v \in \mathbb{R}^{n-1}. \tag{2}$$

Remark 2.2 The elements of the vector j of dimension $z - n + 1$ are referred to as *loop branch currents*, the elements of v of dimension $n - 1$ as *node voltages* or *potentials*.

Electronic devices are often described mathematically by lumped models, i.e., their static behavior is modelled by a set of equations representing the terminal currents, voltage drops, charges associated with terminals and magnetic fluxes associated with branches.

Depending on the physical behavior, the device constitutive equations are described in *admittance form* by

$$f(i,u,t) = i + g(u) + \frac{\mathrm{d}q(u)}{\mathrm{d}t} + b(t) = 0, \quad f : \mathbb{R}^z \times \mathbb{R}^z \times \mathbb{R} \to \mathbb{R}^z, \tag{3}$$

where $q(u)$ represents a vector of *charges*, or by the *impedance form*

$$f(i,u,t) = u + g(i) + \frac{\mathrm{d}\Psi(i)}{\mathrm{d}t} + b(t) = 0, \quad f : \mathbb{R}^z \times \mathbb{R}^z \times \mathbb{R} \to \mathbb{R}^z. \tag{4}$$

In the equation above $\Psi(i)$ is the vector of *magnetic fluxes*. The vector $b(t)$ represents the *stimulus* or *source* vector, which is predetermined. The stimuli are either voltage sources or current sources. Semiconductor devices are typically modelled in the admittance form. However, one encounters often hybrid device equations in a circuit, i.e., partly the circuit's devices are modelled in admittance and partly in impedance form. Such a case usually occurs when ideal voltage sources are used, which can be solely described in impedance form. For the hybrid formulation, it is advantageous to split the vector of branch voltages and terminal currents into two subvectors, i.e., $i = (i_1^T, i_2^T)^T$ and $u = (u_1^T, u_2^T)^T$. The latter form the device constitutive equations in impedance form

$$
\begin{aligned}
f_1(i,u,t) &= i_1 + g_1(u) + h_1(i_2) + \frac{\mathrm{d}q(u)}{\mathrm{d}t} + b_1(t) = 0, \\
f_2(i_2,u,t) &= u_2 + g_2(i_2) + h_2(u) + \frac{\mathrm{d}\Psi(i_2)}{\mathrm{d}t} + b_2(t) = 0,
\end{aligned}
\tag{5}
$$

with $f_1 : \mathbb{R}^z \times \mathbb{R}^z \times \mathbb{R} \to \mathbb{R}^{z_1}$ and $f_2 : \mathbb{R}^{z-z_1} \times \mathbb{R}^z \times \mathbb{R} \to \mathbb{R}^{z-z_1}$. An alternative formulation is the *quasilinear* form of the device constitutive equations, i.e.,

$$
f(i,u,t) = i + g(u) + C(u)\,\frac{\mathrm{d}u}{\mathrm{d}t} + b(t) = 0 \tag{6}
$$

in the case of a device equation given in admittance form with a generally non-constant capacitance matrix $C(u)$. Alternatively, for the impedance form the quasilinear formulation reads as

$$
f(i,u,t) = u + g(i) + L(i)\,\frac{\mathrm{d}i}{\mathrm{d}t} + b(t) = 0, \tag{7}
$$

where $L(i)$ is a generally non-constant inductance matrix. Corresponding formulations hold true for the hybrid form. The capacitance and inductance matrices $C, L \in \mathbb{R}^{z \times z}$ are often singular, resulting in a system of DAEs. As considered below, the quasilinear formulation is critical with regard to charge and flux conservation.

Kirchhoff's voltage and current laws together with the device constitutive equations describe the electronic circuit completely. The solution space of an electronic network is not empty if the system of algebraic equations

$$
f(B\,j, A^T\,v, t) = 0 \tag{8}
$$

has at least one solution, see [21]. If the element equations are given in admittance form (3) one obtains together with (1) and (2)

$$
\begin{aligned}
& A\,i + A\,g(A^T\,v) + A\,\frac{\mathrm{d}q(A^T\,v)}{\mathrm{d}t} + A\,b(t) = 0, \\
\Leftrightarrow \quad & A\,g(A^T\,v) + A\,\frac{\mathrm{d}q(A^T\,v)}{\mathrm{d}t} + A\,b(t) = 0,
\end{aligned}
\tag{9}
$$

where $g, q : \mathbb{R}^{n-1} \to \mathbb{R}^{n-1}$ and $b : \mathbb{R} \to \mathbb{R}^{n-1}$.

The $n-1$ node voltages v yield the vector of branch voltages u by $u = A^T\,v$. Finally, the vector of branch currents i is evaluated from the known branch voltages by (3). The formulation of the circuit equations according to (9) is referred to as *nodal analysis*. Likewise, if the device constitutive equations are given in impedance form (4) the loop currents j are the unknowns of the resulting system

$$
\begin{aligned}
&B^T u + B^T g(B\,j) + B^T \frac{\mathrm{d}\Psi(B\,j)}{\mathrm{d}t} + B^T b(t) = 0, \\
\Leftrightarrow \qquad &B^T g(B\,j) + B^T \frac{\mathrm{d}\Psi(B\,j)}{\mathrm{d}t} + B^T b(t) = 0,
\end{aligned} \tag{10}
$$

where $g, \Psi : \mathbb{R}^{z-n+1} \to \mathbb{R}^{z-n+1}$ and $b : \mathbb{R} \to \mathbb{R}^{z-n+1}$.

After the calculation of the loop currents j the branch currents are uniquely evaluated by $i = B\,j$. Formulating the system of equations according to (10) is referred to as *loop analysis*.

In most practical cases the device equations are given in the hybrid formulation. The formulation most often used in circuit simulators is the *modified nodal analysis* (MNA) by Ho, Ruehli and Brennan [22]. This method, as the name suggests, is essentially the nodal analysis. Nevertheless, it is sufficiently flexible to incorporate devices equations in the impedance form.

The derivation of the MNA starts with the hybrid formulation of the device equations as given in (5). Without loss of generality, the vector i of the branch currents can be subdivided into a vector i_1 representing the branches described in the admittance form and a vector i_2 of the branches in impedance form. Kirchhoff's equations are reformulated into

$$
A\,i = (A_1\,A_2)\begin{pmatrix} i_1 \\ i_2 \end{pmatrix} = 0.
$$

Based on this subdivision, the device equations in hybrid form can be rewritten as

$$
\begin{aligned}
-A_2\,i_2 + A_1\,g_1(A^T v) + A_1\,h_1(i_2) + A_1 \frac{\mathrm{d}q(A^T v)}{\mathrm{d}t} + A_1\,b_1(t) &= 0, \\
A_2^T v + g_2(i_2) + h_2(A^T v) + \frac{\mathrm{d}\Psi(i_2)}{\mathrm{d}t} + b_2(t) &= 0,
\end{aligned} \tag{11}
$$

where $g_1, q : \mathbb{R}^{n-1} \to \mathbb{R}^{n-1}$, $h_1 : \mathbb{R}^{z-z_1} \to \mathbb{R}^{n-1}$, $h_2 : \mathbb{R}^{n-1} \to \mathbb{R}^{z-z_1}$, $g_2, \Psi : \mathbb{R}^{z-z_1} \to \mathbb{R}^{z-z_1}$ and $b_1 : \mathbb{R} \to \mathbb{R}^{n-1}$, $b_2 : \mathbb{R} \to \mathbb{R}^{z-z_1}$.

The MNA is the commonly used technique and employed in a variety of circuit simulators. Unlike the loop analysis the nodal analysis does not require the search for a tree and is therefore easier to code. Moreover, the number of devices in impedance form is smaller than the number of devices in admittance form for nearly all electronic circuits, since semiconductors, both bipolar and MOS devices, are modelled in this form. Therefore, the impedance representation is mainly employed for ideal voltage sources and (coupled) inductors. However, the number of these devices on an integrated circuit is usually small. Hence the number of additional unknowns and additional equations are negligible. Moreover, the MNA improves the flexibility to model the electronic devices drastically. The equations (11) are not formulated directly. Therefore the evaluation of the incidence matrices $A_{1/2}$ is not required. In contrast, there are templates or stamps, for which the matrix structure and right-hand side is known. Each device has its own matrix stamp [23]. Additionally, there are

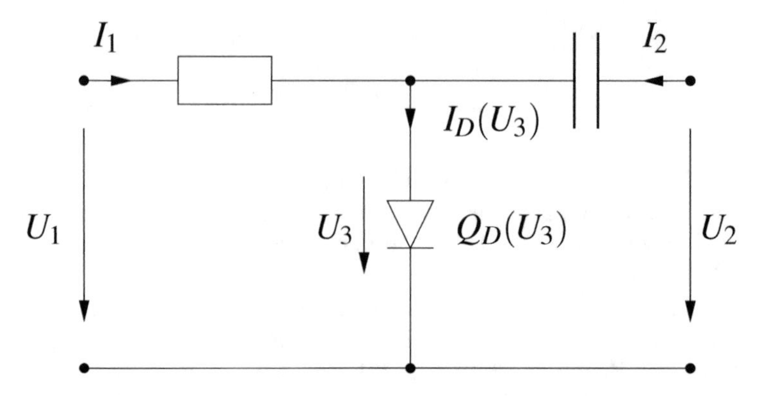

Fig. 1: Circuit example where no terminal charge can be defined [24].

corresponding nonlinear circuit elements. Yet their stamps are formally equivalent to the associated linear counterparts.

The modelling of the device constitutive equations by terminal electric charges and magnetic fluxes associated with branches guarantees the conservation of the total charge and the total flux. This conservation is crucial for, e.g., the design of switched-capacitor circuits, charge pumps, etc. It is also presupposed in device modelling. However, it is not a-priori guaranteed that their existence is physically valid. From measurements one obtains the (partial) derivatives, i.e., the capacitances. Below sufficient conditions for the existence of charges associated with terminals are considered.

Assume that a charge associated with node i, namely $Q_i \in C^2(\mathbb{R})$, exists and let

$$C_{ij} := \frac{\partial Q_i}{\partial u_j} \qquad \text{for } i, j = 1, \ldots, z$$

be the partial derivative w.r.t. the branch voltage u_j. The differential capacity C_{ij} is a measureable quantity. It follows that

$$\frac{\partial C_{ij}}{\partial u_k} = \frac{\partial^2 Q_i}{\partial u_k \partial u_j} = \frac{\partial^2 Q_i}{\partial u_j \partial u_k} = \frac{\partial C_{ik}}{\partial u_j} \qquad \text{for } i, j, k = 1, \ldots, z. \tag{12}$$

If $\frac{\partial C_{ij}}{\partial u_k} \neq \frac{\partial C_{ik}}{\partial u_j}$ holds, the terminal charge assumption is not valid. Modern semiconductor models take into account this problem by introducing extra internal nodes. In [24], the circuit of Fig. 1 is introduced, where no charge representation exists at the nodes 1 and 2. In the same publication, the problem of modelling semiconductor devices by lumped models is deepened. The sum of each column of the capacitance matrix, augmented by the ground node, is zero to guarantee charge neutrality at any point of operation.

All formulations of the network equations represent either systems of ordinary differential equations (ODEs) or systems of differential-algebraic equations (DAEs).

DAEs are classified by their index, see [25]. Tischendorf [26] derived techniques for a topological index calculation.

3 Simulation of Radio Frequency Circuits

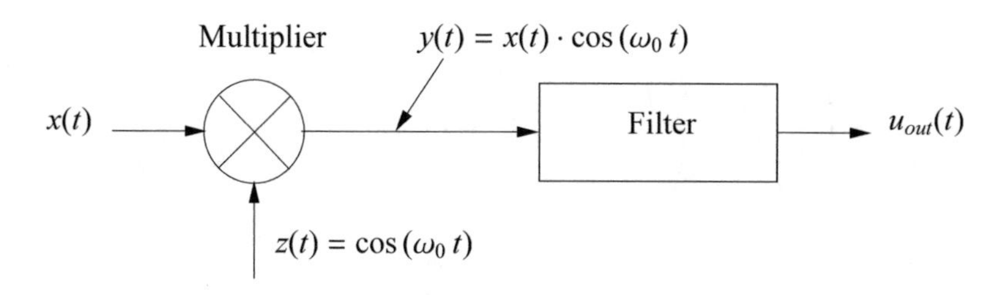

Fig. 2: Block diagram of a mixer circuit.

We consider the block diagram of a mixer circuit depicted in Fig. 2. This configuration represents the classical analog front-end at the transmitter (Tx) side. Such a circuit is typically described by a system of (ordinary) DAEs as discussed in Section 2. The input signal x is referred to as the baseband signal or signal envelope. It is usually generated digitally and later converted to the analog domain by an Analog to Digital Converter (ADC). The essential information is coded inside the signal, e.g. by Orthogonal Frequency Division Multiplexing (OFDM) or other state-of-the-art modulation schemes. Fig. 3 shows schematically the waveforms and spectra of the signals in the Tx chain. The bandwidth $f_{\max}$ of the baseband signal is practically orders of magnitude smaller than the modulation frequency or modulation carrier f_0. The output signal u_{out} is a modulated or bandpass signal, also referred to as RF signal, which gets further amplified and transmitted via the antenna. Due to the sampling theorem, the signal representation of the bandpass signal requires a sampling rate $R > 2 f_0$. Since the signals are locally approximated by algebraic polynomials, the sampling rate must be even significantly higher in circuit simulators.

Such a circuit is often simulated by the so-called two-tone test. Therein, the input signal x is not an arbitrarily modulated baseband signal but a sinusoidal waveform. Fig. 4 depicts the output voltage of a mixer circuit, which exhibits the form

$$u_{\text{out}}(t) \approx U_0 \sin(\omega_0 t) \sin(\omega_1 t) \quad \text{with } \omega_0 \gg \omega_1 \tag{13}$$

in the steady state. The waveform u_{out} can be easily represented by a small amount of Fourier coefficients with two fundamental angular frequencies ω_0 and ω_1. Fig. 5 illustrates the sparse discrete spectrum of the output signal exemplarily. The Harmonic Balance (HB) method captures such a signal waveform by a truncated Fourier series. If a circuit behaves nearly linear, then a small number of Fourier coefficients already

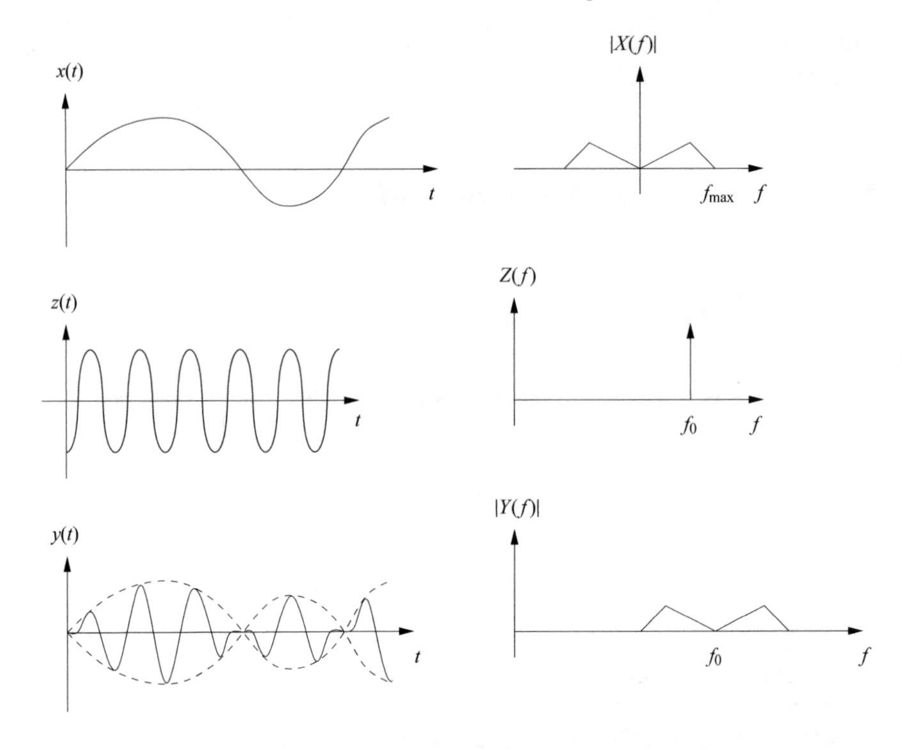

Fig. 3: Amplitude modulation: signal representation in the time domain (top) and frequency domain (bottom).

yields a sufficiently accurate approximation of the signals. Hence HB provides an efficient representation of the signals. However, modern circuit designs behave quasi-digitally or switching-like, because these circuits require less power. Therefore, HB is not well suited for up-to-date RF circuitry. Consequently, one resorts to traditional representations by algebraic polynomials or novel techniques such as spline-wavelet approximations [27, 28, 29, 30, 31].

The multirate technique can be considered as a method to exploit the sparsity of the spectra by decoupling the baseband signal from the carrier signal. Fig. 6 exhibits the same signal (13) in a plane by introducing two different time scales t_1, t_2 for capturing the slow time scale envelope and the RF oscillation.

Representing the quasiperiodic time domain signal (13) by a Fourier series, the argument $\omega_1 t$ is formally replaced by $\omega_1 t_1$ and $\omega_0 t$ by $\omega_0 t_2$, respectively. Hence, the original signal (13) is obtained on the diagonal, where $t_1 = t_2 = t$. The two-dimensional waveform reads as

$$\widehat{u}_{\text{out}}(t_1, t_2) \approx U_0 \sin(\omega_0 t_2) \cdot \sin(\omega_1 t_1). \tag{14}$$

Obviously, it holds that $u_{\text{out}}(t) = \widehat{u}_{\text{out}}(t,t)$.

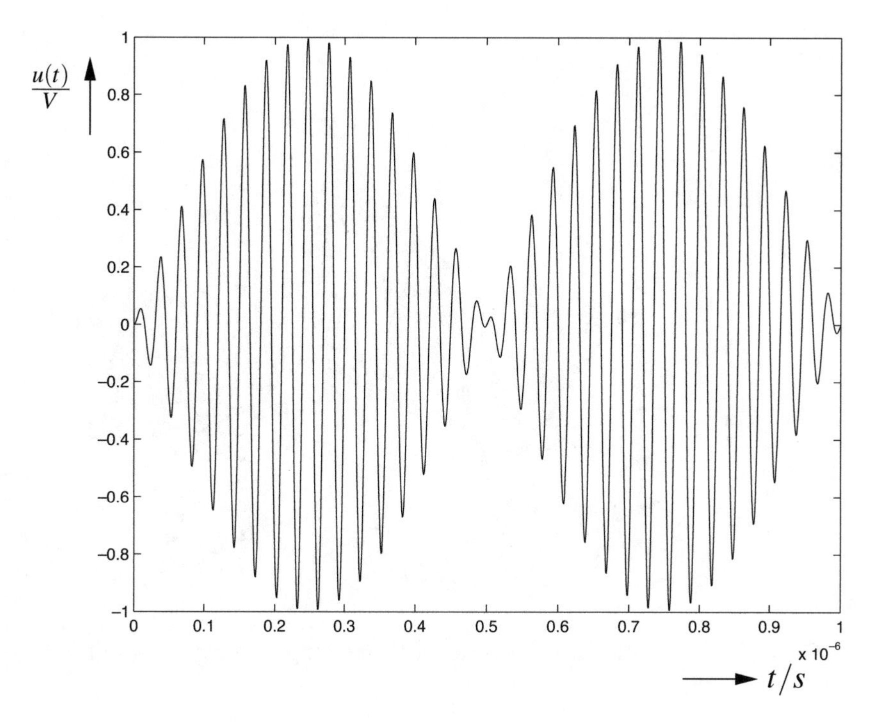

Fig. 4: Typical multitone signal waveform.

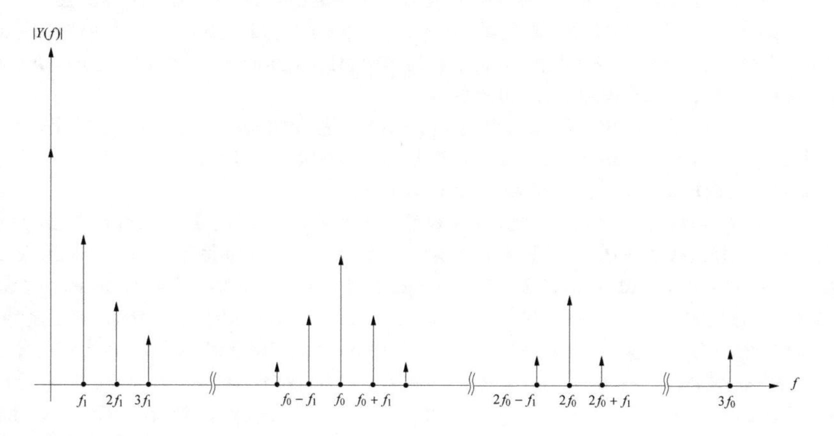

Fig. 5: Typical spectrum of a mixer circuit.

By construction, the two-dimensional steady state (14) is periodic along the axes t_1 and t_2. Moreover, the signal waveform is smooth in both variables. The different time scales of the solution are reflected by the different scales of the axes. The

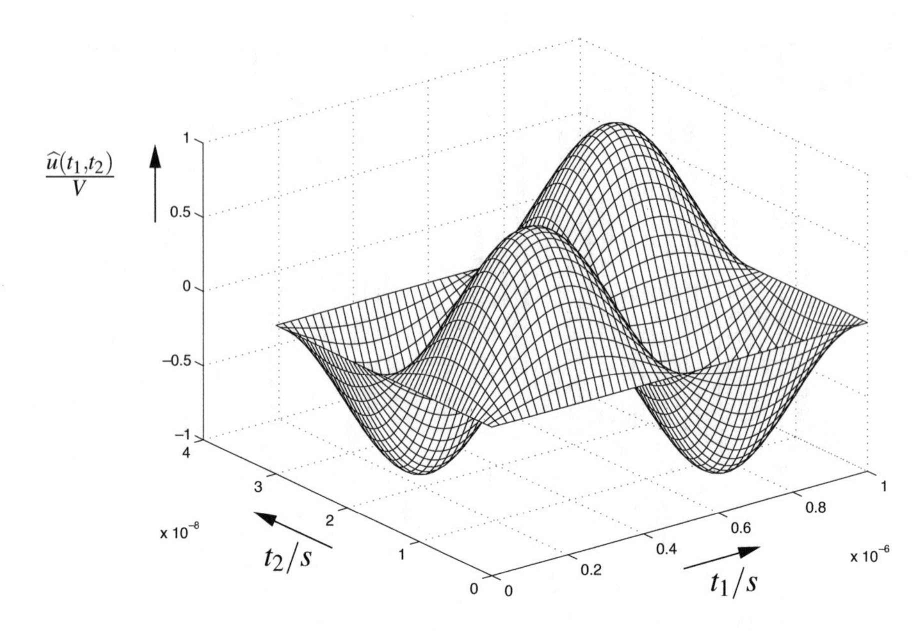

Fig. 6: Two-dimensional representation of the output voltage of a mixer circuit.

discretization of the signal can be realized by choosing different mesh spacings along the coordinate axis. There is no limitation anymore from the sampling theorem. The number of mesh points for representing the signal becomes practically independent of the involved fundamental frequencies.

An alternative interpretation can be given in the frequency domain: the discrete or line spectrum as depicted in Fig. 5. It is clustered around integer multiples of the oscillator frequency with huge gaps in between.

A widely known multirate method is the technique of the Equivalent Complex Baseband (ECB), see e.g. [32], which is familiar to communication engineers. The modulated signals are treated by their baseband counterparts. This is done by two steps: firstly the analytical signal x^+ is generated and secondly the analytical signal is modulated by f_0 or, vice versa in the frequency domain shifted to baseband. The generation of the analytical signal requires a Hilbert transformation or a phase shift of 90°, i.e., $x^+(t) = x(t) + \mathrm{j}\,\mathscr{H}(x(t))$ with $\mathrm{j} = \sqrt{-1}$. There are several devices, which approximate a Hilbert transformation, such as quadrature couplers and Schiffman phase shifters. Moreover, linear time invariant (LTI) systems such as filters and their impulse responses can be treated in the same manner, as well as white Gaussian noise. Therefore, the ECB method became a standard approach for designing and simulating whole communication systems on a system level. However, this technique has its limits when nonlinear effects such as intermodulation distortion come into play.

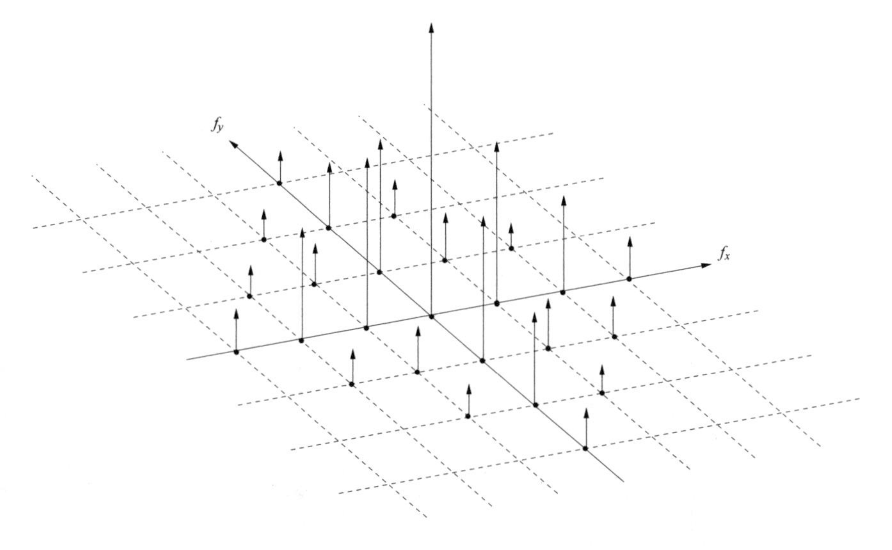

Fig. 7: Demonstration of the embedding technique in the frequency domain.

To generalize the ECB method to nonlinear circuits and systems described by a system of (ordinary) DAEs, one follows the concept of embedding the solution of an initial or boundary value problem into a solution on a domain with different time scales or large anisotropy of the axes scales. This technique will be analyzed further in the next section. It will be shown that solving the partial DAEs

$$f\left(\widehat{x}, \left(\omega_0 \frac{\partial}{\partial t_1} + \omega_1 \frac{\partial}{\partial t_2}\right) \widehat{x}, \widehat{b}(t_1, t_2)\right) = 0 \tag{15}$$

instead for the original (ordinary) DAE $f(x, \dot{x}, b(t)) = 0$, see Section 2, is easier. The system (15) is equipped with 2π-periodic boundary conditions in t_1, t_2, using appropriate time scales. The partial DAE (15) is of hyperbolic type, see [33]. The special signal $x(t) = \widehat{x}(\omega_0 t, \omega_1 t)$ is the solution of the underlying (ordinary) DAE, i.e. obtained along a specific characteristic curve of the partial DAE. Thus the two-dimensional spectrum is more compact.

The frequency domain interpretation of the multirate technique is shown in Fig. 7: a rearrangement into a plane leads to an artificial two-dimensional spectrum such that the frequencies along the axes have equidistant spacings.

In contrast to the periodic steady state problem, another simulation task consists in an initial value problem or transient response of an RF circuit, which is considered next. Exemplarily, a Pierce quartz crystal oscillator is examined as depicted in Fig. 8. Due to the high quality factor (Q-factor) of the quartz crystal (xtal) the initial transient response is by orders of magnitude larger than the period of oscillation. The oscillation frequency of the Pierce oscillator is about 2 MHz and mainly governed by

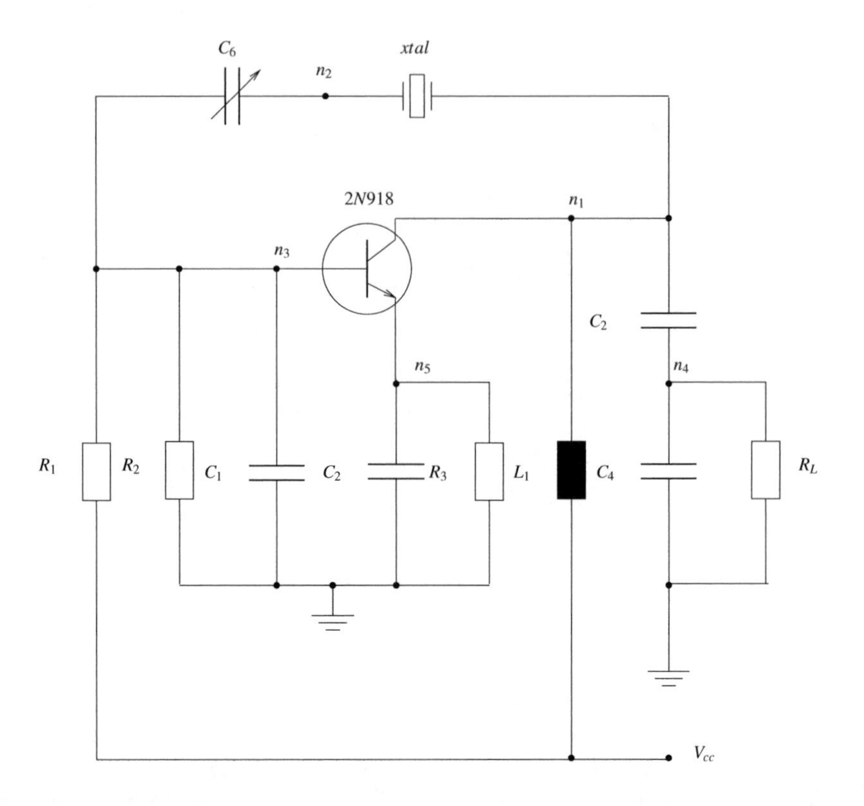

Fig. 8: Schematic of a 2 MHz Pierce oscillator.

the xtal. Fig. 9 shows the simulated initial transient response of the Pierce oscillator, wherein each line represents an oscillation. The envelope is nearly constant within one oscillation. The solution can be represented by a Fourier expansion of the oscillation with slowly time-varying Fourier coefficients

$$x(t) = \sum_{k=-K}^{K} X_k(t) \cdot \exp\left(\mathrm{j}\, k\, \omega(t)\, t\right)$$

with $X_{-k} = X_k^*$. This means that the time constants of the waveforms $X_k(t)$, $\omega(t)$ are orders of magnitude larger than the oscillation period $T(t) = \frac{2\pi}{\omega(t)}$. The waveforms $X_k(t)$ for $k \in \mathbb{Z}$ are addressed as the envelopes of the signal. By the formal replacement

$$\widehat{x}(\tau, t_1) := \sum_{k=-K}^{K} X_k(\tau) \cdot \exp\left(\mathrm{j}\, k\, \omega(\tau)\, t_1\right)$$

one obtains a two-dimensional signal with periodic boundary conditions in t_1. Therefore it is sufficient to evaluate the waveform on a small domain or strip

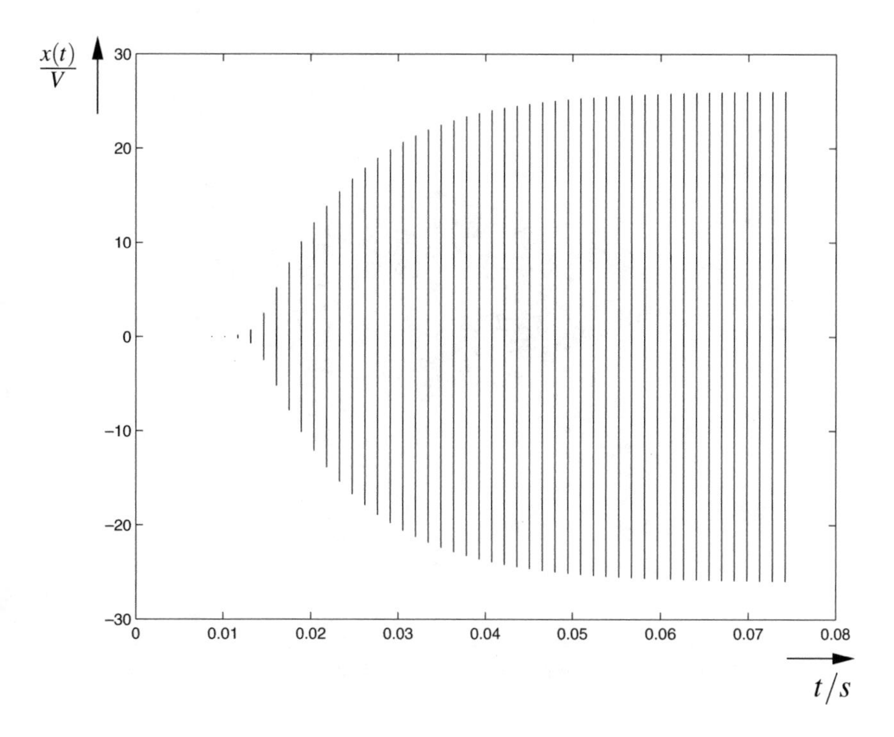

Fig. 9: Initial transient response of a Pierce quartz crystal oscillator.

$$\Omega = \left\{ (\tau, t_1) \mid \tau \geq 0,\ 0 \leq t_1 \leq \tfrac{2\pi}{\omega(\tau)} \right\}.$$

Again the numerical problem due to the sampling theorem is circumvented by introducing two distinct time scales, one for the fast oscillation and one for the slow envelope. As shown below, instead of solving the autonomous (ordinary) DAEs $f(x, \dot{x}) = 0$, the partial DAEs

$$f\left(\widehat{x}, \left(\frac{\partial}{\partial \tau} + \frac{\mathrm{d}(\omega(\tau)\,\tau)}{\mathrm{d}\tau} \cdot \frac{\partial}{\partial t_1}\right) \widehat{x}\right) = 0$$

are solved with mixed initial-boundary value conditions, i.e. a boundary value problem is considered w.r.t. t_1 and an initial value problem is given in τ.

Fig. 10 illustrates the solution obtained by the envelope method. This multiscale signal exhibits a relatively simple behavior. Consequently, the computation is feasible with large step sizes along both axes.

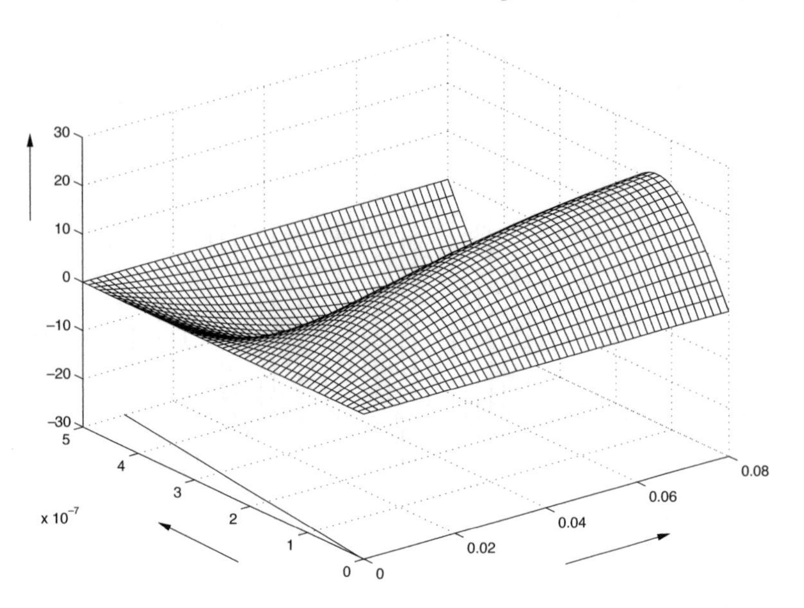

Fig. 10: Sketch of the envelope method. The wanted solution is obtained along the drawn curve through the origin.

4 The Embedding Technique

In Section 2, the MNA has been derived as means to model electronic networks by system of (ordinary) DAEs. Let the system be uniquely solvable for consistent initial conditions. In the following, the notation of the DAE

$$f(x(t),\dot{x}(t),b(t)) = i(x(t)) + \frac{\mathrm{d}}{\mathrm{d}t}q(x(t)) + b(t) = 0, \quad x(0) = x_0, \tag{16}$$

is adopted, where $x \in \mathbb{R}^N$, $i, q : \mathbb{R}^N \to \mathbb{R}^N$ and $b : \mathbb{R} \to \mathbb{R}^N$. It holds that $\frac{\mathrm{d}q(x)}{\mathrm{d}t} = \frac{\partial q}{\partial x}\dot{x}$. Furthermore, for ease of presentation the two-tone case is considered, since generalizations are straightforward and can be found, e.g., in [17].

It is presupposed that the (ordinary) DAE (16) has a unique solution for each member of a *family of consistent initial conditions* parameterized by $\Theta \in \mathbb{R}$

$$f(x_\Theta(t),\dot{x}_\Theta(t),b_\Theta(t)) = i(x_\Theta(t)) + \frac{\mathrm{d}}{\mathrm{d}t}q(x_\Theta(t)) + b_\Theta(t) = 0, \quad x_\Theta(0) = x_0(\Theta). \tag{17}$$

In the system (17), $x_0 : \mathbb{R} \to \mathbb{R}^N$ and $b_\Theta : \mathbb{R} \to \mathbb{R}^N$ are periodic functions of the parameter Θ with normalized period 2π.

The domain $\Omega \subset \mathbb{R}^2$ with

$$\Omega := \{(\tau, t_1) \,|\, \tau \geq 0, t_1 \in [0, 2\pi]\} \tag{18}$$

represents the domain of dependence (Fig. 12). A system of partial DAEs of the form

$$\widehat{f}(\widehat{x}(\tau,t_1),\nabla\widehat{x}(\tau,t_1),\tau,t_1) = i(\widehat{x}(\tau,t_1)) + \mathrm{D}q(\widehat{x}(\tau,t_1)) + \widehat{b}(\tau,t_1) = 0 \qquad (19)$$

is given with initial conditions

$$\widehat{x}(0,t_1) = x_0(t_1) \qquad (20)$$

and 2π-periodic boundary conditions in t_1, i.e.

$$\widehat{x}(\tau,0) = \widehat{x}(\tau,2\pi) \qquad (21)$$

on the domain (18). Note that the initial conditions coincide with the initial conditions (20) of the family of (ordinary) DAEs (17). The symbol ∇ denotes the gradient, i.e., the first-order partial derivatives $\nabla\widehat{x} = (\frac{\partial\widehat{x}}{\partial\tau},\frac{\partial\widehat{x}}{\partial t_1})$. The differential operator D is defined by

$$\mathrm{D} := \frac{\partial}{\partial\tau} + \frac{\mathrm{d}(\tau\cdot\omega_1(\tau))}{\mathrm{d}\tau}\cdot\frac{\partial}{\partial t_1} \qquad (22)$$

with the frequency function $\omega_1 \in C^1(\mathbb{R})$. The stimulus $\widehat{b}$ satisfies

$$\widehat{b}(t,\,\omega_1(t)t+\Theta) = b_\Theta(t). \qquad (23)$$

The relation between the family of initial conditions (17) and the system of partial DAEs (19) is given by the following theorem.

Theorem 4.1 *[17, 34] Let the operator* D *be defined by (22) and let the stimulus vector satisfy (23). The initial-boundary value problem* (20), (21) *for the partial DAE (19) has a solution iff the initial value problem (17) has a solution* x_Θ *for each parameter* Θ*. The relation*

$$x_\Theta(t) = \widehat{x}(t,\,\omega_1(t)t+\Theta) \qquad (24)$$

holds between the periodically extended PDAE solution and the solution of the (ordinary) DAE with the family of initial conditions $x_0(\Theta)$*.*

Remark 4.1 The determination of a suitable frequency function $\omega_1(\tau)\in C^1$ requires a-priori knowledge about the physical background of the DAE (16). This is not a subject of the formulation of the partial DAE, see [17]. Methods for estimating the instantaneous frequency can be found in [17, 35, 36, 37].

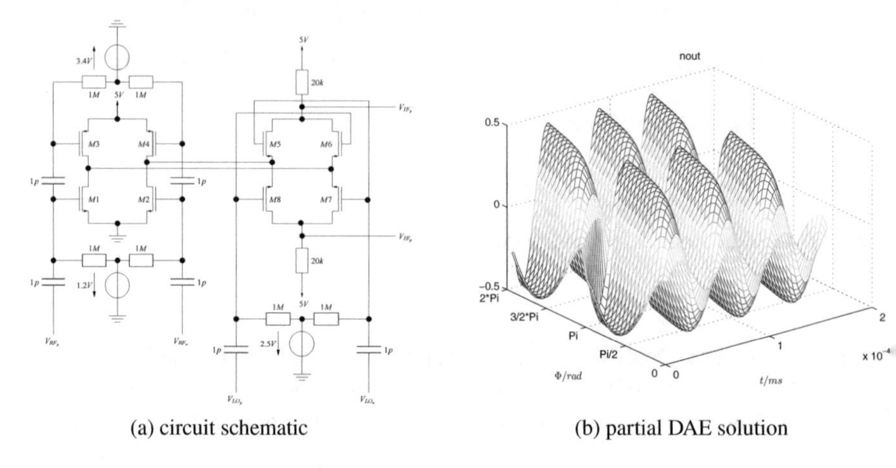

(a) circuit schematic (b) partial DAE solution

Fig. 11: Folded mixer circuit.

In the following, the theorem is illustrated by several examples.

Example 4.1 (Two-tone HB [14, 38, 39]) In the case of two-tone HB we assume that the circuit operates in steady state, that is, the waveforms in (19) are assumed periodic within the τ scale.

The fundamental frequencies are typically given by the signal sources. The corresponding partial DAE takes the form

$$\widehat{f}(\widehat{x}(\tau,t_1),\nabla\widehat{x}(\tau,t_1),\tau,t_1) = \\ i(\widehat{x}(\tau,t_1)) + \omega_0 \frac{\partial}{\partial \tau} q(\widehat{x}(\tau,t_1)) + \omega_1 \frac{\partial}{\partial t_1} q(\widehat{x}(\tau,t_1)) + \widehat{b}(\tau,t_1) = 0 \tag{25}$$

The Harmonic Balance method for multitone signals employs a Ritz-Galerkin method for discretizing (25) based on trigonometric basis functions. Alternatively, spline basis functions, wavelets as well as finite difference schemes are suitable for discretizing the partial differential equations (25), cf. [27].

In Fig. 11a a folded mixer circuit is depicted with differential wiring. One can see the differential inputs of the radio frequency signal (V_{RF}) and the inputs of the local oscillator (V_{LO}). The steady state solution is depicted in Fig. 11b. One can see the sharp transients caused by the switching of the MOS transistors. Therefore, for the numerical solution a spline-wavelet approximation of the signal has been employed [27].

Example 4.2 [34, 40] Let Ω be again the domain (18). The partial DAE is given by

$$\widehat{f}(\widehat{x}(\tau,t_1),\nabla\widehat{x}(\tau,t_1),\tau,t_1) \\ = i(\widehat{x}(\tau,t_1)) + \left(\frac{\partial}{\partial \tau} + \frac{\mathrm{d}(\tau\cdot\omega_1(\tau))}{\mathrm{d}\tau}\frac{\partial}{\partial t_1}\right) q(\widehat{x}(\tau,t_1)) + \widehat{b}(\tau,t_1) = 0. \tag{26}$$

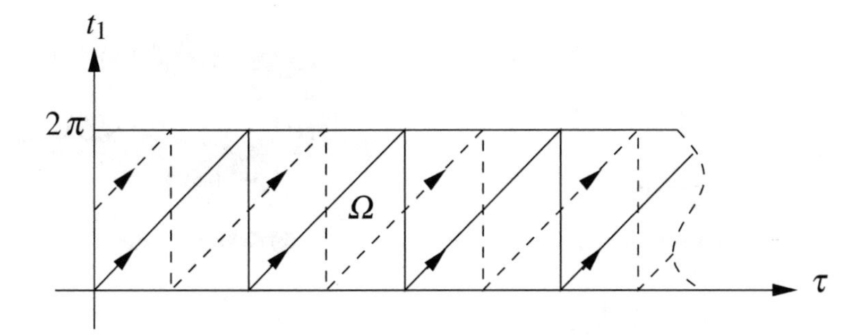

Fig. 12: Graphical illustration of the embedding technique with two characteristic curves sketched.

For a constant stimulus $\widehat{b}(\tau, t_1) \equiv b_0$ the underlying (ordinary) DAE is autonomous. In this case, the frequency function $\omega_1(\tau)$ is unknown and not determined. Consequently, the partial DAE (26) represents the multirate model of a free running oscillator such as the Pierce quartz crystal oscillator in Fig. 8. For the special case $\omega_1(\tau) = \omega_0$ one obtains as a special case the non-autonomous envelope method, see [41, 42].

An example is the simulation of a phase locked loop (PLL) circuit depicted in Fig. 13a as a block diagram. The input signal is a frequency modulated signal of the form

$$s(t) = \sin\left(2\pi f_1 t + \tfrac{\Delta F}{f_2} \sin(2\pi f_2 t)\right).$$

The PLL consists mainly of a voltage controlled oscillator (VCO), a frequency divider, phase comparator and a loop (low-pass) filter. Especially the frequency divider caused severe problems in numerical simulation [43]. In Fig. 13b the feedback and in Fig. 13c the control signals are depicted. The Fig. 13d depicts the estimated instantaneous frequency which is in excellent agreement with the instantaneous frequency of the input signal.

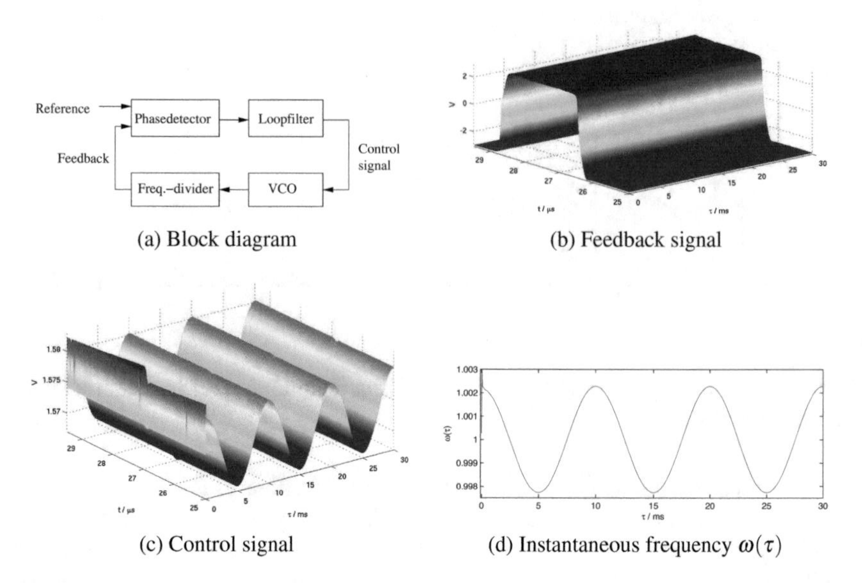

(a) Block diagram (b) Feedback signal

(c) Control signal (d) Instantaneous frequency $\omega(\tau)$

Fig. 13: Phase Locked Loop.

Acknowledgements The authors are indepted to P. Oswald, Bell Laboratories, for fruitful discussions on the multirate PDE approach. This project has been co-financed by the European Union using financial means of the European Regional Development Fund (INTERREG) for sustainable cross boarder cooperation. Further information on INTERREG Austria-Czech Republic is available at url: https://www.at-cz.eu/at.

References

[1] Nagel, L.W.: SPICE2: A simulation program with integrated circuit emphasis. Ph.D. thesis (1975)
[2] Weeks, W.T., Jimenez, A.J., Mahoney, G.W., Metha, D., Qassemzadeh, H., Scott, T.R.: Algorithms for ASTAP - A Network-Analysis Program. IEEE Trans. Circuit Theory **CT-20**, 628–634 (1973)
[3] Kundert, K.S., Sangiovanni-Vincentelli, A.: SPECTRE: A frequency-domain simulator for nonlinear circuits. Tech. rep., University of California, Berkeley (1988)
[4] Rizzoli, V., Cecchetti, C., Lipparini, A., Mastri, F.: General-purpose harmonic balance analysis of nonlinear microwave circuits under multitone excitation. IEEE Trans. on MTT **36**(12), 1650–1660 (1988)
[5] Feldmann, P., Roychowdhury, J.: Computation of circuit waveform envelopes using an efficient, matrix-decomposed harmonic balance algorithm. In: Proc. ICCAD-96, pp. 295–300 (1996)

[6] Rösch, M.: Schnelle Simulation des stationären Verhaltens nichtlinearer Schaltungen. Ph.D. thesis, TU München (1992)
[7] Lelarasmee, E., Ruehli, A., Sangiovanni-Vincentelli, A.: The waveform relaxation method for time-domain analysis of large scale integrated circuits. Computer-Aided Design of Integrated Circuits and Systems, IEEE Transactions on **1**(3), 131–145 (1982). DOI 10.1109/TCAD.1982.1270004
[8] Anheier, W., Laur, R.: Entwicklung von Simulationsverfahren für VLSI-Systeme (Timing Simulation). Forschungsbericht T 84-297, Bundesministerium für Forschung und Technologie, Bonn (1984)
[9] Telichevesky, R., Kundert, K.S., White, J.K.: Efficient Steady-State Analysis based on Matrix-Free Krylov-Subspace Methods. In: Proc. 32nd Design Automation Conference, pp. 480–484. San Francisco (1995)
[10] Feldmann, P., Freund, R.W.: Efficient linear circuit analysis by Pade approximation via the Lanczos process. IEEE Transactions on Computer-Aided Design of Integrated Circuits and Systems **14**(5), 639–649 (1995)
[11] Melville, R., Trajkovic, L., Fang, S.C., Watson, L.: Artificial parameter homotopy methods for the dc operating point problem. Computer-Aided Design of Integrated Circuits and Systems, IEEE Transactions on **12**(6), 861–877 (1993). DOI 10.1109/43.229761
[12] Günther, M., Feldmann, U.: CAD-based electric-circuit modeling in industry II. Impact of circuit configurations and parameters. Surv. Math. Ind. **8**, 131–157 (1999)
[13] Boche, H., Tampubolon, E.: On the existence of the band-limited interpolation of non-band-limited signals. In: 2016 24th European Signal Processing Conference (EUSIPCO), pp. 428–432 (2016). DOI 10.1109/EUSIPCO.2016.7760284
[14] Brachtendorf, H.G., Welsch, G., Laur, R., Bunse-Gerstner, A.: Numerical steady state analysis of electronic circuits driven by multi-tone signals. Electronic Engineering **79**(2), 103–112 (1996)
[15] Kugelmann, B., Pulch, R.: Existence and uniqueness of optimal solutions for multirate partial differential algebraic equations. Appl. Numer. Math. **97**, 69–87 (2015)
[16] Pulch, R., Günther, M., Knorr, S.: Multirate partial differential algebraic equations for simulating radio frequency signals. Euro. Jnl. of Applied Mathematics **18**, 709–743 (2007)
[17] Brachtendorf, H.G.: Theorie und Analyse von autonomen und quasiperiodisch angeregten elektrischen Netzwerken. Eine algorithmisch orientierte Betrachtung. Universität Bremen (2001). Habilitationsschrift
[18] Pels, A., Gyselinck, J., Sabariego, R.V., Schöps, S.: Efficient simulation of DC-DC switch-mode power converters by multirate partial differential equations. IEEE Journal on Multiscale and Multiphysics Computational Techniques **4**, 64–75 (2019). DOI 10.1109/JMMCT.2018.2888900
[19] Pels, A., Gyselinck, J., Sabariego, R.V., Schöps, S.: Solving nonlinear circuits with pulsed excitation by multirate partial differential equations. IEEE Transactions on Magnetics **54**(3), 1–4 (2018). DOI 10.1109/TMAG.2017.2759701

[20] Balabanian, N., Bickart, T.A., Seshu, S.: Electrical Network Theory. John Wiley (1969)
[21] Mathis, W.: Theorie nichtlinearer Netzwerke. Springer, Berlin (1987)
[22] Ho, C.W., Ruehli, A.E., Brennan, P.A.: The modified nodal approach to network analysis. IEEE Trans. Circuits and Systems **CAS-22**, 504–509 (1975)
[23] Singhal, K., Vlach, J.: Computer Methods for Circuit Analysis and Design. Springer-Verlag (1993)
[24] Dirks, H.K.: Kapazitätskoeffizienten nichtlinearer dissipativer Systeme. Technische Hochschule Aachen (1988). Habilitationsschrift
[25] Hairer, E., Wanner, G.: Solving Ordinary Differential Equations II: Stiff and Differential-Algebraic Problems. Springer Series in Computational Mathematics. Springer (2010)
[26] Tischendorf, C.: Topological index calculation of differential-algebraic equations in circuit simulation. Surv. Math. Ind. **8**, 187–199 (1999)
[27] Bittner, K., Brachtendorf, H.: Fast algorithms for grid adaptation using non-uniform biorthogonal spline wavelets. SIAM J. Scient. Computing **37**(2), 283–304 (2015)
[28] Dautbegovic, E., Condon, M., Brennan, C.: An efficient nonlinear circuit simulation technique. Microwave Theory and Techniques, IEEE Transactions on **53**(2), 548 –555 (2005). DOI 10.1109/TMTT.2004.840627
[29] Brachtendorf, H., Bunse-Gerstner, A., Lang, B., Lampe, S.: Steady state analysis of electronic circuits by cubic and exponential splines. Electrical Engineering (Archiv für Elektrotechnik) **91**, 287–299 (2009)
[30] Christoffersen, C.E., Steer, M.B.: Comparison of wavelet- and time-marching-based microwave circuit transient analyses (2001)
[31] Soveiko, N., Nakhla, M.: Wavelet harmonic balance. Microwave and Wireless Components Letters, IEEE **13**(6), 232–234 (2003). DOI 10.1109/LMWC.2003.814600
[32] Proakis, J.G.: Digital Communications. Electrical engineering. McGraw-Hill (2001)
[33] Pulch, R.: PDAE Methoden zur numerischen Simulation quasiperiodischer Grenzzyklen von Oszillatorschaltungen. Ph.D. thesis, Technische Universität München (2003)
[34] Brachtendorf, H.G., Laur, R.: Transient simulation of oscillators. Tech. Rep. 1131G0-980410-09TM, Bell-Laboratories (1998)
[35] Bittner, K., Brachtendorf, H.G.: Optimal frequency sweep method in multi-rate circuit simulation. COMPEL **33**(4), 1189–1197 (2014)
[36] Houben, S.: Circuits in motion: the numerical simulation of electrical oscillators. Ph.D. thesis, Eindhoven University of Technology (2003)
[37] Pulch, R.: Variational methods for solving warped multirate partial differential algebraic equations. SIAM J. Scient. Computing **31**(2), 1016–1034 (2008)
[38] Brachtendorf, H.G.: Simulation des eingeschwungenen Verhaltens elektronischer Schaltungen. Shaker, Aachen (1994)
[39] Brachtendorf, H.G., Welsch, G., Laur, R.: A novel time-frequency method for the simulation of the steady state of circuits driven by multi-tone signals. In:

Proc. IEEE Int. Symp. on Circuits and Systems, pp. 1508–1511. Hongkong (1997)

[40] Brachtendorf, H.G.: On the relation of certain classes of ordinary differential algebraic equations with partial differential algebraic equations. Tech. Rep. 1131G0-971114-19TM, Bell-Laboratories (1997)

[41] Roychowdhury, J.: Efficient methods for simulating highly nonlinear multirate circuits. In: Proc. of the 34th IEEE Design Automation Conference, pp. 269–274 (1997)

[42] Ngoya, E., Larchevèque, R.: Envelope transient analysis: A new method for the transient and steady state analysis of microwave communication circuit and systems. In: Proc. IEEE MTT-S Int. Microwave Symp., pp. 1365–1368. San Francisco (1996)

[43] Bittner, K., Brachtendorf, H.G.: Latency exploitation in wavelet-based multirate circuit simulation. In: Scientific Computing in Electrical Engineering 2014, *Mathematics in Industry*, vol. 23, pp. 13–20. Springer, Berlin Heidelberg (2014)

An Integrated Data-Driven Computational Pipeline with Model Order Reduction for Industrial and Applied Mathematics

Marco Tezzele, Nicola Demo, Andrea Mola, and Gianluigi Rozza

Abstract In this work we present an integrated computational pipeline involving several model order reduction techniques for industrial and applied mathematics, as emerging technology for product and/or process design procedures. Its data-driven nature and its modularity allow an easy integration into existing pipelines. We describe a complete optimization framework with automated geometrical parameterization, reduction of the dimension of the parameter space, and non-intrusive model order reduction such as dynamic mode decomposition and proper orthogonal decomposition with interpolation. Moreover several industrial examples are illustrated.

1 Introduction

A very common problem in the optimization of the design of industrial artifacts is that of finding the shape that minimizes some quantity of interest representing the expected performance. From a mathematical point of view such a problem translates into an optimization problem in which a suitable algorithm makes several queries to a simulation solver allowing for an evaluation of each sample in the design space. This leads to the identification of the optimal solution, possibly subjected to a set of prescribed constraints.

The experience acquired through several industrial projects suggested us that for such pipeline to operate in a robust way in a manufacturing environment, several aspects have to be integrated and developed so as to deal with both complex geometries and solution fields. With the concept of digital twin becoming widespread

---

M. Tezzele e-mail: marco.tezzele@sissa.it · N. Demo e-mail: ndemo@sissa.it · A. Mola e-mail: andrea.mola@sissa.it · G. Rozza e-mail: gianluigi.rozza@sissa.it

Mathematics Area, mathLab, SISSA, International School of Advanced Studies, via Bonomea 265, I-34136 Trieste, Italy

M. Günther, W. Schilders (eds.), *Novel Mathematics Inspired by Industrial Challenges*, Mathematics in Industry 38, https://doi.org/10.1007/978-3-030-96173-2_7

nowadays, we have to be able to pass automatically from industrial CAD geometries to fluid dynamics and structural simulations which allow for virtual performance evaluation. All the steps in the procedure that, moving from a CAD geometry, leads to an optimized shape need to be carefully devised and integrated. First, one has to process the industrial geometry at hand through a suitable shape parameterization strategy which identifies the parameter space describing all possible designs to be investigated. After the generation of a suitable space including all the parameters that satisfy all the structural, geometrical, and regulatory constraints prescribed by the design engineers according to the stakeholders needs, a proper sampling of such space is used to set up a campaign of CFD and structural simulations resulting in the performance evaluation of each shape tested. Usually, a single full order industrial simulation takes days or even weeks, so it is crucial to develop reduced order models (ROMs) so as to speed up the full optimization cycle and make it compatible with the production needs. The computational time of a single sample point evaluation is reduced through different ROM techniques such as dynamic mode decomposition (DMD), and proper orthogonal decomposition (POD), both based on singular value decomposition [7, 8, 45, 47]. In the case of DMD, ROM is used to reconstruct and predict the spatiotemporal dynamics of a high fidelity simulation such that its evolution can be completed at a faster rate. Instead POD-Galerkin or POD with interpolation exploit data on previous simulations, properly stored, to provide accurate surrogate solutions corresponding to untested sample points in the parameter space. In such a way the computational cost of an online optimization cycle can be dropped to hours or even minutes.

In a further post-processing phase, we also apply a reduction of the parameter space exploiting the active subspaces property [12, 62, 64]. Such an analysis allows for the detection of possible redundancies in the chosen parameters, suggesting linear combinations of the original parameters which dominate the system response.

This work aims at presenting and discussing a series of best practice approaches for the application of each of the aforementioned techniques within an industrial optimization framework. Such approaches are the result of the constant involvement of mathLab laboratory of SISSA[1] in industrial projects joining the research efforts of both manufacturing companies and academic institutions. After a brief overview of the overall problem and goals, the contribution is arranged so as to present each of the described industrial numerical pipeline steps in a complete and detailed fashion. We first consider the geometrical treatment of the industrial artifacts shape, then we suggest possible parameter space reduction strategies. We then provide details on data-driven model order reduction methodologies, and finally present a brief summary of numerical results obtained in some applications carried out in the framework of industrial projects in which mathLab is involved.

[1] `www.mathlab.sissa.it`

2 From digital twin to real-time analysis

Nowadays the digital twin is a concept well spread among all the engineering companies and communities. With the exact digital representation of a physical system, it allows, for example, to perform structural and fluid dynamics simulations, to make sensitivity analysis with respect to the parameters, and to optimize the design. The increased number of devices for real-time data acquisition of the physical system makes the digital twin paradigm obsolete, due to its intrinsic static nature.

We are moving every day into a more dynamical representation of the entire system that takes into account more and more real-time data from different sources. This new paradigm, thanks to data-driven models, uncertainty quantification, machine learning algorithms, artificial intelligence and better integration of all the singular computational modules, will provide new capabilities in terms of discovery of hidden correlations, fault detection, predictive maintenance and design optimization. Its goal is to enable delivery of better simulation and modeling results, and thus shorter the product development, the so-called time-to-market, and reduce the product maintenance and potential downtime.

These new needs from the industrial point of view, lead to new mathematical methods and new interdisciplinary pipelines for data acquisition, model order reduction, data elaboration, as well as optimization cycles [4, 17, 25].

We can summarize a modern shape design optimization cycle with the diagram in Figure 1. Usually a CAD file describing the geometry to be optimized is provided by the design team, then we have the structural and CFD teams that provide physical constraints and admissible range for the parameters variation, we have data coming from the experiments on the scale model, and finally we have regulations constraints from the national authorities.

When the inputs of the simulation are set, the output is computed through high fidelity solver and the optimization cycle is closed by validation and control, the imposition of the constraints, and a selection of a new set of parameters. If a complete simulation takes several hours, even days, finding the optimal shape becomes impossible due to the many evaluations of the parametric PDE needed by the optimizer. Here, the model order reduction (MOR) approach allows fast evaluations of the output of interest or derived quantities of the output thus enabling to find the optimum shape in few hours and to test different optimization algorithms. MOR techniques are very versatile, enabling both intrusive and non-intrusive approach with respect to the solver used. If it is a commercial black box it is possible to use numerical methods that work using only precomputed solution fields. While having access to the solver code allows also to reduce the single operators of the PDEs.

After the continued validation and control the optimization cycle ends by providing the final design in a suitable file format that can be analysed by all the interested departments.

In the following we are going to present all the techniques and integrated pipelines developed to accomplish such simulation based design optimization.

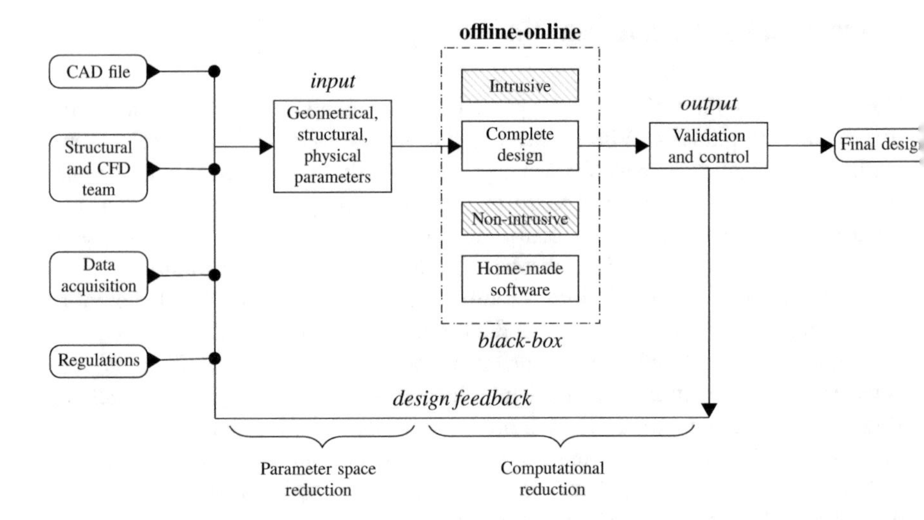

Fig. 1: Complete optimization pipeline involving automatic interface with CAD files, experimental and numerical data acquisition, definition of the parameters constraints, and parameter space dimensionality reduction. Two possible reduced order methods are presented: an intrusive and a non-intrusive approach, allowing the creation of a complete simulation-based design optimization framework.

3 Advanced geometrical parametrization with automatic CAD files interface

The first important step is related to the geometry of the shapes considered. As previously mentioned, one of the aspects typically subjected to optimization in the design of industrial artifacts is in fact their geometry. Finding the shape by maximizing the performance of a certain product or of one of its components is in fact a very common problem in industry. From a mathematical standpoint, such class of problems is obviously formalized as an optimization problem, which consists in the identification of the point of a suitable parameter space that maximizes the value of a prescribed performance parameter (or output function). Although the mathematical algorithms carrying out such task are commonly well assessed, the mathematical formalization of the problem requires that the shape modification can be recast as the corresponding variation of a certain set of parameters. The latter operation, which somehow translates the properties of the geometrical shapes into a set of numbers handled by the optimization algorithms, is usually referred to as *shape parametrization*. In the most common practice the shape parametrization is a rather delicate part of the overall design process. In fact, as the shape to be studied can be specified in several different file formats or analytical descriptions, a unified approach for shape parametrization algorithms has currently not been established. In the present section we will describe and discuss the state of the art of shape parametrization techniques, and present examples of their application to the geometry of different industrial artifacts.

A first shape parametrization algorithm which has been devised so as to be applied to arbitrarily shaped geometries, is the free form deformation (FFD) [36, 56, 58]. FFD consists basically in three different steps, as depicted in Figure 2. The first step is that of mapping the physical domain into a reference one. In the second step, a lattice of points is built in such reference domain, and some points are moved to deform the lattice. Since the lattice points represent the control points of a series of shape functions (typically Bernstein polynomials), the displacement of such points is propagated to compute the deformation of all the points of the reference domain. In the third, final step, the deformed reference domain is mapped back into the physical one, to obtain the resulting morphed geometry.

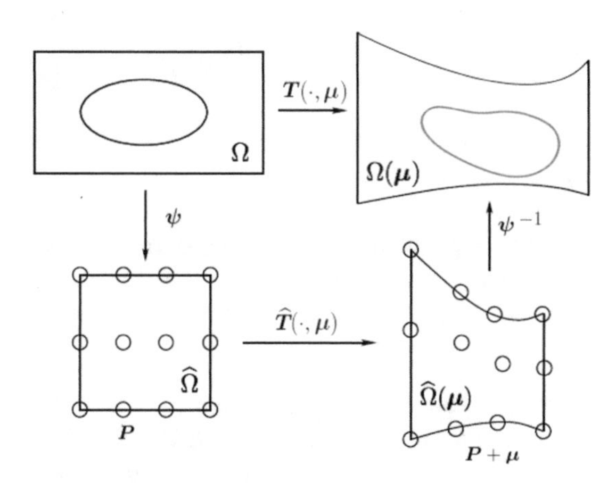

Fig. 2: Sketch of the three maps composing the FFD map construction. ψ maps the physical space to the reference one, then $\widehat{T}$ deforms the entire geometry according to the displacements of the lattice control points, and finally ψ^{-1} maps back the reference domain to the original one.

So, the displacements of one or more of the control points in the lattice are the parameters of the FFD map, which is the composition of the three maps described. As both the number of points in the lattice and the number of control points displaced to generate the deformation are flexible, the FFD map can be built with an arbitrary number of parameters. Thus, a very useful feature of FFD, is that it allows for the generation of global deformations even when few parameters are considered.

Since FFD is able to define a displacement law for each point of the 3D space contained in the control points lattice, it can be quite naturally applied to all the geometries that are specified through surface triangulations, surface grids or even volumetric grids. As a first example, in Figure 3 we present the application of FFD to an STL triangulation, which is a very common output format for CAD modeling tools. The shape deformation illustrated in the picture has been carried out making use of the PyGeM [3] Python package, which has been developed to be directly interfaced with the industrial geometry files and to deform them, so as to cut the communication times between the company simulation team and design team. In this application, the STL triangulation is imported and the coordinates of the nodes in the triangulation are modified according to the FFD map generated through the user specified lattice (included in the pictures).

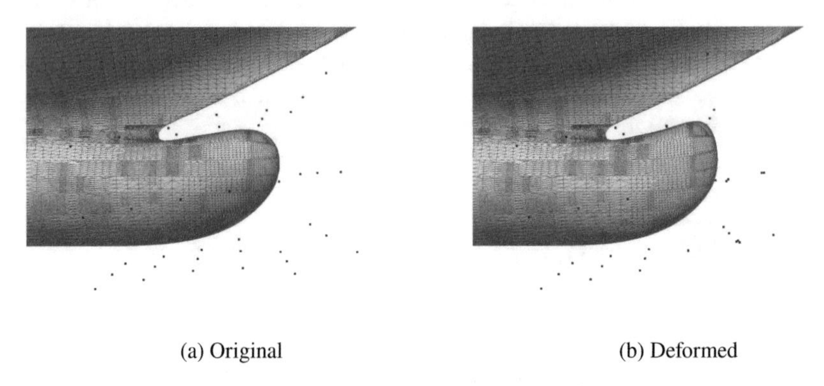

(a) Original (b) Deformed

Fig. 3: An example of the application of FFD to an STL triangulation. Plot (a) shows the FFD lattice over one side of the bulbous bow of a ship. Plot (b) depicts a deformed configuration of the same hull, along with the displaced FFD lattice control points.

The versatility of FFD can be further exploited to deform CAD surfaces that are generated as the collection of parametric patches. In such case, the desired deformation is obtained applying the specified FFD map to the control points of the NURBS and B-Splines surfaces of each patch composing the CAD model. The result in this case will also be a deformed geometry which follows the FFD deformation law specified by the user. Figure 4 presents a real world application of FFD to CAD parametric surfaces. The renderings show the original bulbous bow of the DTMB-5415 US Navy Combatant hull (which is a common benchmark for the CFD simulations community), and one of its several deformations generated to carry out the campaign of fluid dynamic simulations discussed in [18]. The PyGeM capabilities allow for importing the CAD geometry (in IGES or STEP format), extract and modify the control points of its surfaces and curves, and export the result into new CAD files.

Along with FFD, the PyGeM package implements other morphing techniques: the deformation based on radial basis function (RBF) interpolation [9, 38, 40], and the inverse distance weighting (IDW) interpolation [6, 23, 57, 67]. Yet, there are situations in which the shape to be deformed is already been engineered in a specific way, and general purpose deformation algorithms like FFD and the ones just mentioned would not preserve some critical characteristics of the geometry. A rather striking example of this is represented by the deformation of propeller blades illustrated in Figure 5. The shape of a propeller blade is in fact generated in

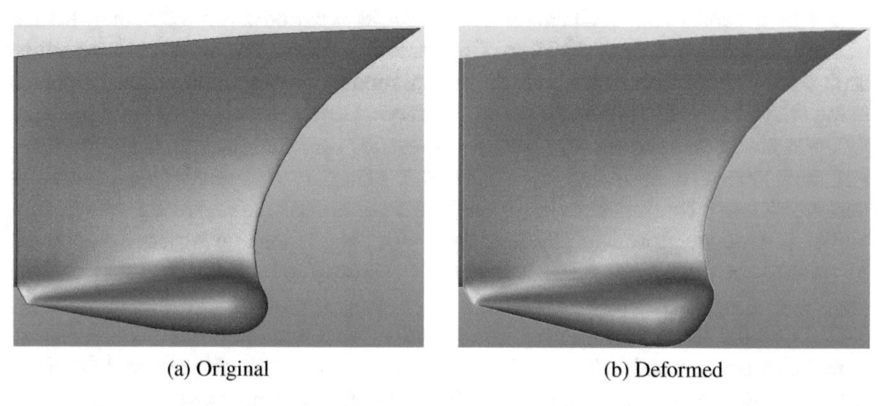

(a) Original (b) Deformed

Fig. 4: An example of the application of FFD to an IGES CAD geometry containing parametric surfaces. Plot (a) shows the original shape of the DTMB-5415 US Navy Combatant hull bulb. Plot (b) presents a deformed configuration of the same bulb.

a bottom-up fashion, first defining a set of sections represented by airfoils, and then properly placing each section in the three dimensional space. Since the aerodynamic feature of each airfoil section are known to the engineers which have selected them, any deformation that alters the sectional shape of the blade will not lead to an acceptable geometry. Thus, for such highly engineered shapes *ad hoc* solutions have to be generated. In most cases such solutions exploit the very algorithms used by the engineers to generate the artifacts in the first place, introducing parameters in one or more points of the generation procedure so as to create a class of deformed shapes. Also in this case, once the shape has been properly deformed (or re-generated), it has to be saved in the proper CAD format (IGES, STEP or STL) to be handled by the CFD or design team.

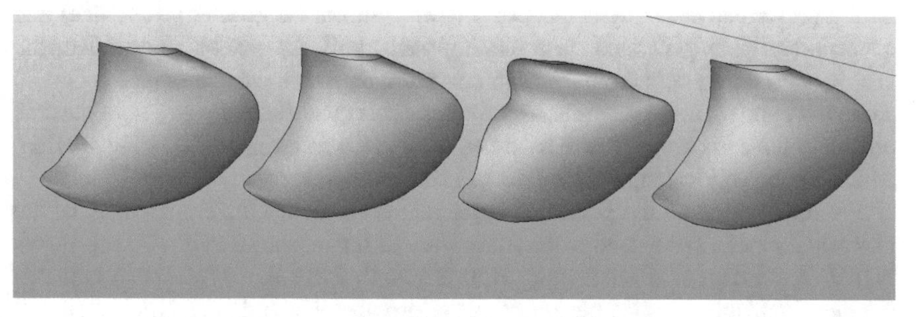

Fig. 5: An example of the deformation of an engineered propeller blade shape. The picture shows four deformed configurations of a PPTC benchmark propeller blade, in which the sectional properties of the blade have been kept untouched, while modifying the pitch, rake and skew distributions. The deformations are performed with the BladeX Python package [2].

As mentioned, in the CAD data structures surfaces can be described both by means of a triangulation and as a collection of parametric patches. Yet, most CAD modeling tools used by engineers to generate the virtual model of any object designed, operate using NURBS and B-Splines curves and surfaces. On the other hand, several tools for CFD or structural analysis only handle geometries specified through triangulations. For such reason, a series of algorithms that generate triangulations on arbitrary parametric surfaces has been implemented in the last years. Among others, the ability to produce closed — or *water tight* — triangulations on surfaces composed of possibly unconnected faces is a crucial feature for both CFD and structural engineering applications. In fact, the neighboring parametric patches composing a CAD model are generally only connected up to a specified tolerance. Thus, it is not infrequent that ideally continuous surfaces present small gaps and overlaps between each patch composing them. To avoid the problem, most CAD modeling tools retain several logical information to complement the geometric data and indicate which patches should be considered as neighbors. Yet, converting to vendor-neutral file formats such as IGES of STEP that allow the digital exchange of information among CAD systems will cause in most case the loss of topological information on neighboring patches. This is often a problem in the numerical analysis community, in which geometries are in most cases obtained by third parties, and in which water tight surfaces are in needed to define (and confine) the three dimensional domains considered in the simulations. So, a possible strategy to avoid a surface triangulation that depends on the local patches parametrization, and suffers from parametrization jumps and gaps, is to create new nodes not in the surface parametric space, but in the physical three dimensional space. Since such new nodes will not be initially located onto the CAD surface, a series of surface and curves projectors are used to make sure that the new grid points are properly placed onto the surface in a way that is completely independent of the parametrization. Along with the projectors, presented in [16], the work in [39] describes an algorithm which allows for the hierarchical refinement of an initial blocking made of quadrilateral cells. Across each level of refinement, the cells located in the highest curvature regions are refined, until a prescribed accuracy is reached. Figure 6 shows the geometry (left side) and quadrilateral grid generated on a planing yacht hull. As can be appreciated, the grid is consistently refined in the high curvature regions located around the double chine line, and on the spray rails. Finally, once this quadrilateral water tight mesh is obtained, the cells are split into triangles to obtain a water tight STL triangulation, which can be an ideal input for numerical analysis software.

We remark that all the presented applications exemplify the employment of the numerical pipeline proposed in the framework of the industrial POR-FESR projects SOPHYA "Seakeeping Of Planing Hull YAchts", PRELICA "Advanced methodologies for hydro-acoustic design of naval propulsion", and FSE HEaD "Higher Education and Development" programme founded by European Social Fund, in which mathLab laboratory has been involved in the last years.

Fig. 6: Water tight quadrilateral mesh generated on the water non tight IGES surface of a planing yacht hull.

4 Parameter space dimensionality reduction

After all the contributions from the different teams, the number of parameters could be too big for a reasonable optimization cycle in terms of computational time. In other cases, even if the parameters are not too many, there could be some of them dependent on the others. To overcome this problem it is possible to reduce the dimensionality of the parameter space by finding an active subspace (AS) [12] of the target functions. This technique ascertains whether the output of interest can be approximated by a function depending by linear combinations of all the original parameters. Its application has been proven successful in several parametrized engineering models [13, 18, 26, 64].

Now we briefly review the process of finding active subspaces of a scalar function f representing the output of interest, and depending on the inputs $\boldsymbol{\mu} \in \mathbb{R}^m$. Let us assume $f : \mathbb{R}^m \to \mathbb{R}$ is a scalar function continuous and differentiable in the support of a probability density function $\rho : \mathbb{R}^m \to \mathbb{R}^+$. We assume f with continuous and square-integrable (with respect to the measure induced by ρ) derivatives. We define the active subspaces of the pair (f, ρ) as the eigenspaces of the covariance matrix associated to the gradients $\nabla_{\boldsymbol{\mu}} f$. The elements of this matrix, the so-called uncentered covariance matrix of the gradients of f, denoted by $\mathbf{C}$, are the average products of partial derivatives of the simulations' input/output map, i.e.:

$$\mathbf{C} = \mathbb{E}[\nabla_{\boldsymbol{\mu}} f \nabla_{\boldsymbol{\mu}} f^T] = \int_{\mathbb{D}} (\nabla_{\boldsymbol{\mu}} f)(\nabla_{\boldsymbol{\mu}} f)^T \rho \, d\boldsymbol{\mu},$$

where $\mathbb{E}[\cdot]$ is the expected value. The matrix $\mathbf{C}$ is symmetric and positive semidefinite, so it admits a real eigenvalue decomposition $\mathbf{C} = \mathbf{W} \blacksquare \mathbf{W}^T$, where $\mathbf{W}$ is a $m \times m$ orthogonal matrix of eigenvectors, and $\blacksquare$ is the diagonal matrix of the eigenvalues, which are non-negative, arranged in descending order.

Now we select the first M eigenvectors, for some $M < m$, forming a lower dimensional parameter subspace. We underline that, on average, low eigenvalues suggest the corresponding vector is in the nullspace of the covariance matrix. So we can construct an approximation of f by taking the eigenvectors corresponding to the most energetic eigenvalues. Let us partition $\blacksquare$ and $\mathbf{W}$ as follows:

$$\blacksquare = \begin{bmatrix} \blacksquare_1 & \\ & \blacksquare_2 \end{bmatrix}, \qquad \mathbf{W} = [\mathbf{W}_1 \quad \mathbf{W}_2],$$

where $\blacksquare_1 = \mathrm{diag}(\lambda_1, \dots, \lambda_M)$, and $\mathbf{W}_1$ contains the first M eigenvectors. The range of $\mathbf{W}_1$ is the active subspace, while the inactive subspace is the range of $\mathbf{W}_2$. By projecting the full parameter space onto the active subspace we approximate the behaviour of the target function with respect to the new reduced parameters.

The active variable $\boldsymbol{\mu}_M$, and the inactive one $\boldsymbol{\eta}$, are obtained from the input parameters as follows:

$$\boldsymbol{\mu}_M = \mathbf{W}_1^T \boldsymbol{\mu} \in \mathbb{R}^M, \qquad \boldsymbol{\eta} = \mathbf{W}_2^T \boldsymbol{\mu} \in \mathbb{R}^{m-M}.$$

That means that we can express any point in the parameter space $\boldsymbol{\mu} \in \mathbb{R}^m$ in terms of $\boldsymbol{\mu}_M$ and $\boldsymbol{\eta}$ as:

$$\boldsymbol{\mu} = \mathbf{W}\mathbf{W}^T \boldsymbol{\mu} = \mathbf{W}_1 \mathbf{W}_1^T \boldsymbol{\mu} + \mathbf{W}_2 \mathbf{W}_2^T \boldsymbol{\mu} = \mathbf{W}_1 \boldsymbol{\mu}_M + \mathbf{W}_2 \boldsymbol{\eta}.$$

So we can rewrite f as $f(\boldsymbol{\mu}) = f(\mathbf{W}_1 \boldsymbol{\mu}_M + \mathbf{W}_2 \boldsymbol{\eta})$, and construct a surrogate quantity of interest g using only the active variable $\boldsymbol{\mu}_M$

$$f(\boldsymbol{\mu}) \approx g(\mathbf{W}_1^T \boldsymbol{\mu}) = g(\boldsymbol{\mu}_M).$$

Active subspaces can also be seen in the more general context of ridge approximation [14, 33, 42].

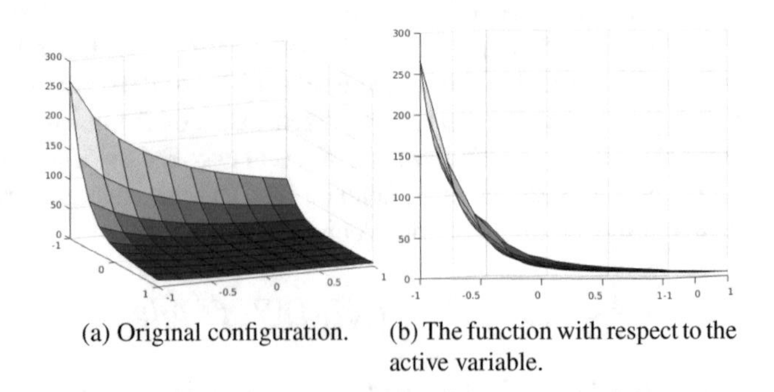

(a) Original configuration. (b) The function with respect to the active variable.

Fig. 7: Example of a scalar output function depending on two parameters. After a proper rotation of the whole domain it is possible to highlight the behaviour of the quantity of interest along the active direction.

From a practical point of view, expressing a target function with respect to its active variable means to rescale the parameter space on the origin and then rotate it so as to unveil the low rank behaviour of the function. In Figure 7 an example involving a bivariate scalar function is depicted. This approach can be viewed as a preprocessing step in the optimization cycle that helps both in reducing the number of the parameters and to increase the accuracy of a further reduction of the model as proven in [62] for the computation of the pressure drop in an occluded carotid artery using active subspace and POD-Galerkin methods.

Other approaches focus only on the shape parameters. To retain a significant geometric variance while reducing the number of geometrical parameters there exist nonlinear methods such as, among others, Kernel PCA [54, 55], Local PCA [32], and particular neural network such as Auto Encoders and Deep Auto Encoders [31]. For a comprehensive comparison among them we refer to [15], where the methods are demonstrated for the design-space dimensionality reduction of a destroyer hull. For a comparison between 13 different nonlinear techniques see [66].

5 Data driven model order reduction

In the *big data* era, data-driven models is becoming more and more popular in order to extract as much information as possible from all the data acquired during the physical experiments and the simulations. We mention also *uncertainty quantification* and ROM algorithms modified "ad hoc" [11]. Also in model order reduction community, several techniques have been developed to face industrial problems with a non-intrusive approach.

5.1 Dynamic mode decomposition

Dynamic mode decomposition (DMD) has emerged as a powerful tool for analyzing the dynamics of nonlinear systems, and for postprocessing spatio-temporal data in fluid mechanics [52, 53, 61]. It was developed by Schmid in [51], and it is an equation-free algorithm, and it does not make any assumptions about the underlying system. DMD allows to describe a non-linear time-dependent system as linear combination of few main structures evolving linearly in time. Many variants of the DMD were developed in the last years like forward backward DMD, compressed DMD [21], multiresolution DMD [35], higher order DMD [37], and DMD with control [43] among others. For a complete review refer to [34, 65], while for an implementation of them we refer to the Python package called PyDMD [20]. Lots of these variants arose to solve particular industrial problems such as streaming DMD [27] that are able to feed the classical algorithm with new real-time data coming from sensors, and do not require storage of past data, and they prove useful for real-time PIV or smoke/dye visualizations. In presence of very large dataset for complex industrial model the DMD modes are computed via randomized methods [22]. We cite also a new paradigm for data-driven modeling that simultaneously learns the dynamics and estimates the measurement noise at each observation that uses deep learning and DMD for signal-noise decomposition [46].

Now we present a brief overview of the standard algorithm. Let us consider m vectors, equispaced in time, representing the state of our system, also called *snapshots*: $\{\boldsymbol{x}_i\}_{i=1}^m$. The idea is that there exists a linear operator $\mathbf{A}$ that approximates the nonlinear dynamics of $\boldsymbol{x}(t)$, i.e. $\boldsymbol{x}_{k+1} = \mathbf{A}\boldsymbol{x}_k$. Without explicitly computing the operator $\mathbf{A}$ we seek to approximate its eigenvectors and eigenvalues, and we call them DMD modes and eigenvalues. First of all we arrange the snapshots in two matrices $\mathbf{X}$ and $\mathbf{Y}$ so as each column of the latter contains the state vector at the next timestep of the one in the corresponding $\mathbf{X}$ column, as follows

$$\mathbf{X} = \begin{bmatrix} x_1^1 & x_2^1 & \cdots & x_{m-1}^1 \\ x_1^2 & x_2^2 & \cdots & x_{m-1}^2 \\ \vdots & \vdots & \ddots & \vdots \\ x_1^n & x_2^n & \cdots & x_{m-1}^n \end{bmatrix}, \qquad \mathbf{Y} = \begin{bmatrix} x_2^1 & x_3^1 & \cdots & x_m^1 \\ x_2^2 & x_3^2 & \cdots & x_m^2 \\ \vdots & \vdots & \ddots & \vdots \\ x_2^n & x_3^n & \cdots & x_m^n \end{bmatrix}.$$

We are looking for $\mathbf{A}$ such that $\mathbf{Y} \approx \mathbf{A}\mathbf{X}$. The best-fit $\mathbf{A}$ matrix is given by $\mathbf{A} = \mathbf{Y}\mathbf{X}^{\dagger}$, where the symbol † represents the Moore-Penrose pseudo-inverse.

The DMD algorithm projects the data onto a low-rank subspace defined by the POD modes, that are the first r left-singular vectors of the matrix $\mathbf{X}$. We compute them via truncated singular value decomposition as $\mathbf{X} \approx \mathbf{U}_r \boldsymbol{\Sigma}_r \mathbf{V}_r^*$. The unitary matrix $\mathbf{U}_r$ contains the first r modes. So we can express the reduced operator $\widetilde{\mathbf{A}} \in \mathbb{C}^{r \times r}$ as

$$\widetilde{\mathbf{A}} = \mathbf{U}_r^* \mathbf{A} \mathbf{U}_r = \mathbf{U}_r^* \mathbf{Y} \mathbf{X}^{\dagger} \mathbf{U}_r = \mathbf{U}_r^* \mathbf{Y} \mathbf{V}_r \boldsymbol{\Sigma}_r^{-1} \mathbf{U}_r^* \mathbf{U}_r = \mathbf{U}_r^* \mathbf{Y} \mathbf{V}_r \boldsymbol{\Sigma}_r^{-1},$$

avoiding the computation of the high-dimensional operator $\mathbf{A}$. $\widetilde{\mathbf{A}}$ defines the linear evolution of the low-dimensional model $\widetilde{\boldsymbol{x}}_{k+1} = \widetilde{\mathbf{A}}\widetilde{\boldsymbol{x}}_k$, where $\widetilde{\boldsymbol{x}}_k \in \mathbb{R}^r$ is the low-rank approximated state. The high-dimensional state $\boldsymbol{x}_k$ can then be easily computed as $\boldsymbol{x}_k = \mathbf{U}_r \widetilde{\boldsymbol{x}}_k$.

Exploiting the eigendecomposition of $\widetilde{\mathbf{A}}$, that is $\widetilde{\mathbf{A}}\mathbf{W} = \mathbf{W}\boldsymbol{\Lambda}$, we can reconstruct the eigenvectors and eigenvalues of the matrix $\mathbf{A}$. The elements in $\boldsymbol{\Lambda}$ correspond to the nonzero eigenvalues of $\mathbf{A}$, while the eigenvectors of $\mathbf{A}$ can be computed in two ways. The first one is by projecting the low-rank approximation $\mathbf{W}$ on the high-dimensional space: $\boldsymbol{\Phi} = \mathbf{U}_r \mathbf{W}$. We call the eigenvectors $\boldsymbol{\Phi}$ the *projected* DMD modes. The other possibility is the so called *exact* DMD modes [65], that are the real eigenvectors of $\mathbf{A}$, and are computed as $\boldsymbol{\Phi} = \mathbf{Y}\mathbf{V}_r \boldsymbol{\Sigma}_r^{-1} \mathbf{W}$.

DMD has also been successfully used to accelerate the computation of the total drag resistance of a hull advancing in calm water [17, 18, 63]. This responded to the industrial needs of a rapid creation of the offline dataset. We decided not only to identify the approximated dynamics of the system but also to predict its evolution in order to achieve the regime state using only few snapshots, as we show in the example reported in Figure 8.

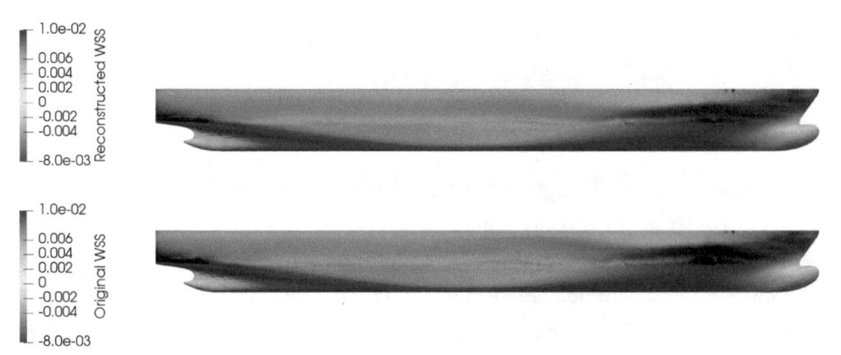

Fig. 8: Example of DMD application for wall shear stress prevision. In the top image, we show the wall shear stress along the x direction, at time $t = 50s$, computed using full-order solver. In the bottom, we show the wall shear stress along x direction reconstructed at time $t = 50s$ using 30 snapshots equispaced in the temporal window $[1, 30]$.

5.2 Proper orthogonal decomposition with interpolation

Proper orthogonal decomposition with interpolation (PODI) is an equation-free model order reduction technique providing a fast approximation of the solution of a parametric PDE. The key idea is to approximate the solution manifold by interpolating a finite set of high-fidelity snapshots, computed for some chosen parameters. Since interpolation of high dimensional data can be very expensive, we need reduced order modelling for a real-time evaluation of the solution for the new parameters.

This method consists in two logical phases: in the *offline* one, the high-fidelity solutions of a finite set of deformed configurations are computed and stored into the matrix $\mathbf{S}$ such that:

$$\mathbf{S} = \begin{bmatrix} s_1 \; s_2 \; \dots \; s_m \end{bmatrix}, \quad s_i \in \mathbb{R}^n \quad \text{for}\, i = 1, 2, \dots, m.$$

The basis spanning the low dimensional space is computed applying the singular value decomposition on the snapshots matrix:

$$\mathbf{S} = \mathbf{U}\blacksquare\mathbf{V}^*,$$

where $\mathbf{U} \in \mathbb{C}^{n \times m}$ refers to the matrix whose columns are the left singular vectors — the so called *POD modes* — of the snapshots matrix. We project the high-fidelity solutions onto the low-rank space, so they are represented as linear combination of the modes and the coefficients of this combination are called modal coefficients.

In the *online* phase the modal coefficients are interpolated and finally, for any new parameter, the solution of the parametric PDE is approximated. This method has the great benefit of being based only on the system output, but the accuracy of the approximation depends on the chosen interpolation method. The algorithm has been implemented in an open source Python package called EZyRB [19]. For deeper details about the PODI, we recommend [10, 44, 49].

6 Simulation-based design optimization framework

As previously stated, a shape optimization pipeline is usually composed by three fundamental ingredients: a deformation technique to construct the set of admissible shape, an objective function, and an optimal strategy to converge to the optimal shape with the lowest number of evaluations. Depending on the studied physical phenomena, the entire process can be very long: many complex problems, as for example conductivity, diffusion and fluid dynamic, are described through partial differential equations (PDEs). The numerical solution of such equations is usually expensive from the computational viewpoint. Moreover, in an optimization scenario, these equations have to be solved at each iteration, making the computational cost unaffordable for many applications, especially in the industrial sectors where a high responsiveness is requested to reduce the time-to-market. The model order reduction (MOR) offers the possibility to efficiently compute the solution of parametric PDEs, drastically reducing the computational effort. We exploited MOR techniques to design an innovative shape optimization pipeline which fits the industrial needs, primarily in terms of efficiency, reliability and modularity. The key idea of this optimization procedure is to collect the solutions, or the output of interest, from the full-order model for a finite set of parameters, then combining these solutions for a fast evaluation of the solution for any new iteration of the optimization algorithm.

In the first step, the deformed shapes are created from the initial geometry by using a combination of parameters. There are many possible techniques to choose from, as presented in Section 3. The important aspect is that given a set of parameters, the software is able to generate a new deformed geometry.

The parameter space is sampled and the system configurations so-created are evaluated using the high-fidelity numerical method. The pipeline relies only on the system outputs, without requiring information about the physical system, making all the procedure independent from the high-fidelity solver. Especially in an industrial context, this guarantees a great plus, allowing to adopt any solver — also commercial — within the pipeline. Further, the non-intrusive approach preserves the industrial know-how and reduces the complexity in the implementation phase.

We use two different data-driven model order reduction methods to accelerate the optimization. With the dynamic mode decomposition described in Section 5.1 we can simulate the physical problem at hand for a shorter temporal window using the computational expensive full-order solver and apply the DMD on the produced output to predict the solution/output of interest at regime. The second model order reduction technique adopted in the pipeline is the PODI, discusses in Section 5.2. Thanks to this method, we have the possibility to approximate in a real-time context the solution of parametric PDEs, combining several pre-computed snapshots. We adopt PODI in order to deal only with the output data of the high-fidelity solver,

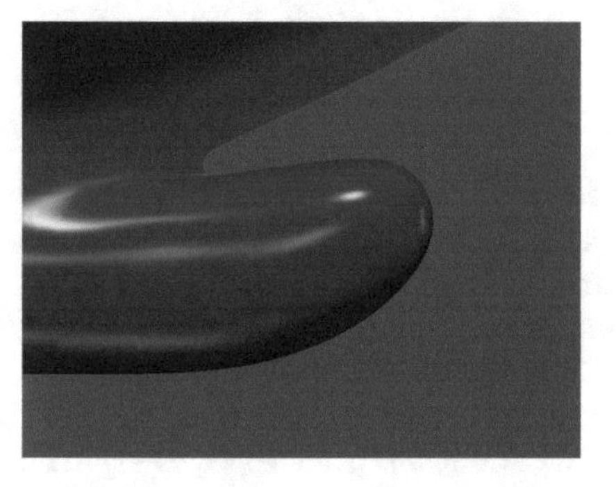
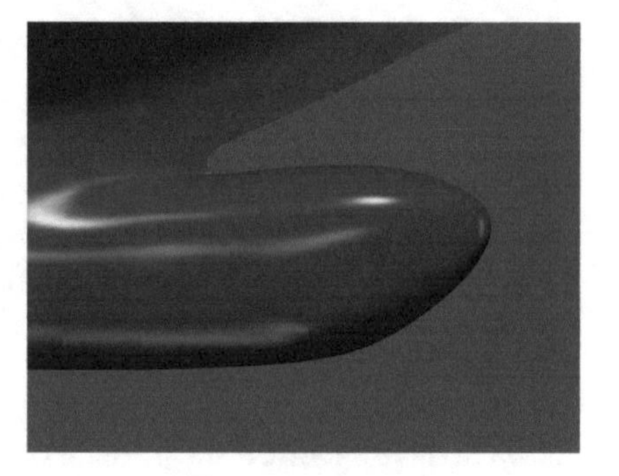

Fig. 9: Example of the shape optimization pipeline applied on naval hull: the original shape (left) and the optimized shape (right).

thus let the pipeline be independent from the used full-order model. To increase the accuracy of the reduced order model, an intrusive approach can be adopted. For an exhaustive discussion on the intrusive model order reduction, we suggest [59, 60] due to the implementation of MOR methods in a finite volume (FV) framework, the nowadays industrial standard for many fluid dynamics applications. For an overview on projection-based ROMs and the effort in increasing the Reynolds number see [5, 29], and for the joint use of such methods and uncertainty quantification strategies based on non intrusive polynomial chaos see [30]. Another possibility is to link together the isogeometric analysis with MOR into a complete parametric design pipeline from CAD to accurate and efficient numerical simulation [24, 48]. For a complete discussion about ROM for parametric PDEs, we recommend [28].

The optimization algorithm relies so on the reduced order model: since the online phase returns the approximated solutions in a quasi real-time scenario, the optimization algorithm lasts minutes or hours to reach the optimal shape, also if thousands of iterations are needed. The computational cost of the procedure is due to the creation of the solutions database. Thanks to MOR, we have also the possibility to run and tests many different optimization algorithms, avoiding any further high-fidelity simulations. Moreover, the solutions database can be enriched to increase the accuracy of the reduced order model. Examples of optimization procedure involving MOR techniques applied into naval and aerodynamics fields are respectively [17, 50]. Figure 9 shows the results of the application of the shape optimization system on the bulbous bow of a cruise ship. This achievement has been developed in the framework of a regional European Social Fund project from Regione Friuli Venezia-Giulia: HEaD in collaboration with Fincantieri - Cantieri Navali Italiani S.p.A..

7 Conclusions and perspectives

Industrial computational needs are every day more and more demanding in terms of computational time, reliability, error certification, data-assimilation, robustness, and easiness of use. In this work we presented several model order reduction and shape parameterization techniques to solve industrial and applied mathematics problems.

More has to be made to integrate real-time data-assimilation, machine learning and prediction, but we are moving along this horizon and MOR will play a crucial role to tackle many complexities arising from complex industrial artifacts management. A step in this direction is the planned webserver ARGOS [1], developed by mathLab group at SISSA that will make possible the exploitation of reduced order models to a vast category of people working in design, structural, and CFD teams. Through specific web applications the user will be able to solve many industrial and biomedical problems without the need of being an expert in numerical analysis and scientific computing. Figure 10 depicts some of the possible applications that are currently being developed.

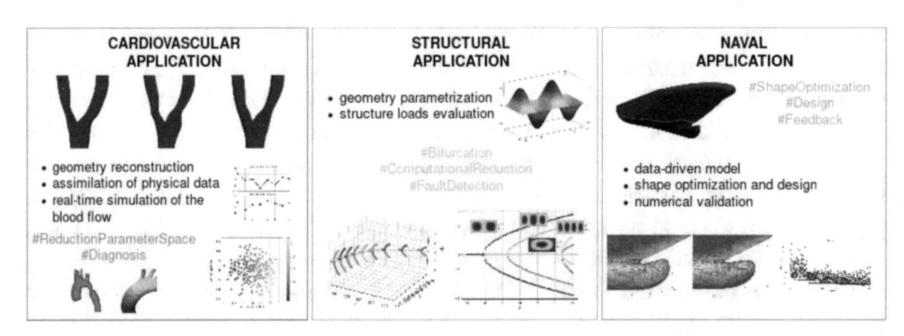

Fig. 10: Possible extension of the presented pipeline with different goal and application fields. From cardiovascular problems like real-time blood flow simulation, to structure load analysis and identification, as well as naval applications.

We also cite the Artificial Student "Artie" [41] that accepts problem statements posed in natural language, and solves numerically some PDEs problems, that will help both students, the scientific staff, and engineers in general. Moreover we want to highlight the effort of the Italian government in the technology transfer thanks to the institution of several competence centers connecting research facilities, university, and companies in the framework of Industry 4.0. Similar initiatives are undergoing in many other European countries (France, Germany, UK, . . .).

Acknowledgment

We acknowledge the scientific collaboration within SISSA mathLab group and the support provided by Dr Francesco Ballarin and Mr Federico Pichi for casting future extensions.

Competing Interests

None

Ethics approval and consent to participate

Not applicable.

Availability of data and materials

Open source software (PyGeM, EZyRB, PyDMD, BladeX) cited in this work is available on the website `https://mathlab.sissa.it/cse-software`

Funding acknowledgements

This work was partially performed in the context of the project SOPHYA - "Seakeeping Of Planing Hull YAchts" and of the project PRELICA - "Advanced methodologies for hydro-acoustic design of naval propulsion", both supported by Regione FVG, POR-FESR 2014-2020, Piano Operativo Regionale Fondo Europeo per lo Sviluppo Regionale, partially funded by the project HEaD, "Higher Education and Development" in collaboration with Fincantieri, supported by Regione FVG, European Social Fund FSE 2014-2020, and partially supported by European Union Funding for Research and Innovation — Horizon 2020 Program — in the framework of European Research Council Executive Agency: H2020 ERC CoG 2015 AROMA-CFD project 681447 "Advanced Reduced Order Methods with Applications in Computational Fluid Dynamics" P.I. Gianluigi Rozza.

References

[1] ARGOS: Advanced Reduced order modelinG Online computational web server for parametric Systems. Available at: `http://argos.sissa.it/`.

[2] BladeX: Python Package for Blade Deformation. Available at: `https://github.com/mathLab/BladeX`.

[3] PyGeM: Python Geometrical Morphing. Available at: `https://github.com/mathLab/PyGeM`.

[4] E. Abisset-Chavanne, J. L. Duval, E. Cueto, and F. Chinesta. Model and system learners, optimal process constructors and kinetic theory-based goal-oriented design: A new paradigm in materials and processes informatics. In *AIP Conference Proceedings*, volume 1960, page 090004. AIP Publishing, 2018.

[5] F. Ballarin, and A. Manzoni, A. Quarteroni, G. Rozza. Supremizer stabilization of POD–Galerkin approximation of parametrized steady incompressible Navier–Stokes equations. International Journal for Numerical Methods in Engineering 102(5): 1136–1161 (2015).

[6] F. Ballarin, A. D'Amario, S. Perotto, and G. Rozza. A POD-Selective Inverse Distance Weighting method for fast parametrized shape morphing. *Int. J. Num. Meth. Eng.*, in press, 2018.

[7] P. Benner, A. Cohen, M. Ohlberger, and K. Willcox. *Model reduction and approximation: theory and algorithms*, volume 15. SIAM review, 2017.

[8] P. Benner, M. Ohlberger, A. Patera, G. Rozza, and K. Urban. *Model Reduction of Parametrized Systems*, volume 17. Springer, 2017.

[9] M. D. Buhmann. *Radial basis functions: theory and implementations*, volume 12. Cambridge University Press, 2003.

[10] T. Bui-Thanh, M. Damodaran, and K. Willcox. Proper orthogonal decomposition extensions for parametric applications in compressible aerodynamics. In *21st AIAA Applied Aerodynamics Conference*, page 4213, 2003.

[11] P. Chen, A. Quarteroni, and G. Rozza. Reduced Basis Methods for Uncertainty Quantification. SIAM/ASA Journal on Uncertainty Quantification **5** (2017), 813–869.

[12] P. G. Constantine. *Active subspaces: Emerging ideas for dimension reduction in parameter studies*, volume 2. SIAM, 2015.

[13] P. G. Constantine and A. Doostan. Time-dependent global sensitivity analysis with active subspaces for a lithium ion battery model. *Statistical Analysis and Data Mining: The ASA Data Science Journal*, 10(5):243–262, 2017.

[14] P. G. Constantine, A. Eftekhari, and R. Ward. A near-stationary subspace for ridge approximation. *arXiv preprint arXiv:1606.01929*, 2016.

[15] D. D'Agostino, A. Serani, E. F. Campana, and M. Diez. Nonlinear Methods for Design-Space Dimensionality Reduction in Shape Optimization. In *International Workshop on Machine Learning, Optimization, and Big Data*, pages 121–132. Springer, 2017.

[16] F. Dassi, A. Mola, and H. Si. Curvature-adapted remeshing of cad surfaces. *Engineering with Computers*, 34(3):565–576, 2018.

[17] N. Demo, M. Tezzele, G. Gustin, G. Lavini, and G. Rozza. Shape optimization by means of proper orthogonal decomposition and dynamic mode decomposition. In *Technology and Science for the Ships of the Future: Proceedings of NAV 2018: 19th International Conference on Ship & Maritime Research*, pages 212–219. IOS Press, 2018.

[18] N. Demo, M. Tezzele, A. Mola, and G. Rozza. An efficient shape parametrisation by free-form deformation enhanced by active subspace for hull hydrodynamic ship design problems in open source environment. In *The 28th International Ocean and Polar Engineering Conference*, 2018.

[19] N. Demo, M. Tezzele, and G. Rozza. EZyRB: Easy Reduced Basis method. *The Journal of Open Source Software*, 3(24):661, 2018.

[20] N. Demo, M. Tezzele, and G. Rozza. PyDMD: Python Dynamic Mode Decomposition. *The Journal of Open Source Software*, 3(22):530, 2018.

[21] N. B. Erichson, S. L. Brunton, and J. N. Kutz. Compressed dynamic mode decomposition for background modeling. *Journal of Real-Time Image Processing*, pages 1–14, 2016.

[22] N. B. Erichson and C. Donovan. Randomized low-rank dynamic mode decomposition for motion detection. *Computer Vision and Image Understanding*, 146:40–50, 2016.

[23] D. Forti and G. Rozza. Efficient geometrical parametrisation techniques of interfaces for reduced-order modelling: application to fluid–structure interaction coupling problems. *International Journal of Computational Fluid Dynamics*, 28(3-4):158–169, 2014.

[24] F. Garotta, N. Demo, M. Tezzele, M. Carraturo, A. Reali, and G. Rozza. Reduced Order Isogeometric Analysis Approach for PDEs in Parametrized Domains. Submitted for publication.

[25] C. Ghnatios, F. Masson, A. Huerta, A. Leygue, E. Cueto, and F. Chinesta. Proper generalized decomposition based dynamic data-driven control of thermal processes. *Computer Methods in Applied Mechanics and Engineering*, 213:29–41, 2012.

[26] Z. Grey and P. Constantine. Active subspaces of airfoil shape parameterizations. In *58th AIAA/ASCE/AHS/ASC Structures, Structural Dynamics, and Materials Conference*, page 0507, 2017.

[27] M. S. Hemati, M. O. Williams, and C. W. Rowley. Dynamic mode decomposition for large and streaming datasets. *Physics of Fluids*, 26(11):111701, 2014.

[28] J. S. Hesthaven, G. Rozza, B. Stamm. Certified Reduced Basis Methods for Parametrized Partial Differential Equations. Springer Briefs in Mathematics (2015).

[29] S. Hijazi, S. Ali, G. Stabile, F. Ballarin, and G. Rozza. The Effort of Increasing Reynolds Number in Projection-Based Reduced Order Methods: from Laminar to Turbulent Flows. In press, FEF special volume 2017 (`https://arxiv.org/abs/arXiv:1807.11370`).

[30] S. Hijazi, G. Stabile, A. Mola, and G. Rozza. Non-Intrusive Polynomial Chaos Method Applied to Full-Order and Reduced Problems in Computational Fluid Dynamics: a Comparison and Perspectives. Submitted for publication.

[31] G. E. Hinton and R. R. Salakhutdinov. Reducing the dimensionality of data with neural networks. *Science*, 313(5786):504–507, 2006.

[32] N. Kambhatla and T. K. Leen. Dimension reduction by local principal component analysis. *Neural computation*, 9(7):1493–1516, 1997.

[33] S. Keiper. Analysis of generalized ridge functions in high dimensions. In *Sampling Theory and Applications (SampTA), 2015 International Conference on*, pages 259–263. IEEE, 2015.

[34] J. N. Kutz, S. L. Brunton, B. W. Brunton, and J. L. Proctor. *Dynamic mode decomposition: data-driven modeling of complex systems*, volume 149. SIAM, 2016.

[35] J. N. Kutz, X. Fu, and S. L. Brunton. Multiresolution dynamic mode decomposition. *SIAM Journal on Applied Dynamical Systems*, 15(2):713–735, 2016.

[36] T. Lassila and G. Rozza. Parametric free-form shape design with PDE models and reduced basis method. *Computer Methods in Applied Mechanics and Engineering*, 199(23–24):1583–1592, 2010.

[37] S. Le Clainche and J. M. Vega. Higher order dynamic mode decomposition. *SIAM Journal on Applied Dynamical Systems*, 16(2):882–925, 2017.

[38] A. Manzoni, A. Quarteroni, and G. Rozza. Model reduction techniques for fast blood flow simulation in parametrized geometries. *International journal for numerical methods in biomedical engineering*, 28(6-7):604–625, 2012.

[39] A. Mola, L. Heltai, and A. DeSimone. A fully nonlinear potential model for ship hydrodynamics directly interfaced with CAD data structures. In *24th Int. Ocean Polar Eng. Conf.*, 2014.

[40] A. Morris, C. Allen, and T. Rendall. CFD-based optimization of aerofoils using radial basis functions for domain element parameterization and mesh deformation. *International Journal for Numerical Methods in Fluids*, 58(8):827–860, 2008.

[41] A. Patera. Project Artie: An Artificial Student for Disciplines Informed by Partial Differential Equations. arXiv:1809.06637 (2018)

[42] A. Pinkus. *Ridge functions*, volume 205. Cambridge University Press, 2015.

[43] J. L. Proctor, S. L. Brunton, and J. N. Kutz. Dynamic mode decomposition with control. *SIAM Journal on Applied Dynamical Systems*, 15(1):142–161, 2016.

[44] M. Ripepi, M. Verveld, N. Karcher, T. Franz, M. Abu-Zurayk, S. Görtz, and T. Kier. Reduced-order models for aerodynamic applications, loads and MDO. *CEAS Aeronautical Journal*, 9(1):171–193, 2018.

[45] G. Rozza, M. H. Malik, N. Demo, M. Tezzele, M. Girfoglio, G. Stabile, and A. Mola. Advances in Reduced Order Methods for Parametric Industrial Problems in Computational Fluid Dynamics. ECCOMAS proceedings, Glasgow, 2018.

[46] S. H. Rudy, J. N. Kutz, and S. L. Brunton. Deep learning of dynamics and signal-noise decomposition with time-stepping constraints. *arXiv preprint arXiv:1808.02578*, 2018.
[47] F. Salmoiraghi, F. Ballarin, G. Corsi, A. Mola, M. Tezzele, and G. Rozza. Advances in geometrical parametrization and reduced order models and methods for computational fluid dynamics problems in applied sciences and engineering: overview and perspectives. ECCOMAS proceedings, Crete, 2016.
[48] F. Salmoiraghi, F. Ballarin, L. Heltai, and G. Rozza. Isogeometric analysis-based reduced order modelling for incompressible linear viscous flows in parametrized shapes. *Advanced Modeling and Simulation in Engineering Sciences*, 3(1):21, 2016.
[49] F. Salmoiraghi, A. Scardigli, H. Telib, and G. Rozza. Free-form deformation, mesh morphing and reduced-order methods: enablers for efficient aerodynamic shape optimisation. *International Journal of Computational Fluid Dynamics*, 0(0):1–15, 2018.
[50] A. Scardigli, R. Arpa, A. Chiarini, and H. Telib. Enabling of Large Scale Aerodynamic Shape Optimization Through POD-Based Reduced-Order Modeling and Free Form Deformation. In *Advances in Evolutionary and Deterministic Methods for Design, Optimization and Control in Engineering and Sciences*, pages 49–63. Springer, 2019.
[51] P. J. Schmid. Dynamic mode decomposition of numerical and experimental data. *Journal of fluid mechanics*, 656:5–28, 2010.
[52] P. J. Schmid. Application of the dynamic mode decomposition to experimental data. *Experiments in fluids*, 50(4):1123–1130, 2011.
[53] P. J. Schmid, L. Li, M. P. Juniper, and O. Pust. Applications of the dynamic mode decomposition. *Theoretical and Computational Fluid Dynamics*, 25(1-4):249–259, 2011.
[54] B. Schölkopf, A. Smola, and K.-R. Müller. Kernel principal component analysis. In *International Conference on Artificial Neural Networks*, pages 583–588. Springer, 1997.
[55] B. Schölkopf, A. Smola, and K.-R. Müller. Nonlinear component analysis as a kernel eigenvalue problem. *Neural computation*, 10(5):1299–1319, 1998.
[56] T. Sederberg and S. Parry. Free-Form Deformation of solid geometric models. In *Proceedings of SIGGRAPH - Special Interest Group on GRAPHics and Interactive Techniques*, pages 151–159. SIGGRAPH, 1986.
[57] D. Shepard. A two-dimensional interpolation function for irregularly-spaced data. In *Proceedings-1968 ACM National Conference*, pages 517–524. ACM, 1968.
[58] D. Sieger, S. Menzel, and M. Botsch. On shape deformation techniques for simulation-based design optimization. In *New Challenges in Grid Generation and Adaptivity for Scientific Computing*, pages 281–303. Springer, 2015.
[59] G. Stabile, S. Hijazi, A. Mola, S. Lorenzi, G. Rozza. POD-Galerkin reduced order methods for CFD using Finite Volume Discretisation: vortex shedding around a circular cylinder. Communications in Applied and Industrial Mathematics 8(1): 210–236 (2017).

[60] G. Stabile and G. Rozza. Finite volume POD-Galerkin stabilised reduced order methods for the parametrised incompressible Navier–Stokes equations. Computers & Fluids 173: 273–284 (2018).

[61] P. Stegeman, A. Ooi, and J. Soria. Proper orthogonal decomposition and dynamic mode decomposition of under-expanded free-jets with varying nozzle pressure ratios. In *Instability and Control of Massively Separated Flows*, pages 85–90. Springer, 2015.

[62] M. Tezzele, F. Ballarin, and G. Rozza. Combined parameter and model reduction of cardiovascular problems by means of active subspaces and POD-Galerkin methods. In *Mathematical and Numerical Modeling of the Cardiovascular System and Applications*. SEMA SIMAI Springer Series 16, 2018.

[63] M. Tezzele, N. Demo, M. Gadalla, A. Mola, and G. Rozza. Model order reduction by means of active subspaces and dynamic mode decomposition for parametric hull shape design hydrodynamics. In *Technology and Science for the Ships of the Future: Proceedings of NAV 2018: 19th International Conference on Ship & Maritime Research*, pages 569–576. IOS Press, 2018.

[64] M. Tezzele, F. Salmoiraghi, A. Mola, and G. Rozza. Dimension reduction in heterogeneous parametric spaces with application to naval engineering shape design problems. *Advanced Modeling and Simulation in Engineering Sciences*, 5(1):25, Sep 2018.

[65] J. H. Tu, C. W. Rowley, D. M. Luchtenburg, S. L. Brunton, and J. N. Kutz. On dynamic mode decomposition: theory and applications. *Journal of Computational Dynamics*, 1(2):391–421, 2014.

[66] L. Van Der Maaten, E. Postma, and J. Van den Herik. Dimensionality reduction: a comparative review. *J Mach Learn Res*, 10:66–71, 2009.

[67] J. Witteveen and H. Bijl. Explicit mesh deformation using Inverse Distance Weighting interpolation. In *19th AIAA Computational Fluid Dynamics*. AIAA, 2009.

From Rotating Fluid Masses and Ziegler's Paradox to Pontryagin- and Krein Spaces and Bifurcation Theory

Oleg N. Kirillov and Ferdinand Verhulst

Abstract Four classical systems, the Kelvin gyrostat, the Maclaurin spheroids, the Brouwer rotating saddle, and the Ziegler pendulum have directly inspired development of the theory of Pontryagin and Krein spaces with indefinite metric and singularity theory as independent mathematical topics, not to mention stability theory and nonlinear dynamics. From industrial applications in shipbuilding, turbomachinery, and artillery to fundamental problems of astrophysics, such as asteroseismology and gravitational radiation — that is the range of phenomena involving the Krein collision of eigenvalues, dissipation-induced instabilities, and spectral and geometric singularities on the neutral stability surfaces, such as the famous Whitney's umbrella.

1 Historical background

The purpose of this paper is to show how a curious phenomenon observed in the natural sciences, destabilization by dissipation, was solved by mathematical analysis. After the completion of the analysis, the eigenvalue calculus of matrices which it involved was (together with other applications) an inspiration for the mathematical theory of structural stability of matrices. But the story is even more intriguing. Later it was shown that the bifurcation picture in parameter space is related to a seemingly pure mathematical object in singularity theory, Whitney's umbrella.

In many problems in physics and engineering, scientists found what looked like a counter-intuitive phenomenon: certain systems that without dissipation show stable behaviour became unstable when any form of dissipation was introduced. This can

Oleg N. Kirillov
Northumbria University, NE1 8ST Newcastle upon Tyne, UK, e-mail: oleg.kirillov@northumbria.ac.uk

Ferdinand Verhulst
Mathematisch Instituut, PO Box 80.010, 3508TA Utrecht, Netherlands e-mail: F.Verhulst@uu.nl

M. Günther, W. Schilders (eds.), *Novel Mathematics Inspired by Industrial Challenges*, Mathematics in Industry 38, https://doi.org/10.1007/978-3-030-96173-2_8

be viscosity in fluids and solids, magnetic diffusivity, or losses due to radiation of waves of different nature, including recently detected gravitational waves [1, 3], to name a few. For instance, dissipation-induced modulation instabilities [27] are widely discussed in the context of modern nonlinear optics [100].

The destabilization by dissipation is especially sophisticated when several dissipative mechanisms are acting simultaneously. In this case, "no simple rule for the effect of introducing small viscosity or diffusivity on flows that are neutral in their absence appears to hold" [132] and "the ideal limit with zero dissipation coefficients has essentially nothing to do with the case of small but finite dissipation coefficients" [96].

In hydrodynamics, a classical example is given by secular instability of the Maclaurin spheroids due to both fluid viscosity (Kelvin and Tait, 1879) and gravitational radiation reaction (Chandrasekhar, 1970), where the critical eccentricity of the meridional section of the spheroid depends on the ratio of the two dissipative mechanisms and reaches its maximum, corresponding to the onset of dynamical instability in the ideal system, when this ratio equals 1 [85]. In meteorology this phenomenon is known as the 'Holopäinen instability mechanism' (Holopäinen, 1961) for a baroclinic flow when waves that are linearly stable in the absence of Ekman friction become dissipatively destabilized in its presence, with the result that the location of the curve of marginal stability is displaced by an order one distance in the parameter space, even if the Ekman number is infinitesimally small [49, 143]. For a baroclinic circular vortex with thermal and viscous diffusivities this phenomenon was studied by McIntyre in 1970 [94].

In rotor dynamics, the generic character of the discontinuity of the instability threshold in the zero dissipation limit was noticed by Smith already in 1933 [120]. In mechanical engineering such a phenomenon is called Ziegler's paradox, it was found in the analysis of a double pendulum with a nonconservative positional force with and without damping in 1952 [147, 148]. The importance of solving the Ziegler paradox for mechanics was emphasized by Bolotin [20]: "The greatest theoretical interest is evidently centered in the unique effect of damping in the presence of pseudo-gyroscopic forces, and in particular, in the differences in the results for systems with slight damping which then becomes zero and systems in which damping is absent from the start." Encouraging further research of the destabilization paradox, Bolotin was not aware that the crucial ideas for its explanation were formulated as early as 1956.

Ziegler's paradox was solved in 1956 by an expert in classical geometry and mechanics, Oene Bottema [25]. He formulated the problem of the stability of an equilibrium in two degrees-of-freedom (4 dimensions), allowing for gyroscopic and nonconservative positional forces. The solution by Bottema in the form of concrete analysis was hardly noticed at that time. Google Scholar gives no citation of Bottema's paper in the first 30 years after 1956.

As mentioned above, a new twist to the treatment of the problem came from identifying geometric considerations independently introduced by Whitney in singularity theory with the bifurcation analysis of Bottema. Interestingly Whitney's "umbrella singularity" predated Bottema's analysis, it gives the right geometric picture but it ignores the stability questions of the dynamical context which is the essential question of Ziegler's paradox. Later, in 1971-72, V.I. Arnold showed that the umbrella singularity is generic in parameter families of real matrices. This result links to stability by linearisation of the vector field near equilibrium.

The phenomenon of dissipation-induced instability and the Ziegler paradox raised important questions in mechanics and mathematics. For instance, what is the connection between the conservative and deterministic Hamiltonian systems and real systems involving both dissipation of energy and stochastic effects [79]? Also it added a new bifurcation in the analysis of dynamical systems. An important consequence of the results is that in a large number of problem-fields one can now predict and characterise precisely this type of instability [50, 64, 66, 68, 74, 79, 87, 119].

1.1 Stability of Kelvin's gyrostat and spinning artillery shells filled with liquid

Nov. 18, 1880] *NATURE* 69

ON AN EXPERIMENTAL ILLUSTRATION OF MINIMUM ENERGY[2]

THIS illustration consists of a liquid gyrostat of exactly the same construction as that described and represented by the annexed drawing, repeated from NATURE, February 1, 1877, p. 297, 298, with the difference that the figure of the shell is prolate instead of oblate. The experiment was in fact conducted with the actual apparatus which was exhibited to the British Association at Glasgow in 1876, altered by the substitution of a shell having its equatorial diameter about $\frac{9}{10}$ of its axial diameter, for the shell with axial diameter $\frac{9}{10}$ of equatorial diameter which was used when the apparatus was shown as a successful gyrostat.

[1] In illustration of this see an exhaustive mathematical paper on the values of iron ores, by Prof. A. Habets: *Cuyper's Revue Universelle des Mines* (1877), t. i. p. 504.
[2] By Sir William Thomson, F.R.S. British Association, Swansea, Section A.

The oblate and prolate shells were each of them made from the two hemispheres of sheet copper which plumbers solder together to make their globular floaters. By a little hammering it is easy to alter the hemispheres to the proper shapes to make either the prolate or the oblate figure.

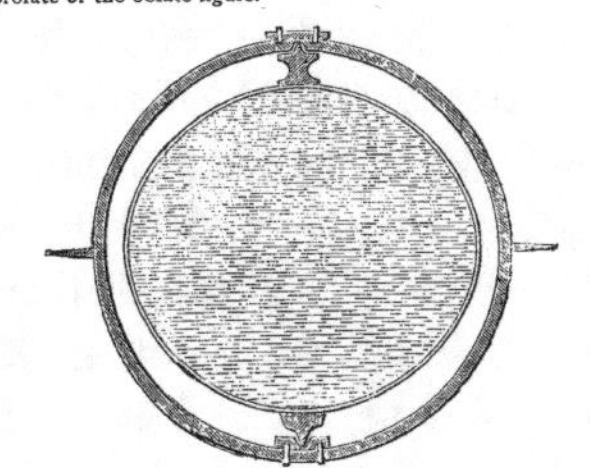

The result of the first trial was literally startling.

The spinning in the case of the oblate shell, as was known from previous experiments, would have given amply sufficient rotation to the contained water to cause the apparatus to act with great firmness like a solid gyrostat. In the first experiment with the oval shell the shell was seen to be rotating with great velocity during the last minute of the spinning; but the moment it was released from the cord, and when, holding the framework in my hands, I commenced carrying it towards the horizontal glass table to test its gyrostatic quality, the framework which I held in my hands gave a violent uncontrollable lurch, and in a few seconds the shell stopped turning. Its utter failure as a gyrostat is precisely what was expected from the theory, and presents a truly wonderful contrast from what is observed with the apparatus and operations in every respect similar, except having an oblate instead of a prolate shell to contain the liquid.

Fig. 1: The Kelvin gyrostat [131].

Already von Laue (1905) [138], Lamb (1908) [77] and Heisenberg (1924) [47] realized that dissipation easily destabilizes waves and modes of negative energy of an ideal system supported by rotating and translating continua [123, 124]. Williamson (1936) proposed normal forms for Hamiltonian systems allowing sorting stable modes of negative and positive energy according to their symplectic sign [89, 141,

142]. However, further generalization — the theory of spaces with indefinite metric, or Krein [75] and Pontryagin [104] spaces, — was directly inspired by the problem of stability of the Kelvin gyrostat [44, 131], having both astrophysical and industrial (and even military) applications.

Kelvin [131] experimentally demonstrated in 1880 that a thin-walled and slightly oblate spheroid completely filled with liquid remains stable if rotated fast enough about a fixed point, which does not happen if the spheroid is slightly prolate, Figure 1. In the same year this observation was confirmed theoretically by Greenhill [44], who found that rotation around the center of gravity of the top in the form of a weightless ellipsoidal shell completely filled with an ideal and incompressible fluid, is unstable when $a < c < 3a$, where c is the length of the semiaxis of the ellipsoid along the axis of rotation and the lengths of the two other semiaxes are equal to a [44].

Quite similarly, bullets and projectiles fired from the rifled weapons can relatively easily be stabilized by rotation, if they are solid inside. In contrast, the shells, containing a liquid substance inside, have a tendency to turn over despite seemingly revolved fast enough to be gyroscopically stabilized. Motivated by such artillery applications, in 1942 Sobolev, then director of the Steklov Mathematical Institute in Moscow, considered stability of a rotating heavy top with a cavity entirely filled with an ideal incompressible fluid [121]—a problem that is directly connected to the classical XIXth century models of astronomical bodies with a crust surrounding a molten core [122].

For simplicity, the solid shell of the top and the domain V occupied by the cavity inside it, can be assumed to have a shape of a solid of revolution. They have a common symmetry axis where the fixed point of the top is located. The velocity profile of the stationary unperturbed motion of the fluid is that of a solid body rotating with the same angular velocity Ω as the shell around the symmetry axis.

Following Sobolev, we denote by M_1 the mass of the shell, M_2 the mass of the fluid, ρ and p the density and the pressure of the fluid, g the gravity acceleration, and l_1 and l_2 the distances from the fixed point to the centers of mass of the shell and the fluid, respectively. The moments of inertia of the shell and the 'frozen' fluid with respect to the symmetry axis are C_1 and C_2, respectively; A_1 (A_2) stands for the moment of inertia of the shell (fluid) with respect to any axis that is orthogonal to the symmetry axis and passes through the fixed point. Let, additionally,

$$L = C_1 + C_2 - A_1 - A_2 - \frac{K}{\Omega^2}, \quad K = g(l_1 M_1 + l_2 M_2). \tag{1}$$

The solenoidal ($\operatorname{div} \mathbf{v} = 0$) velocity field $\mathbf{v}$ of the fluid is assumed to satisfy the no-flow condition on the boundary of the cavity: $\mathbf{v}_n|_{\partial V} = 0$.

Stability of the stationary rotation of the top around its vertically oriented symmetry axis is determined by the system of linear equations derived by Sobolev in the frame (x, y, z) that has its origin at the fixed point of the top and rotates with respect to an inertial frame around the vertical z-axis with the angular velocity of the unperturbed top, Ω. If the real and imaginary part of the complex number Z describe the deviation of the unit vector of the symmetry axis of the top in the coordinates x, y, and z, then these equations are, see e.g. [121, 146]:

$$
\begin{aligned}
\frac{dZ}{dt} &= i\Omega W,\\
(A_1+\rho\kappa^2)\frac{dW}{dt} &= i\Omega LZ+i\Omega(C_1-2A_1+\rho E)W+i\rho\int_V\left(v_x\frac{\partial\chi}{\partial y}-v_y\frac{\partial\chi}{\partial x}\right)dV,\\
\partial_t v_x &= 2\Omega v_y-\rho^{-1}\partial_x p+2i\Omega^2 W\partial_y\overline{\chi},\\
\partial_t v_y &= -2\Omega v_x-\rho^{-1}\partial_y p-2i\Omega^2 W\partial_x\overline{\chi},\\
\partial_t v_z &= -\rho^{-1}\partial_z p,
\end{aligned} \tag{2}
$$

where $2\kappa^2=\int_V|\nabla\chi|^2dV$, $E=i\int_V(\partial_x\overline{\chi}\partial_y\chi-\partial_y\overline{\chi}\partial_x\chi)\,dV$, and the function χ is determined by the conditions

$$
\nabla^2\chi=0,\quad \partial_n\chi|_{\partial V}=z(\cos nx+i\cos ny)-(x+iy)\cos nz, \tag{3}
$$

with n the absolute value of a vector $\mathbf{n}$, normal to the boundary of the cavity.

Sobolev realized that some qualitative conclusions on the stability of the top can be drawn with the use of the bilinear form

$$
Q(R_1,R_2)=L\Omega Z_1\overline{Z}_2+(A_1+\rho\kappa^2)W_1\overline{W}_2+\frac{\rho}{2\Omega^2}\int_V\overline{\mathbf{v}}_2^T\mathbf{v}_1dV \tag{4}
$$

on the elements R_1 and R_2 of the space $\{R\}=\{Z,W,\mathbf{v}\}$. The linear operator B defined by Eqs. (2) that can be written as $\frac{dR}{dt}=iBR$ has all its eigenvalues real when $L>0$, which yields Lyapunov stability of the top. The number of pairs of complex-conjugate eigenvalues of B (counting multiplicities) does not exceed the number of negative squares of the quadratic form $Q(R,R)$, which can be equal only to one when $L<0$. Hence, for $L<0$ an unstable solution $R=e^{i\lambda_0 t}R_0$ can exist with $\mathrm{Im}\lambda_0<0$; all real eigenvalues are simple except for maybe one [73, 121, 146].

In the particular case when the cavity is an ellipsoid of rotation with the semi-axes a, a, and c, the space of the velocity fields of the fluid can be decomposed into a direct sum of subspaces, one of which is finite-dimensional. Only the movements from this subspace interact with the movements of the rigid shell, which yields a finite-dimensional system of ordinary differential equations that describes coupling between the shell and the fluid.

Calculating the moments of inertia of the fluid in the ellipsoidal container

$$
C_2=\frac{8\pi\rho}{15}a^4c,\quad A_2=l_2^2M_2+\frac{4\pi\rho}{15}a^2c(a^2+c^2),
$$

denoting $m=\frac{c^2-a^2}{c^2+a^2}$, and assuming the field $\mathbf{v}=(v_x,v_y,v_z)^T$ in the form

$$
v_x=(z-l_2)a^2m\xi,\quad v_y=-i(z-l_2)a^2m\xi,\quad v_z=-(x-iy)c^2m\xi,
$$

one can eliminate the pressure in Eqs. (2) and obtain the reduced model

$$
\frac{d\mathbf{x}}{dt}=i\Omega\mathbf{A}^{-1}\mathbf{C}\mathbf{x}=i\Omega\mathbf{B}\mathbf{x}, \tag{5}
$$

where $\mathbf{x} = (Z, W, \xi)^T \in \mathbb{C}^3$ and

$$\mathbf{A} = \begin{pmatrix} 1 & 0 & 0 \\ 0 & A_1 + l_2^2 M_2 + \frac{4\pi\rho}{15} a^2 c \frac{(c^2-a^2)^2}{c^2+a^2} & 0 \\ 0 & 0 & c^2 + a^2 \end{pmatrix},$$

$$\mathbf{C} = \begin{pmatrix} 0 & 1 & 0 \\ L & C_1 - 2A_1 - 2l_2^2 M_2 - \frac{8\pi\rho}{15} a^2 c^3 m^2 & -\frac{8\pi\rho}{15} a^4 c^3 m^2 \\ 0 & -2 & -2a^2 \end{pmatrix}. \tag{6}$$

The matrix $\mathbf{B} \neq \mathbf{B}^T$ in Eq. (5) after multiplication by a symmetric matrix

$$\mathbf{G} = \begin{pmatrix} L & 0 & 0 \\ 0 & A_1 + l_2^2 M_2 + \frac{4\pi\rho}{15} a^2 c \frac{(c^2-a^2)^2}{c^2+a^2} & 0 \\ 0 & 0 & \frac{4\pi\rho}{15} a^4 c^3 \frac{(c^2-a^2)^2}{c^2+a^2} \end{pmatrix} \tag{7}$$

yields a Hermitian matrix $\mathbf{GB} = \overline{(\mathbf{GB})}^T$, i.e. $\mathbf{B}$ is a self-adjoint operator in the space $\mathbb{C}^3$ endowed with the metric

$$[\mathbf{u}, \mathbf{u}] := (\mathbf{G}\mathbf{u}, \mathbf{u}) = \overline{\mathbf{u}}^T \mathbf{G}\mathbf{u}, \quad \mathbf{u} \in \mathbb{C}^3, \tag{8}$$

which is *definite* when $L > 0$ and *indefinite* with one negative square when $L < 0$. If λ is an eigenvalue of the matrix $\mathbf{B}$, i.e. $\mathbf{Bu} = \lambda \mathbf{u}$, then $\overline{\mathbf{u}}^T \mathbf{GBu} = \lambda \overline{\mathbf{u}}^T \mathbf{Gu}$. On the other hand, $\overline{\mathbf{u}}^T (\overline{\mathbf{GB}})^T \mathbf{u} = \overline{\lambda}\, \overline{\overline{\mathbf{u}}^T \mathbf{Gu}} = \overline{\lambda}\, \overline{\mathbf{u}}^T \mathbf{Gu}$. Hence,

$$(\lambda - \overline{\lambda}) \overline{\mathbf{u}}^T \mathbf{Gu} = 0,$$

implying $\overline{\mathbf{u}}^T \mathbf{Gu} = 0$ on the eigenvector $\mathbf{u}$ of the complex $\lambda \neq \overline{\lambda}$. For real eigenvalues $\lambda = \overline{\lambda}$ and $\overline{\mathbf{u}}^T \mathbf{Gu} \neq 0$. The sign of the quantity $\overline{\mathbf{u}}^T \mathbf{Gu}$ (or Krein sign) can be different for different real eigenvalues.

For example, when the ellipsoidal shell is massless and the supporting point is at the center of mass of the system, then $A_1 = 0$, $C_1 = 0$, $M_1 = 0$, $l_2 = 0$. The matrix $\mathbf{B}$ has thus one real eigenvalue ($\lambda_1^+ = -1$, $\overline{\mathbf{u}}_1^{+T} \mathbf{Gu}_1^+ > 0$) and the pair of eigenvalues

$$\lambda_2^\pm = -\frac{1}{2} \pm \frac{1}{2} \sqrt{1 + \frac{32\pi\rho}{15} \frac{ca^4}{L}}, \quad L = \frac{4\pi\rho}{15} a^2 c (a^2 - c^2), \tag{9}$$

which are real if $L > 0$ and can be complex if $L < 0$. The latter condition together with the requirement that the radicand in Eq. (9) is negative, reproduces the Greenhill's instability zone: $a < c < 3a$ [44]. With the change of c, the real eigenvalue λ_2^+ with $\overline{\mathbf{u}}_2^{+T} \mathbf{Gu}_2^+ > 0$ collides at $c = 3a$ with the real eigenvalue λ_2^- with $\overline{\mathbf{u}}_2^{-T} \mathbf{Gu}_2^- < 0$ into a real double defective eigenvalue λ_d with the algebraic multiplicity two and geometric multiplicity one. This Krein collision is illustrated in Figure 2. Note that $\overline{\mathbf{u}}_d^T \mathbf{Gu}_d = 0$, where $\mathbf{u}_d$ is the eigenvector at λ_d.

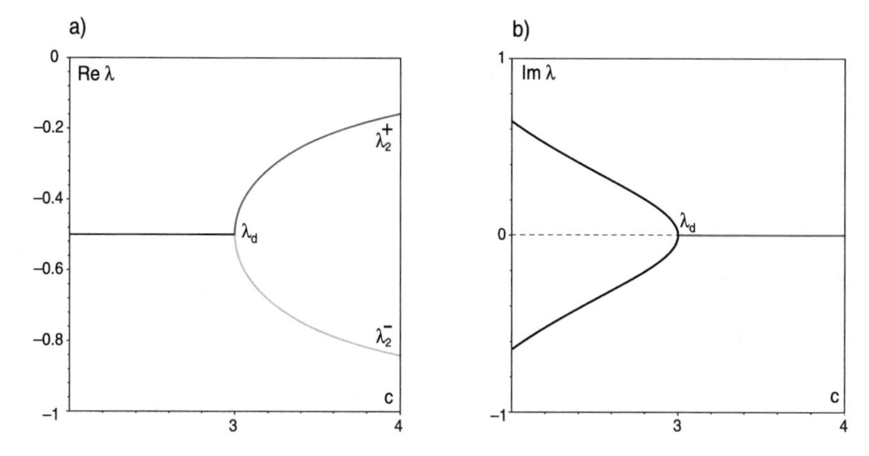

Fig. 2: (a) Simple real eigenvalues (9) of the Sobolev's top in the Greenhill's case for $a = 1$ with (red) $\overline{\mathbf{u}}^T \mathbf{G}\mathbf{u} > 0$ and (green) $\overline{\mathbf{u}}^T \mathbf{G}\mathbf{u} < 0$. (b) At simple complex-conjugate eigenvalues (black) and at the double real eigenvalue λ_d we have $\overline{\mathbf{u}}^T \mathbf{G}\mathbf{u} = 0$.

Therefore, in the case of the ellipsoidal shapes of the shell and the cavity, the Hilbert space $\{R\} = \{Z, W, \mathbf{v}\}$ of the Sobolev's problem endowed with the indefinite metric $(L < 0)$ decomposes into the three-dimensional space of the reduced model (5), where the self-adjoint operator B can have complex eigenvalues and real defective eigenvalues, and a complementary infinite-dimensional space, which is free of these complications. The very idea that the signature of the indefinite metric can serve for counting unstable eigenvalues of an operator that is self-adjoint in a functional space equipped with such a metric, turned out to be a concept of a rather universal character possessing powerful generalizations that were initiated by Pontryagin in 1944 [104] and developed into a general theory of indefinite inner product spaces or Krein spaces [19, 72, 75]. Relation of the Krein sign to the sign of energy or action has made it a popular tool for predicting instabilities in physics [97, 149].

1.2 Secular instability of the Maclaurin spheroids by viscous and radiative losses

It is hard to find a physical application that would stimulate development of mathematics to such an extent as the problem of stability of equilibria of rotating and self-gravitating masses of fluids. Rooted in the Newton and Cassini thoughts on the actual shape of the Earth, the rigorous analysis of this question attracted the best minds of the XVIII-th and XIX-th centuries, from Maclaurin to Riemann, Poincaré and Lyapunov. In fact, modern nonlinear dynamics [5, 53, 76] and Lyapunov stability theory [88] are by-products of the efforts invested in solution of this question of the astrophysical fluid dynamics [22], which experiences a revival nowadays [3] inspired by the recent detection of gravitational waves [1].

We recall that in 1742 Maclaurin established that an oblate spheroid

$$\frac{x^2}{a_1^2}+\frac{y^2}{a_2^2}+\frac{z^2}{a_3^2}=1, \quad a_3<a_2=a_1$$

is a shape of relative equilibrium of a self-gravitating mass of inviscid fluid in a solid-body rotation about the z-axis, provided that the rate of rotation, Ω, is related to the eccentricity $e=\sqrt{1-\frac{a_3^2}{a_1^2}}$ through the formula [91]

$$\Omega^2(e)=2e^{-3}(3-2e^2)\sin^{-1}(e)\sqrt{1-e^2}-6e^{-2}(1-e^2). \tag{10}$$

A century later, Jacobi (1834) has discovered less symmetric shapes of relative equilibria in this problem that are tri-axial ellipsoids

$$\frac{x^2}{a_1^2}+\frac{y^2}{a_2^2}+\frac{z^2}{a_3^2}=1, \quad a_3<a_2<a_1.$$

Later on a student of Jacobi, Meyer (1842) [95], and then Liouville (1851) [86] have shown that the family of Jacobi's ellipsoids has one member in common with the family of Maclaurin's spheroids at $e\approx 0.8127$. The equilibrium with the Meyer-Liouville eccentricity is neutrally stable.

In 1860 Riemann [105] established neutral stability of inviscid Maclaurin's spheroids on the interval of eccentricities $(0<e<0.952..)$. At the Riemann point with the critical eccentricity $e\approx 0.9529$ the Hamilton-Hopf bifurcation sets in and causes dynamical instability with respect to ellipsoidal perturbations beyond this point [38, 113, 114].

A century later Chandrasekhar [31] used a virial theorem to reduce the problem to a finite-dimensional system, which stability is governed by the eigenvalues of the matrix polynomial

$$\mathbf{L}_i(\lambda)=\lambda^2\begin{pmatrix}1&0\\0&1\end{pmatrix}+\lambda\begin{pmatrix}0&-4\Omega\\\Omega&0\end{pmatrix}+\begin{pmatrix}4b-2\Omega^2&0\\0&4b-2\Omega^2\end{pmatrix}, \tag{11}$$

where $\Omega(e)$ is given by the Maclaurin law (10) and $b(e)$ is as follows

$$b=\frac{\sqrt{1-e^2}}{4e^5}\left\{e(3-2e^2)\sqrt{1-e^2}+(4e^2-3)\sin^{-1}(e)\right\}. \tag{12}$$

The eigenvalues of the matrix polynomial (11) are

$$\lambda=\pm\left(i\Omega\pm i\sqrt{4b-\Omega^2}\right). \tag{13}$$

Requiring $\lambda = 0$ we can determine the critical Meyer-Liouville eccentricity by solving with respect to e the equation [31]

$$4b(e) = 2\Omega^2(e).$$

The critical eccentricity at the Riemann point follows from requiring the radicand in (13) to vanish:

$$4b(e) = \Omega^2(e),$$

which is equivalent to the equation

$$e = \sin\left(\frac{e(3+4e^2)\sqrt{1-e^2}}{3+2e^2-4e^4}\right)$$

that has a root $e \approx 0.9529$.

Remarkably, when

$$\Omega^2(e) < 4b(e) < 2\Omega^2(e) \tag{14}$$

both eigenvalues of the stiffness matrix

$$\begin{pmatrix} 4b-2\Omega^2 & 0 \\ 0 & 4b-2\Omega^2 \end{pmatrix}$$

are negative. The number of negative eigenvalues of the matrix of potential forces is known as the Poincaré instability degree. The Poincaré instability degree of the equilibria with the eccentricities (14) is even and equal to 2. Hence, the interval (14) corresponding to $0.812.. < e < 0.952..$, which is stable according to Riemann, is, in fact, the interval of gyroscopic stabilization [78] of the Maclaurin spheroids, Figure 3.

In 1879 Kelvin and Tait [128] realized that viscosity of the fluid can destroy the gyroscopic stabilization of the Maclaurin spheroids: "If there be any viscosity, however slight, in the liquid, the equilibrium [beyond $e \approx 0.8127$] in any case of energy either a minimax or a maximum cannot be secularly stable".

The prediction made by Kelvin and Tait [128] has been rigorously verified only in the XX-th century by Roberts and Stewartson [107]. Using the virial approach by Chandrasekhar, the linear stability problem can be reduced to the study of eigenvalues of the matrix polynomial

$$\mathbf{L}_\nu(\lambda) = \lambda^2\begin{pmatrix} 1 & 0 \\ 0 & 1 \end{pmatrix} + \lambda\begin{pmatrix} 10\mu & -4\Omega \\ \Omega & 10\mu \end{pmatrix} + \begin{pmatrix} 4b-2\Omega^2 & 0 \\ 0 & 4b-2\Omega^2 \end{pmatrix}, \tag{15}$$

where $\mu = \frac{\nu}{a_1^2\sqrt{\pi G\rho}}$, ν is the viscosity of the fluid, G is the universal gravitation constant, and ρ the density of the fluid [31]. The λ and Ω are measured in units of $\sqrt{\pi G\rho}$. The operator $\mathbf{L}_\nu(\lambda)$ differs from the operator of the ideal system, $\mathbf{L}_i(\lambda)$, by the matrix of dissipative forces $10\lambda\mu\mathbf{I}$, where $\mathbf{I}$ is the 2×2 unit matrix.

The characteristic polynomial written for $\mathbf{L}_\nu(\lambda)$ yields the equation governing the growth rates of the ellipsoidal perturbations in the presence of viscosity:

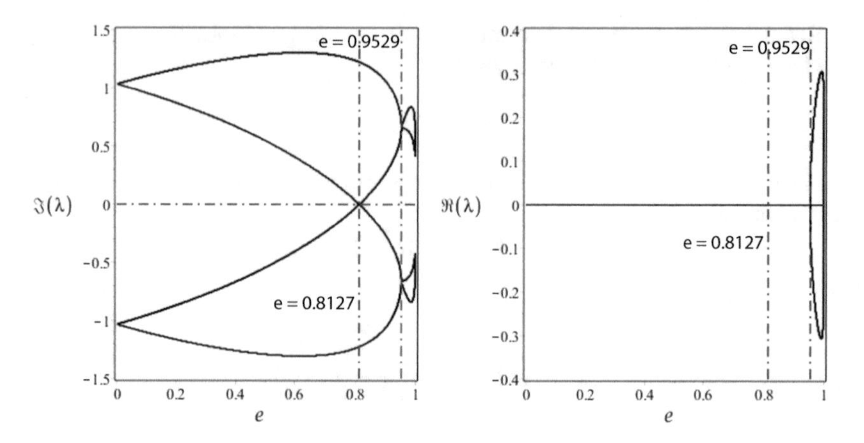

Fig. 3: (Left) Frequencies and (right) growth rates of the eigenvalues of the inviscid eigenvalue problem $\mathbf{L}_i(\lambda)\mathbf{u} = 0$ demonstrating the Hamilton-Hopf bifurcation at the Riemann critical value of the eccentricity, $e \approx 0.9529$ and neutral stability at the Meyer-Liouville point, $e \approx 0.8127$.

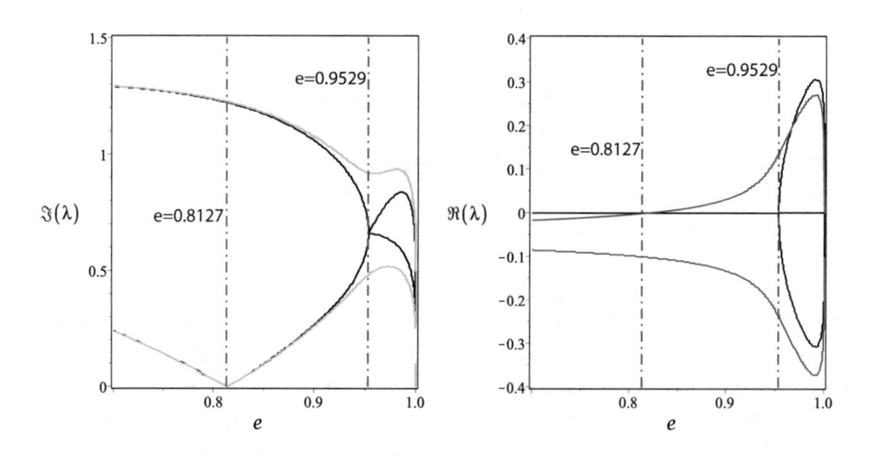

Fig. 4: (Left) Frequencies and (right) growth rates of the (black lines) inviscid Maclaurin spheroids and (green and red lines) viscous ones with $\mu = \frac{\nu}{a_1^2\sqrt{\pi G\rho}} = 0.01$. Viscosity destroys the gyroscopic stabilization of the Maclaurin spheroids on the interval $0.8127.. < e < 0.9529..$, which is stable in the inviscid case [31, 33, 107].

$$25\Omega^2\mu^2 + (\mathrm{Re}\lambda + 5\mu)^2(\Omega^2 - \mathrm{Re}\lambda^2 - 10\mathrm{Re}\lambda\mu - 4b) = 0. \tag{16}$$

The right panel of Figure 4 shows that the growth rates (16) become positive beyond the Meyer-Liouville point. Indeed, assuming $\mathrm{Re}\lambda = 0$ in (16), we reduce it to $50\mu^2(\Omega^2 - 2b) = 0$, meaning that the growth rate vanishes when $\Omega^2 = 2b$ no matter how small the viscosity coefficient μ is. But, as we already know, the equation $\Omega^2(e) = 2b(e)$ determines exactly the Meyer-Liouville point, $e \approx 0.8127$.

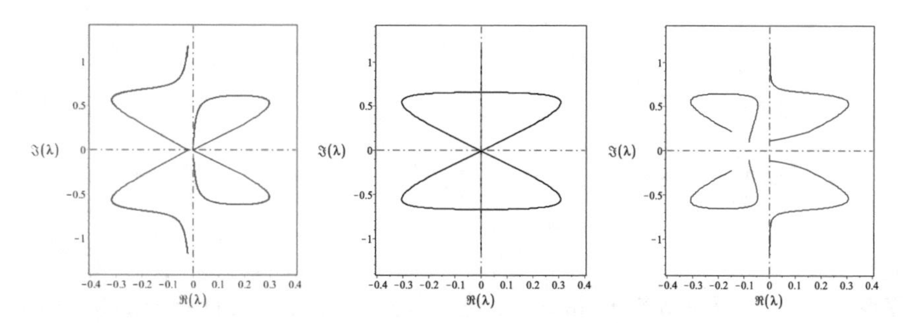

Fig. 5: Paths of the eigenvalues in the complex plane for (left) viscous Maclaurin spheroids with $\mu = \frac{\nu}{a_1^2\sqrt{\pi G\rho}} = 0.002$, (centre) Maclaurin spheroids without dissipation, and (right) inviscid Maclaurin spheroids with radiative losses for $\delta = 0.05$. The collision of two modes of the non-dissipative Hamiltonian system shown in the centre occurs at the Riemann critical value $e \approx 0.9529$. Both types of dissipation destroy the collision and destabilize one of the two interacting modes at the Meyer-Liouville critical value $e \approx 0.8127$.

It turns out, that the critical eccentricity of the viscous Maclaurin spheroid is equal to the Meyer-Liouville value, $e \approx 0.8127$, even in the limit of vanishing viscosity, $\mu \to 0$, and thus does not converge to the inviscid Riemann value $e \approx 0.9529$.

Viscous dissipation destroys the interaction of two modes at the Riemann critical point and destabilizes one of them beyond the Meyer-Liouville point, showing an avoided crossing in the complex plane, Figure 5(left).

Kelvin and Tait [128] hypothesised that the instability, which is stimulated by the presence of viscosity in the fluid, will result in a slow, or *secular*, departure of the system from the unperturbed equilibrium of the Maclaurin family at the Meyer-Liouville point and subsequent evolution along the Jacobi family, as long as the latter is stable [31, 130].

Therefore, a rotating, self-gravitating fluid mass, initially symmetric about the axis of rotation, can undergo an axisymmetric evolution in which it first loses stability to a nonaxisymmetric disturbance, and continues evolving along a non-axisymmetric family toward greater departure from axial symmetry; then it undergoes a further loss of stability to a disturbance tending toward splitting into two parts. Rigorous mathematical treatment of the validity of the fission theory of binary stars by Lyapunov and Poincaré has laid a foundation to modern nonlinear analysis. In particular, it has led Lyapunov to the formulation of a general theory of stability of motion [22].

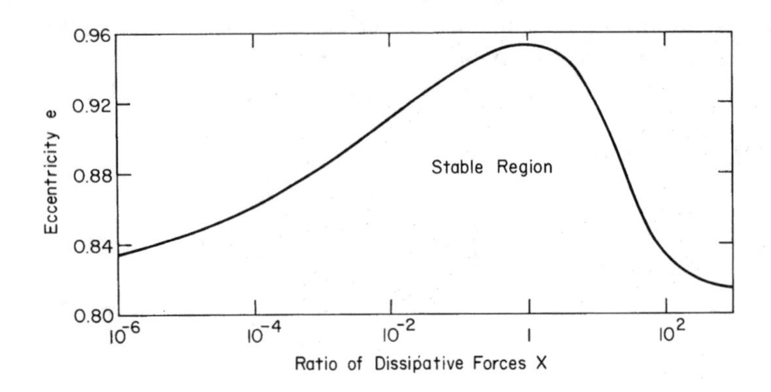

Fig. 6: Critical eccentricity in the limit of vanishing dissipation depends on the damping ratio, X, and attains its maximum (Riemann) value, $e \approx 0.9529$ exactly at $X = 1$. As X tends to zero or infinity, the critical value tends to the Meyer-Liouville value $e \approx 0.8127$, [33, 85].

In 1970 Chandrasekhar [32] demonstrated that there exists another mechanism making Maclaurin spheroids unstable beyond the Meyer-Liouville point of bifurcation, namely, the radiative losses due to emission of gravitational waves. However, the mode that is made unstable by the radiation reaction is not the same one that is made unstable by the viscosity, Figure 5(right).

In the case of the radiative damping mechanism stability is determined by the spectrum of the following matrix polynomial [32]

$$\mathbf{L}_g(\lambda) = \lambda^2 + \lambda(\mathbf{G}+\mathbf{D}) + \mathbf{K} + \mathbf{N}$$

that contains the matrices of gyroscopic, $\mathbf{G}$, damping, $\mathbf{D}$, potential, $\mathbf{K}$, and nonconservative positional, $\mathbf{N}$, forces

$$\mathbf{G} = \frac{5}{2}\begin{pmatrix} 0 & -\Omega \\ \Omega & 0 \end{pmatrix}, \quad \mathbf{D} = \begin{pmatrix} \delta 16\Omega^2(6b-\Omega^2) & -3\Omega/2 \\ -3\Omega/2 & \delta 16\Omega^2(6b-\Omega^2) \end{pmatrix}$$

$$\mathbf{K} = \begin{pmatrix} 4b-\Omega^2 & 0 \\ 0 & 4b-\Omega^2 \end{pmatrix}, \quad \mathbf{N} = \delta\begin{pmatrix} 2q_1 & 2q_2 \\ -q_2/2 & 2q_1 \end{pmatrix},$$

where $\Omega(e)$ and $b(e)$ are given by equations (10) and (12). Explicit expressions for q_1 and q_2 can be found in [32]. The coefficient $\delta = \frac{GMa_1^2(\pi G\rho)^{3/2}}{5c^5}$ is related to gravitational radiation reaction, G is the universal gravitation constant, ρ the density of the fluid, M the mass of the ellipsoid, and c the velocity of light in vacuum.

In 1977 Lindblom and Detweiler [85] studied the combined effects of the gravitational radiation reaction and viscosity on the stability of the Maclaurin spheroids. As we know, each of these dissipative effects induces a secular instability in the Maclaurin sequence past the Meyer-Liouville point of bifurcation. However, when both effects are considered together, the sequence of stable Maclaurin spheroids reaches past the bifurcation point to a new point determined by the *ratio*, $X = \frac{25}{2\Omega_0^4}\frac{\mu}{\delta}$, of the strengths of the viscous and radiation reaction forces, where $\Omega_0 = 0.663490\ldots$.

Figure 6 shows the critical eccentricity as a function of the damping ratio in the limit of vanishing dissipation. This limit coincides with the inviscid Riemann point only at a particular damping ratio. At any other ratio, the critical value is below the Riemann one and tends to the Meyer-Liouville value as this ratio tends either to zero or infinity. Lindblom and Detweiler [85] correctly attributed the cancellation of the secular instabilities to the fact that viscous dissipation and radiation reaction cause different modes to become unstable, see Figure 5. In fact, the mode destabilized by the fluid viscosity is the prograde moving spherical harmonic that appears to be retrograde in the frame rotating with the fluid mass and the mode destabilized by the radiative losses is the retrograde moving spherical harmonic when it appears to be prograde in the inertial frame [3]. It is known [77, 97] that to excite the positive energy mode one must provide additional energy to the mode, while to excite the negative energy mode one must extract energy from the mode [35, 109]. The latter can be done by dissipation and the former by the non-conservative positional forces [62, 63, 66]. Both are present in the model by Lindblom and Detweiler [85].

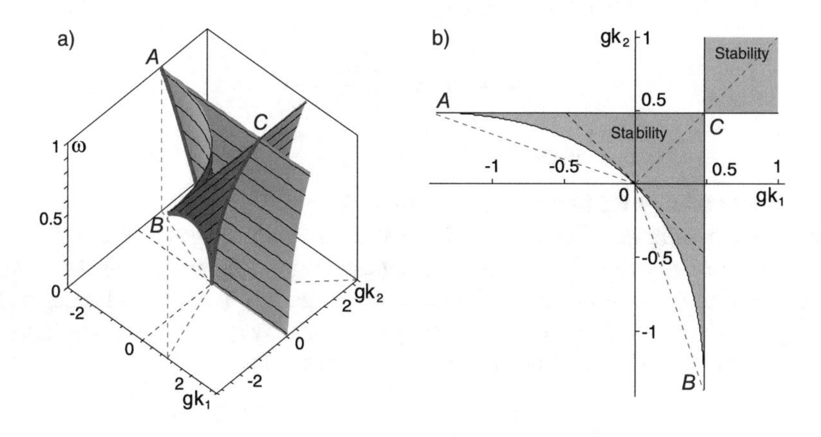

Fig. 7: (a) Stability domain (18) of the Brouwer's rotating vessel and (b) its cross-section at $\omega = 0.7$ [65, 66].

1.3 Brouwer's rotating vessel

In 1918, Brouwer [28] published a simple model for the motion of a point mass near the bottom of a rotating vessel. For an English translation see [29] pp. 665-675. The shape of the vessel is described by a surface S, rotating around a vertical axis with constant angular velocity ω. With the equilibrium chosen on the vertical axis at $(x,y) = (0,0)$ on S, the linearized equations of motion without dissipation are

$$\begin{aligned} \ddot{x} - 2\omega\dot{y} + (gk_1 - \omega^2)x &= 0, \\ \ddot{y} + 2\omega\dot{x} + (gk_2 - \omega^2)y &= 0. \end{aligned} \tag{17}$$

The constants k_1 and k_2 are the x,y-curvatures of S at $(0,0)$, g is the gravitational constant. Suppose that $k_1 \geq k_2$.

Stability and instability without friction
Assuming there is no damping, there are the following three cases [28, 29]:

- $0 < k_2 < k_1$ (single-well at equilibrium).
  Stability iff $0 < \omega^2 < gk_2$ (slow rotation) or $\omega^2 > gk_1$ (fast rotation).
- $k_2 < 0$ and $k_1 > 0$, $k_1 > -k_2$ (saddle at equilibrium).
  Stability iff $\omega^2 > gk_1$.
- $k_2 < 0$ and $k_1 > 0$, $k_1 < -k_2$ (saddle).
  If $3k_1 + k_2 < 0$: instability.
  If $3k_1 + k_2 > 0$: stability if [28, 29]

$$gk_1 < \omega^2 < -\frac{g}{8}\frac{(k_1 - k_2)^2}{(k_1 + k_2)}. \tag{18}$$

- $k_1 < 0$, $k_2 < 0$: instability.

Stability of triangular libration points L_4 and L_5
Brouwer's rotating vessel model includes both a well with two positive curvatures k_1 and k_2 and a saddle with the curvatures of opposite signs. Remarkably, the latter case describes stability of triangular libration points L_4 and L_5 (discovered by Lagrange in 1772) in the restricted circular three-body problem of celestial mechanics. Indeed, the linearized equation for this problem is (17) where $\omega = 1$,

$$\begin{aligned} gk_1 &= -\frac{1}{2} + \frac{3}{2}\sqrt{1 - 3\mu(1-\mu)} \geq \frac{1}{4}, \\ gk_2 &= -\frac{1}{2} - \frac{3}{2}\sqrt{1 - 3\mu(1-\mu)} \leq -\frac{5}{4}, \end{aligned} \tag{19}$$

$\mu = \frac{m_1}{m_1+m_2}$, and m_1 and m_2 are the masses of the two most massive bodies (in comparison with each of them the mass of the third body is assumed to be negligible) [2, 39, 106]. Since k_1 and k_2 are of opposite signs for all $0 \leq \mu \leq 1$, the linear stability of the triangular Lagrange equilibriums is determined by the stability conditions (18) for the rotating saddle [12]. Note that the coefficients (19) satisfy the constraint

$gk_1 + gk_2 = -1$. Intervals of intersection of this line with the narrow corners of the curvilinear triangle in Figure 7b correspond to the stable Lagrange points. After substitution of (19) into (7), we reproduce the classical condition for their linear stability first established by Gascheau in 1843 [39]

$$(gk_1 - gk_2)^2 - 8 = 1 - 27\mu(1-\mu) > 0.$$

Stability in a rotating saddle potential is a subject of current active discussion in respect with the particle trapping [12, 13, 65, 129]. An effect of slow precession of trajectories of the trapped particles in a rotating saddle potential [70, 118] has recently inspired new works leading to the improvement of traditional averaging methods [34, 84, 145].

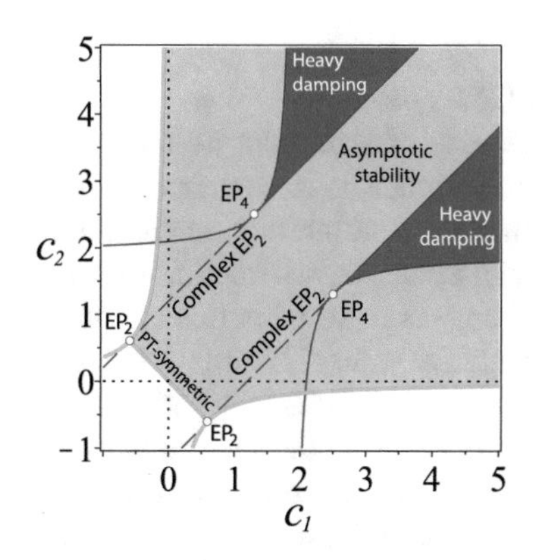

Fig. 8: Domain of asymptotic stability of the damped Brouwer's rotating vessel in the plane of the damping coefficients for $gk_1 = 1$, $gk_2 = 1$, and $\omega = 0.3$ [66, 67].

Destabilization by friction

Adding constant (Coulomb) damping, Brouwer [28] finds a number of instability cases. The equations of motion become in this case:

$$\ddot{x} - 2\omega\dot{y} + c_1\dot{x} + (gk_1 - \omega^2)x = 0, \tag{20}$$

$$\ddot{y} + 2\omega\dot{x} + c_2\dot{y} + (gk_2 - \omega^2)y = 0. \tag{21}$$

The friction constants c_1, c_2 are positive. The characteristic (eigenvalue) equation takes the form:

$$\lambda^4 + a_1\lambda^3 + a_2\lambda^2 + a_3\lambda + a_4 = 0,$$

with $a_1 = c_1 + c_2$,
$a_2 = g(k_1 + k_2) + 2\omega^2 + c_1c_2$
$a_3 = c_1(gk_2 - \omega^2) + c_2(gk_1 - \omega^2)$,
$a_4 = (gk_1 - \omega^2)(gk_2 - \omega^2)$.

There are two cases that drastically change the stability (see also Bottema [26] and the recent works [23, 136]) :

- $0 < k_2 < k_1$ (single-well).
  Stability iff $0 < \omega^2 < gk_2$.
  The fast rotation branch $\omega^2 > gk_1$ has vanished.
- $k_2 < 0$ and $k_1 > 0$, $k_1 < -k_2$ (saddle).
  The requirement $a_4 > 0$ produces $0 < gk_1 < \omega$. This is not compatible with $a_3 < 0$ so a saddle is always unstable with any size of positive damping.

Brouwer studied this model probably to use in his lectures. In a correspondence with O. Blumenthal and G. Hamel he asked whether the results of the calculations were known; see [29] pp. 677-686. Hamel confirmed that the results were correct and surprising, but there is no reference to older literature in this correspondence. See also [26].

Indefinite damping and PT-symmetry
Brouwer's model for the case of a rotating well with two positive curvatures has a direct relation to rotordynamics as it contains the Jeffcott rotor model, see for instance [65]. Dissipation induces instabilities in this model at high speeds of rotation ω [20, 35, 120], of course, under the assumption that the damping coefficients are both positive. But what happens if we relax this constraint and extend the space of damping parameters to negative values? It turns out that at low speeds ω the domain of asymptotic stability spreads to the area of negative damping, Figure 8. Even more, a part of the neutral stability curve belongs to the line $c_1 + c_2 = 0$ where one of the damping coefficients is positive and the other one is negative. On this line the system is invariant under time and parity reversion transformation and is therefore PT-symmetric. Its eigenvalues are imaginary in spite of the presence of the loss (positive damping) and gain (negative damping) in the system. PT-symmetric systems with the indefinite damping can easily be realized in the laboratory experiments [112]. Recent study [150] shows their connection to complex G-Hamiltonian systems [68, 144].

2 Ziegler's paradox

We already know that Greenhill's analysis of stability of the Kelvin gyrostat [44] inspired the works of Sobolev [121] and Pontryagin [104] which have led to the development of the theory of spaces with indefinite inner product [19, 75]. Another work of Greenhill [45] ultimately brought about the famous Ziegler paradox [147]. As Gladwell remarked in his historical account of the genesis of the field of nonconservative stability [40], "It was Greenhill who started the trouble though he never knew it."

Motivated by the problem of buckling of propeller-shafts of steamers Greenhill (1883) analyzed stability of an elastic shaft of a circular cross-section under the action of a compressive force and an axial torque [45]. He managed to find the critical torque that causes buckling of the shaft for a number of boundary conditions. For the clamped-free and the clamped-hinged shaft loaded by an axial torque the question remained open until Nicolai in 1928 reconsidered a variant of the clamped-hinged boundary conditions, in which the axial torque is replaced with the *follower* torque [98]. The vector of the latter is directed along the tangent to the deformed axis of the shaft at the end point [40].

Nicolai had established that no nontrivial equilibrium configuration of the shaft exists different from the rectilinear one, meaning stability for all magnitudes of the follower torque. Being unsatisfied with this overoptimistic result, he assumed that the equilibrium method does not work properly in the case of the follower torque. He decided to study small oscillations of the shaft about its rectilinear configuration using what is now known as the Lyapunov stability theory [88] that, in particular, can predict instability via eigenvalues of the linearized problem.

Surprisingly, it turned out that there exist eigenvalues with positive real parts (instability) for all magnitudes of the torque, meaning that the critical value of the follower torque for an elastic shaft of a circular cross-section is actually zero. Because of its unusual behavior, this instability phenomenon received a name "Nicolai's paradox" [40, 87, 98].

In 1951-56 Hans Ziegler of the ETH Zürich re-considered the five original Greenhill problems with the Lyapunov approach and found that the shaft is unstable in cases of the clamped-free and the clamped-hinged boundary conditions for all values of the axial torque, just as in Nicolai's problem with the follower torque [148]. Moreover, the non-self-adjoint boundary eigenvalue problem for the Greenhill's shaft with the axial torque turned out to be a Hermitian adjoint of the non-self-adjoint boundary eigenvalue problem for the Greenhill's shaft with the follower torque [20].

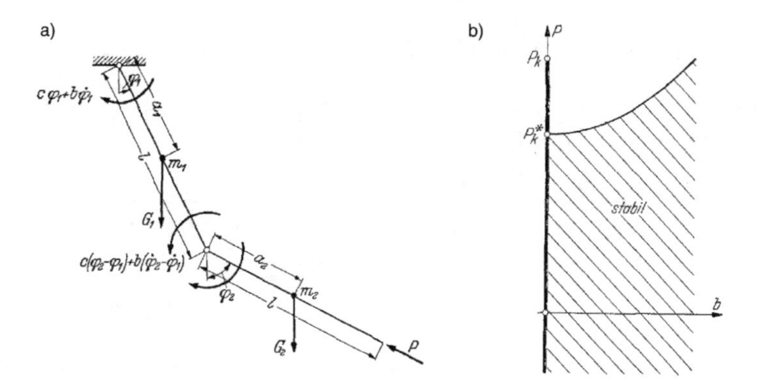

Fig. 9: Original drawings from Ziegler's work of 1952 [147]: (a) double linked pendulum under the follower load P, (b) (bold line) stability interval of the undamped pendulum and (shaded area) the domain of asymptotic stability of the damped pendulum with equal coefficients of dissipation in both joints. If $b = 0$ we have no dissipation and the system is marginally stable for $P < P_k$.

In 1952, inspired by the paradoxes of Greenhill-Nicolai follower torque problems, Ziegler introduced the notion of the *follower force* and published a paper [147] that became widely known in the engineering community, in particular among those interested in theoretical mechanics. It was followed by a second paper [148] that added more details.

Ziegler considered a double pendulum consisting of two rigid rods of length l each. The pendulum is attached at one of the endpoints and can swing freely in a vertical plane; see Figure 9. The angular deflections with respect to the vertical are denoted by ϕ_1, ϕ_2, two masses m_1 and m_2 resulting in the external forces G_1 and G_2 are concentrated at the distances a_1 and a_2 from the joints. At the joints we have elastic restoring forces of the form $c\phi_1$, $c(\phi_2 - \phi_1)$ and internal damping torques

$$b_1 \frac{d\phi_1}{dt}, \quad b_2 \left(\frac{d\phi_2}{dt} - \frac{d\phi_1}{dt} \right).$$

So if $b_1 = b_2 = 0$ we have no dissipation. We impose a follower force P on the lowest hanging rod, see Figure 9. We consider only the quadratic terms of kinetic and potential energy. With these assumptions the kinetic energy T of the system is:

$$T = \frac{1}{2} \left[(m_1 a_1^2 + m_2 l^2) \dot{\phi_1}^2 + 2 m_2 l a_2 \dot{\phi}_1 \dot{\phi}_2 + m_2 a_2^2 \dot{\phi_2}^2 \right]. \tag{22}$$

A dot denotes differentiation with respect to time t. The potential energy V reads:

$$V = \frac{1}{2} \left[(G_1 a_1 + G_2 l + 2c) \phi_1^2 - 2c \phi_1 \phi_2 + (G_2 a_2 + c) \phi_2^2 \right]. \tag{23}$$

This leads to the generalised dissipative and non-conservative forces Q_1, Q_2:

$$Q_1 = Pl(\phi_1 - \phi_2) - ((b_1 + b_2)\dot{\phi}_1 - b_2 \dot{\phi}_2), \; Q_2 = b_2(\dot{\phi}_1 - \dot{\phi}_2). \tag{24}$$

Writing the Lagrange's equations of motion $\dot{L}_{\dot{\varphi}_i} - L_{\varphi_i} = Q_i$, where $L = T - V$ and assuming $G_1 = 0$ and $G_2 = 0$ for simplicity, we find

$$\begin{pmatrix} m_1 a_1^2 + m_2 l^2 & m_2 l a_2 \\ m_2 l a_2 & m_2 a_2^2 \end{pmatrix} \begin{pmatrix} \ddot{\varphi}_1 \\ \ddot{\varphi}_2 \end{pmatrix} + \begin{pmatrix} b_1 + b_2 & -b_2 \\ -b_2 & b_2 \end{pmatrix} \begin{pmatrix} \dot{\varphi}_1 \\ \dot{\varphi}_2 \end{pmatrix} + \begin{pmatrix} -Pl + 2c & Pl - c \\ -c & c \end{pmatrix} \begin{pmatrix} \varphi_1 \\ \varphi_2 \end{pmatrix} = 0. \tag{25}$$

The stability analysis of equilibrium follows the standard procedure. With the substitution $\varphi_i = A_i \exp(\lambda t)$, equation (25) yields a 4-dimensional eigenvalue problem with respect to the spectral parameter λ.

Putting $m_1 = 2m$, $m_2 = m$, $a_1 = a_2 = l$, $b_1 = b_2 = b$ and assuming that internal damping is absent ($b = 0$), Ziegler found from the characteristic equation that the vertical equilibrium position of the pendulum looses its stability when the magnitude of the follower force exceeds the critical value P_k, where

$$P_k = \left(\frac{7}{2} - \sqrt{2}\right)\frac{c}{l} \simeq 2.086\frac{c}{l}. \tag{26}$$

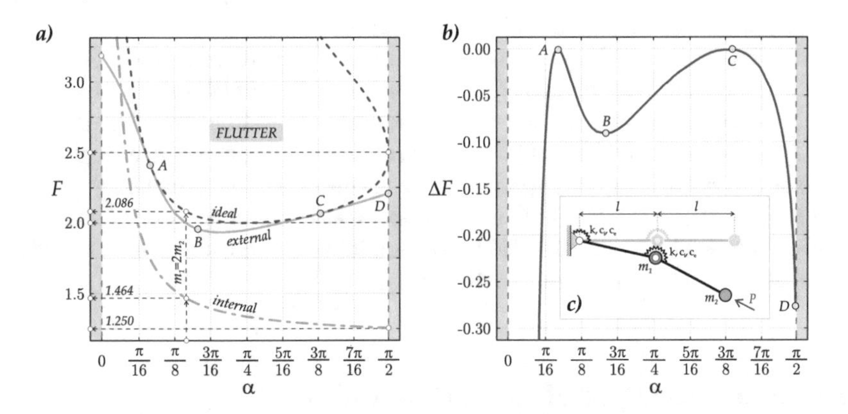

Fig. 10: (a) The (dimensionless) follower force F, shown as a function of the (transformed via $\cot\alpha = m_1/m_2$) mass ratio α, represents the flutter domain of (dashed/red line) the undamped, or 'ideal', Ziegler pendulum [111] and the flutter boundary of the dissipative system in the limit of vanishing (dot-dashed/green line) internal and (continuous/blue line) external damping. (b) Discrepancy ΔF between the critical flutter load for the ideal Ziegler pendulum and for the same structure calculated in the limit of vanishing external damping. The discrepancy quantifies the Ziegler's paradox due to external (air drag) damping [133].

In the presence of damping ($b > 0$) the Routh-Hurwitz condition yields the new critical follower load that depends on the square of the damping coefficient b

$$P_k(b) = \frac{41}{28}\frac{c}{l} + \frac{1}{2}\frac{b^2}{ml^3}. \tag{27}$$

Ziegler found that the domain of asymptotic stability for the damped pendulum is given by the inequalities $P < P_k(b)$ and $b > 0$ and he plotted it against the stability interval $P < P_k$ of the undamped system, Figure 9(b). Surprisingly, the limit of the critical load $P_k(b)$ when b tends to zero turned out to be significantly lower than the critical load of the undamped system

$$P_k^* = \lim_{b\to 0} P_k(b) = \frac{41}{28}\frac{c}{l} \simeq 1.464\frac{c}{l} < P_k. \tag{28}$$

Note that in the original work of Ziegler, formula (27) contains a misprint which yields linear dependency of the critical follower load on the damping coefficient b. Nevertheless, the domain of asymptotic stability found in [147] and reproduced in Figure 9(b), is correct.

Ziegler limited his original calculation to a particular mass distribution, $m_1 = 2m_2$, and took into account only internal damping in the joints, neglecting, e.g, the air drag (an external damping). Later studies confirmed that the Ziegler paradox is a generic phenomenon and exists at all mass distributions, both for internal and external damping [15, 16, 125, 133], see Figure 10.

After invention of robust methods of practical realization of follower forces [14] the Ziegler paradox was immediately observed in the recent laboratory experiments [15, 16]. Nowadays follower forces find new applications in cytosceletal dynamics [9, 37] and acoustics of friction [48]. In general, the interest to mathematical models involving nonconservative positional forces (known also as circulatory [148] or curl [10] forces) is growing both in traditional areas such as energy harvesting and fluid-structure interactions [93, 102] and in rapidly emerging new research fields of optomechanics [126] and light robotics [101].

3 Bottema's analysis of Ziegler's paradox

In 1956, in the journal 'Indagationes Mathematicae', there appeared an article by Oene Bottema (1901-1992) [25], then Rector Magnificus of the Technical University of Delft and an expert in classical geometry and mechanics, that outstripped later findings for decades. Bottema's work on stability in 1955 [24] can be seen as an introduction, it was directly motivated by Ziegler's paradox. However, instead of examining the particular model of Ziegler, he studied in [25] a much more general class of non-conservative systems.

Following [24, 25], we consider a holonomic scleronomic linear system with two degrees of freedom, of which the coordinates x and y are chosen in such a way that the kinetic energy is $T = (\dot{x}^2 + \dot{y}^2)/2$. Hence the Lagrange equations of small oscillations near the equilibrium configuration $x = y = 0$ are as follows

$$\begin{aligned} \ddot{x} + a_{11}x + a_{12}y + b_{11}\dot{x} + b_{12}\dot{y} &= 0, \\ \ddot{y} + a_{21}x + a_{22}y + b_{21}\dot{x} + b_{22}\dot{y} &= 0, \end{aligned} \tag{29}$$

where a_{ij} and b_{ij} are constants, $\mathbf{A} := (a_{ij})$ is the matrix of the forces depending on the coordinates, $\mathbf{B} := (b_{ij})$ of those depending on the velocities. If $\mathbf{A}$ is symmetrical and while disregarding the damping associated with the matrix $\mathbf{B}$, there exists a potential energy function $V = (a_{11}x^2 + 2a_{12}xy + a_{22}y^2)/2$, if it is antisymmetrical, the forces are circulatory. When the matrix $\mathbf{B}$ is symmetrical, we have a non-gyroscopic damping force, which is positive when the dissipative function $(b_{11}x^2 + 2b_{12}xy + b_{22}y^2)/2$ is positive definite. If $\mathbf{B}$ is antisymmetrical the forces depending on the velocities are purely gyroscopic.

The matrices $\mathbf{A}$ and $\mathbf{B}$ can both be written uniquely as the sum of symmetrical and antisymmetrical parts: $\mathbf{A} = \mathbf{K} + \mathbf{N}$ and $\mathbf{B} = \mathbf{D} + \mathbf{G}$, where

$$\mathbf{K} = \begin{pmatrix} k_{11} & k_{12} \\ k_{21} & k_{22} \end{pmatrix}, \quad \mathbf{N} = \begin{pmatrix} 0 & \nu \\ -\nu & 0 \end{pmatrix}, \quad \mathbf{D} = \begin{pmatrix} d_{11} & d_{12} \\ d_{21} & d_{22} \end{pmatrix}, \quad \mathbf{G} = \begin{pmatrix} 0 & \Omega \\ -\Omega & 0 \end{pmatrix}, \quad (30)$$

with $k_{11} = a_{11}, k_{22} = a_{22}, k_{12} = k_{21} = (a_{12} + a_{21})/2, \nu = (a_{12} - a_{21})/2$ and $d_{11} = b_{11}$, $d_{22} = b_{22}$, $d_{12} = d_{21} = (b_{12} + b_{21})/2$, $\Omega = (b_{12} - b_{21})/2$.

Disregarding damping, the system (29) has a potential energy function when $\nu = 0$, it is purely circulatory for $k_{11} = k_{12} = k_{22} = 0$, it is non-gyroscopic for $\Omega = 0$, and has no damping when $d_{11} = d_{12} = d_{22} = 0$. If damping exists, we suppose in this section that it is positive.

In order to solve the linear stability problem for equations (29) we put $x = C_1 \exp(\lambda t)$, $y = C_2 \exp(\lambda t)$ and obtain the characteristic equation for the frequencies of the small oscillations around equilibrium

$$Q := \lambda^4 + a_1 \lambda^3 + a_2 \lambda^2 + a_3 \lambda + a_4 = 0, \quad (31)$$

where [55, 56, 61]

$$a_1 = \mathrm{tr}\mathbf{D}, \quad a_2 = \mathrm{tr}\mathbf{K} + \det\mathbf{D} + \Omega^2, \quad a_3 = \mathrm{tr}\mathbf{K}\mathrm{tr}\mathbf{D} - \mathrm{tr}\mathbf{K}\mathbf{D} + 2\Omega\nu, \quad a_4 = \det\mathbf{K} + \nu^2. \quad (32)$$

For the equilibrium to be stable all roots of the characteristic equation (31) must be semi-simple and have real parts which are non-positive.

It is always possible to write, in at least one way, the left hand-side as the product of two quadratic forms with real coefficients, $Q = (\lambda^2 + p_1\lambda + q_1)(\lambda^2 + p_2\lambda + q_2)$. Hence

$$a_1 = p_1 + p_2, \quad a_2 = p_1 p_2 + q_1 + q_2, \quad a_3 = p_1 q_2 + p_2 q_1, \quad a_4 = q_1 q_2. \quad (33)$$

For all the roots of the equation (31) to be in the left side of the complex plane (L) it is obviously necessary and sufficient that p_i and q_i are positive. Therefore in view of (33) we have: a necessary condition for the roots of $Q = 0$ having negative real parts is $a_i > 0$ $(i = 1, 2, 3, 4)$. This system of conditions however is not sufficient, as the example $(\lambda^2 - \lambda + 2)(\lambda^2 + 2\lambda + 3) = \lambda^4 + \lambda^3 + 3\lambda^2 + \lambda + 6$ shows. But if $a_i > 0$ it is not possible that either one root of three roots lies in L (for then $a_4 \leq 0$); it is also impossible that no root is in it (for, then $a_4 \leq 0$). Hence if $a_i > 0$ at least two roots are in L; the other ones are either both in L, or both on the imaginary axis, or both in R. In order to distinguish between these cases we deduce the condition for two roots being on the imaginary axis. If μi ($\mu \neq 0$ is real) is a root, then $\mu^4 - a_2\mu^2 + a_4 = 0$ and $-a_1\mu^2 + a_3 = 0$. Hence $H := a_1^2 a_4 + a_3^2 - a_1 a_2 a_3 = 0$. Now by means of (33) we have

$$H = -p_1 p_2 (a_1 a_3 + (q_1 - q_2)^2). \quad (34)$$

In view of $a_1 > 0, a_3 > 0$ the second factor is positive; furthermore $a_1 = p_1 + p_2 > 0$, hence p_1 and p_2 cannot both be negative. Therefore $H < 0$ implies $p_1 > 0$, $p_2 > 0$, for $H = 0$ we have either $p_1 = 0$ or $p_2 = 0$ (and not both, because $a_3 > 0$), for $H > 0$ p_1 and p_2 have different signs. We see from the decomposition of the polynomial (31) that all its roots are in L if p_1 and p_2 are positive.

Hence: a set of necessary and sufficient conditions for all roots of (31) to be on the left hand-side of the complex plane is

$$a_i > 0 \ \ (i = 1,2,3,4), \quad H < 0. \tag{35}$$

We now proceed to the cases where all roots have non-positive real parts, so that they lie either in L or on the imaginary axis. If three roots are in L and one on the imaginary axis, this root must be $\lambda = 0$. Reasoning along the same lines as before we find that necessary and sufficient conditions for this are $a_i > 0 \ \ (i = 1,2,3)$, $a_4 = 0$, and $H < 0$. If two roots are in L and two (different) roots on the imaginary axis we have $p_1 > 0$, $q_1 > 0$, $p_2 = 0$, $q_2 > 0$ and the conditions are $a_i > 0 \ \ (i = 1,2,3,4)$ and $H = 0$. If one root is in L and three are on the imaginary axis, then $p_1 > 0$, $q_1 = 0$, $p_2 = 0$, $q_2 > 0$ and the conditions are $a_i > 0 \ \ (i = 1,2,3)$, $a_4 = 0$, and $H = 0$.

The obtained conditions are border cases of (35). This does not occur with the last type we have to consider: all roots are on the imaginary axis. We now have $p_1 = 0$, $p_2 = 0$, $q_1 > 0$, $q_2 > 0$. Hence $a_2 > 0$, $a_4 > 0$, $a_1 = a_3 = 0$ and therefore $H = 0$. This set of relations is necessary, but not sufficient, as the example $Q = \lambda^4 + 6\lambda^2 + 25 = 0$ (which has two roots in L and two in the righthand side of the complex plane (R)) shows. The proof given above is not valid because as seen from (35), $H = 0$ does not imply now $p_1 p_2 = 0$, the second factor being zero for $a_1 a_3 = 0$ and $q_1 = q_2$. The condition can of course easily be given; the equation (31) is $\lambda^4 + a_2\lambda^2 + a_4 = 0$ and therefore it reads $a_2 > 0$, $a_4 > 0$, $a_2^2 > 4a_4$.

Summing up we have: all roots of (31) (assumed to be different) have non-positive real parts if and only if one of the two following sets of conditions is satisfied [25]

$$\begin{aligned} A: &\quad a_1 > 0,\ a_2 > 0,\ a_3 > 0,\ a_4 \geq 0,\ a_2 \geq \frac{a_1^2 a_4 + a_3^2}{a_1 a_3}, \\ B: &\quad a_1 = 0,\ a_2 > 0,\ a_3 = 0,\ a_4 > 0,\ a_2 > 2\sqrt{a_4}. \end{aligned} \tag{36}$$

Note that a_1 represents the damping coefficients b_{11} and b_{22} in the system. One could expect B to be a limit of A, so that for $a_1 \to 0$, $a_3 \to 0$ the set A would continuously tend to B. *That is not the case.*

Remark first of all that the roots of (31) never lie outside R if $a_1 = 0$, $a_3 \neq 0$ (or $a_1 \neq 0$, $a_3 = 0$). Furthermore, if A is satisfied and we take $a_1 = \varepsilon b_1$, $a_3 = \varepsilon b_3$, where b_1 and b_3 are fixed and $\varepsilon \to 0$, the last condition of A reads ($\varepsilon \neq 0$)

$$a_2 > \frac{b_1^2 a_4 + b_3^2}{b_1 b_3} = g_1$$

while for $\varepsilon = 0$ we have

$$a_2 > 2\sqrt{a_4} = g_2.$$

Obviously we have [25]

$$g_1 - g_2 = \frac{(b_1\sqrt{a_4} - b_3)^2}{b_1 b_3}$$

so that $(g_1 > g_2)$ but for $b_3 = b_1\sqrt{a_4}$. That means that in all cases where $b_3 \neq b_1\sqrt{a_4}$ we have a discontinuity in our stability condition.

In 1987, Leipholz remarked in his monograph on stability theory [81] that "Independent works of Bottema and Bolotin for *second-order systems* have shown that in the non-conservative case and for different damping coefficients the stability condition is discontinuous with respect to the undamped case." However, Leipholz did not mention that, in contrast to Bolotin, Bottema illustrated the phenomenon of the discontinuity in a remarkable *geometric diagram*, first published in [25] and reproduced in Figure 11.

Following Bottema [25] we substitute in (31) $\lambda = c\mu$, where c is the positive fourth root of $a_4 > 0$. The new equation reads $P := \mu^4 + b_1\mu^3 + b_2\mu^2 + b_3\mu + 1 = 0$, where $b_i = a_i/c^i$ $(i = 1,2,3,4)$. If we substitute $a_i = c^i b_i$ in A and B we get the same condition as when we write b_i for a_i, which was to be expected, because if the roots of (31) are outside R, those of $P = 0$ are also outside R and inversely. We can therefore restrict ourselves to the case $a_4 = 1$, so that we have only three parameters a_1, a_2, a_3. We take them as coordinates in an orthogonal coordinate system.

The condition $H = 0$ or

$$a_1a_2a_3 = a_1^2 + a_3^2 \tag{37}$$

is the equation of a surface V of the third degree, which we have to consider for $a_1 \geq 0$, $a_3 \geq 0$, Figure 11. Obviously V is a *ruled surface*, the line $a_3 = ma_1$, $a_2 = m + 1/m$ $(0 < m < \infty)$ being on V. The line is parallel to the $0a_1a_3$-plane and intersects the a_2-axis in $a_1 = a_3 = 0$, $a_2 = m + 1/m \geq 2$. The a_2-axis is the double line of V, $a_2 > 2$ being its active part. Two generators pass through each point of it; they coincide for $a_2 = 2$ $(m = 1)$, and for $a_2 \to \infty$ their directions tend to those of the a_1 and a_3-axis $(m = 0, m = \infty)$. The conditions A and B express that the image point (a_1, a_2, a_3) lies on V or above V. The point $(0,2,0)$ is on V, but if we go to the a_2-axis along the line $a_3 = ma_1$ the coordinate a_2 has the limit $m + 1/m$, which is > 2 but for $m = 1$.

Note that we started off with 8 parameters in Eq. (29), but that the surface V bounding the stability domain is described by 3 parameters. It is described by a map of E^2 into E^3 as in Whitney's papers [139, 140]. Explicitly, a transformation of (19) to (2) is given by

$$a_1 = \frac{1}{2}y_3 + w, \quad a_2 = 2 + y_2, \quad a_3 = -\frac{1}{2}y_3 + w$$

with $w^2 = \frac{1}{4}y_3^2 + y_1y_2$.

Returning to the non-conservative system (29) $(\nu \neq 0)$, with damping, but without gyroscopic forces, so $\Omega = 0$, and assuming as in [24] that $k_{12} = 0$, $k_{11} > 0$, and $k_{22} > 0$ (a similar setting but with $d_{12} = 0$ and $k_{12} \neq 0$ was considered later by Bolotin in [20]), we find that the condition for stability $H \leq 0$ reads

$$\begin{aligned} \nu^2 < {} & \frac{(k_{11} - k_{22})^2}{4} \\ & - \frac{(d_{11} - d_{22})^2(k_{11} - k_{22})^2 - 4(k_{11}d_{22} + k_{22}d_{11})(d_{11}d_{22} - d_{12}^2)(d_{11} + d_{22})}{4(d_{11} + d_{22})^2}. \end{aligned} \tag{38}$$

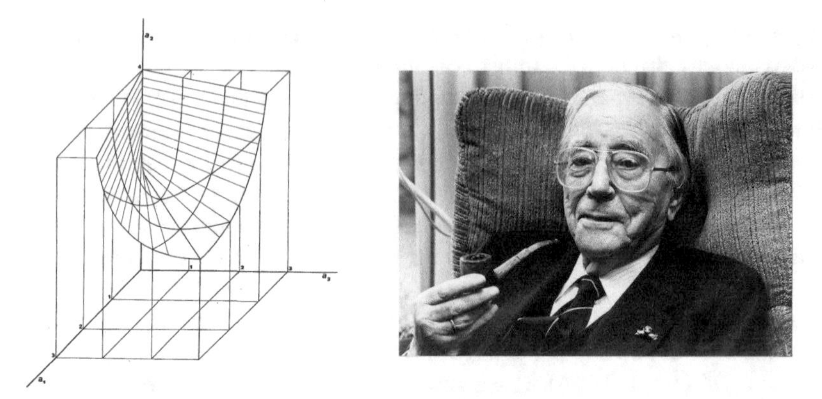

Fig. 11: Original drawing (left) from the 1956 work [25] of Oene Bottema (right), showing the domain of the asymptotic stability of the real polynomial of fourth order and of the two-dimensional non-conservative system with Whitney's umbrella singularity discussed in the sequel. The ruled surface (called V in the text) is given by equation (37).

Suppose now that the damping force decreases in a uniform way, so we put $d_{11} = \varepsilon d'_{11}$, $d_{12} = \varepsilon d'_{12}$, $d_{22} = \varepsilon d'_{22}$, where d_{11}, d_{12}, d_{22} are constants and $\varepsilon \to 0$. Then, for the inequality (38) we get

$$\nu^2 < \nu_{cr}^2 := \frac{(k_{11}-k_{22})^2}{4} - \frac{(d'_{11}-d'_{22})^2(k_{11}-k_{22})^2}{4(d'_{11}+d'_{22})^2}. \tag{39}$$

But if there is no damping, we have to make use of condition B, which gives

$$\nu^2 < {\nu_0}^2 := \frac{(k_{11}-k_{22})^2}{4} = \left(\frac{\mathrm{tr}\mathbf{K}}{2}\right)^2 - \det\mathbf{K}. \tag{40}$$

Obviously

$${\nu_0}^2 - \nu_{cr}^2 = \frac{(d'_{11}-d'_{22})^2(k_{11}-k_{22})^2}{4(d'_{11}+d'_{22})^2} = \left[\frac{2\mathrm{tr}\mathbf{KD} - \mathrm{tr}\mathbf{K}\mathrm{tr}\mathbf{D}}{2\mathrm{tr}\mathbf{D}}\right]^2 \geq 0, \tag{41}$$

where the expressions written in terms of the invariants of the matrices involved [61] are valid also without the restrictions on the matrices $\mathbf{D}$ and $\mathbf{K}$ that were adopted in [20, 24]. For the values of $\frac{2\mathrm{tr}\mathbf{KD}-\mathrm{tr}\mathbf{K}\mathrm{tr}\mathbf{D}}{2\mathrm{tr}\mathbf{D}}$ which are small with respect to ν_0 we can approximately write [57, 58]

$$\nu_{cr} \simeq \nu_0 - \frac{1}{2\nu_0}\left[\frac{2\mathrm{tr}\mathbf{KD} - \mathrm{tr}\mathbf{K}\mathrm{tr}\mathbf{D}}{2\mathrm{tr}\mathbf{D}}\right]^2. \tag{42}$$

If $\mathbf{D}$ depends on two parameters, say δ_1 and δ_2, then (42) has a canonical form (44) for the Whitney's umbrella in the $(\delta_1, \delta_2, \nu)$-space. Due to discontinuity existing for $2\text{tr}\mathbf{KD} - \text{tr}\mathbf{K}\text{tr}\mathbf{D} \neq 0$ the equilibrium may be stable if there is no damping, but unstable if there is damping, however small it may be. We observe also that the critical non-conservative parameter, ν_{cr}, depends on the *ratio* of the damping coefficients and thus is strongly sensitive to the distribution of damping similarly to how it happens in other applications, including the viscous Chandrasekhar-Friedman-Schutz (CFS) instability of the Maclaurin spheroids [85].

The analytical approximations of the type (42) for the onset of the flutter instability in the general finite-dimensional and infinite-dimensional cases were obtained for the first time in the works [55, 56, 57, 58, 60, 62] as a result of further development of the sensitivity analysis of simple and multiple eigenvalues in multiparameter families of non-self-adjoint operators. The previous important works include [4, 11, 17, 35, 38, 46, 54, 90, 99, 113, 114, 115, 116, 117]. Recent results on the perturbation analysis of dissipation-induced instabilities and the destabilization paradox are summarized in the works [66], [87] and [36].

4 An umbrella without dynamics

Part of global analysis, a topic of pure mathematics, is concerned with singularity theory, which deals with the geometric characterisation and classification of singularities (stationary points) of vector fields. In dynamics these singularities are recognised as equilibria of dynamical systems. Well-known representatives of this singularity school are René Thom [127] and Christopher Zeeman. Among pure mathematicians they were exceptional as they promoted singularity theory as useful for real-life problems in biology, the social sciences and physics. Unfortunately their approach gave singularity theory a bad name as in their examples they used geometric methods without explaining a possible relation between realistic vector fields and dynamics. It makes little sense to describe equilibria and transitions (bifurcations) between equilibria without discussing actual causes that are tied in with dynamical processes and corresponding equations of motion. We want to stress here that notwithstanding the lack of dynamics the geometry of singularities as an ingredient of dynamical systems theory can be very useful.

Before Ziegler's results a geometric result in singularity theory was obtained (1943-44) by Hassler Whitney. This result turned out to be an excellent complement to Bottema's analytic approach. In his paper [139], Whitney described singularities of maps of a differential n-manifold into E^m with $m = 2n - 1$. It turns out that in this case a special kind of singularity plays a prominent role. Later, the local geometric structure of the manifold near the singularity has been aptly called 'Whitney's umbrella'. In Figure 12 we reproduce a sketch of the singular surface.

The paper [139] contains two main theorems. Consider the C^k map $f : E^n \mapsto E^m$ with $m = 2n - 1$.

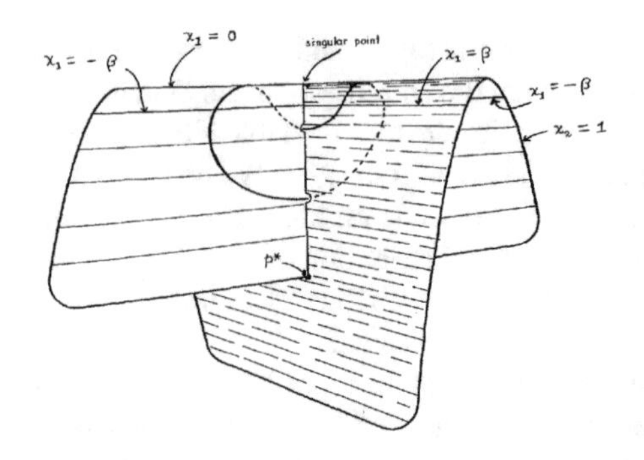

Fig. 12: Whitney's umbrella, lowest dimensional case with 3 parameters [139, 140].

1. The map f can be altered slightly, forming f^*, for which the singular points are isolated. For each such an isolated singular point p, a technical regularity condition C is valid which relates to the map f^* of the independent vectors near p and of the differentials, the vectors in tangent space.
2. Consider the map f^* which satisfies condition C. Then we can choose coordinates $x = (x_1, x_2, \cdots, x_n)$ in a neighborhood of p and coordinates $y = (y_1, y_2, \cdots, y_m)$ (with $m = 2n - 1$) in a neighborhood of $y = f(p)$ such that in a neighborhood of $f^*(p)$ we have exactly

$$\begin{aligned} y_1 &= x_1^2, \\ y_i &= x_i, \ i = 2, \cdots, n, \\ y_{n+i-1} &= x_1 x_i, \ i = 2, \cdots, n. \end{aligned}$$

If for instance $n = 2$, $m = 3$, the simplest interesting case, we have near the origin

$$y_1 = x_1^2, \ y_2 = x_2, \ y_3 = x_1 x_2, \tag{43}$$

so that $y_1 \geq 0$ and on eliminating x_1 and x_2:

$$y_1 y_2^2 - y_3^2 = 0. \tag{44}$$

Starting on the y_2-axis for $y_1 = y_3 = 0$, the surface widens up for increasing values of y_1. For each y_2, the cross-section is a parabola; as y_2 passes through 0, the parabola degenerates to a half-ray, and opens out again (with sense reversed); see Figure 12.

Note that because of the C^k assumption for the differentiable map f, the analysis is delicate. There is a considerable simplification of the treatment if the map is analytical.

The analysis of singularities of functions and maps is a fundamental ingredient for bifurcation studies of differential equations. After the pioneering work of Hassler Whitney and Marston Morse, it has become a huge research field, both in theoretical investigations and in applications. We can not even present a summary of this field here, so we restrict ourselves to citing a number of survey texts and discussing a few key concepts and examples. In particular we mention [5, 6, 7, 8, 41, 42, 43, 137]. A monograph relating bifurcation theory with normal forms and numerics is [76].

5 Hopf bifurcation near 1:1 resonance and structural stability

A study of the stability of equilibria of dynamical systems will usually involve the analysis of matrices obtained by linearisation of the equations of motion in a neighbourhood of the equilibria. This triggered off the study of *structural stability of matrices* as an independent topic in singularity theory [6, 7].

More explicitly, consider a dynamical system described by the autonomous ODE

$$\dot{\mathbf{x}} = \mathbf{f}(\mathbf{x}, \mathbf{p}),\ \mathbf{x} \in \mathbb{R}^n,\ \mathbf{f} : \mathbb{R}^n \mapsto \mathbb{R}^n,$$

where $\mathbf{p} \in \mathbb{R}^k$ is a vector of parameters. An equilibrium $\mathbf{x}_0$ of the system arises if $\mathbf{f}(\mathbf{x}_0, \mathbf{p}) = \mathbf{0}$. With a little smoothness of the map $\mathbf{f}$ we can linearise near $\mathbf{x}_0$ so that we can write

$$\dot{\mathbf{x}} = \mathbf{A}_\mathbf{p}(\mathbf{x} - \mathbf{x}_0) + \mathbf{g}(\mathbf{x}, \mathbf{p}) \tag{45}$$

with $\mathbf{A}_\mathbf{p}$ a constant $n \times n-$ matrix, $\mathbf{g}(\mathbf{x}, \mathbf{p})$ contains higher-order terms only. In other words

$$\lim_{\mathbf{x} \to \mathbf{x}_0} \frac{\|\mathbf{g}(\mathbf{x}, \mathbf{p})\|}{\|\mathbf{x} - \mathbf{x}_0\|} = 0,$$

$\mathbf{g}(\mathbf{x}, \mathbf{p})$ is tangent to the linear map in $\mathbf{x}_0$. The properties of the matrix $\mathbf{A}_\mathbf{p}$ determine in a large number of cases the local behavior of the dynamical system.

Suppose that for $\mathbf{p} = \mathbf{0}$, $\mathbf{A}_\mathbf{0}$ has two equal non-zero imaginary eigenvalues and their complex conjugates, $\pm i\omega, \omega > 0$, and no other eigenvalues with zero real part. This equilibrium is called a $1:1$ resonant double Hopf point [43]. (Similarly, in a Hopf point the matrix of linearization has a single conjugate pair of imaginary eigenvalues $\pm i\omega$ and in a double Hopf point there are two distinct such pairs: $\pm i\omega_1$, $\pm i\omega_2$ [43].) Then, without loss of generality, we may assume that the system (45) has been already reduced to a centre manifold of dimension $n = 4$. Considering further a generic case of double non-semisimple eigenvalues with geometric multiplicity 1, after a linear change of coordinates and re-scaling time to get $\omega = 1$, we can transform $\mathbf{A}_\mathbf{0}$ to [137]

$$\begin{pmatrix} 0 & -1 & 1 & 0 \\ 1 & 0 & 0 & 1 \\ 0 & 0 & 0 & -1 \\ 0 & 0 & 1 & 0 \end{pmatrix}. \tag{46}$$

Setting $z_1 = \Delta x_1 + i\Delta x_2$ and $z_2 = \Delta x_3 + i\Delta x_4$, where $i = \sqrt{-1}$ and $\Delta\mathbf{x} = \mathbf{x} - \mathbf{x}_0$, and assuming $\mathbf{A}_\mathbf{0}$ to be in the form (46) we re-write (45) at $\mathbf{p} = \mathbf{0}$ in the complex form [137]

$$\begin{pmatrix} \dot{z}_1 \\ \dot{z}_2 \end{pmatrix} = \begin{pmatrix} i & 1 \\ 0 & i \end{pmatrix} \begin{pmatrix} z_1 \\ z_2 \end{pmatrix} + \widetilde{\mathbf{g}}(z_1, z_2, \bar{z}_1, \bar{z}_2). \tag{47}$$

The second pair of equations governing the conjugates $\bar{z}_1$, $\bar{z}_2$ is omitted here for simplicity.

Arnold [6, 7] has proven that a universal unfolding of the linear vector field with the matrix

$$\begin{pmatrix} i & 1 \\ 0 & i \end{pmatrix}$$

is given by the three-parameter family of complex matrices

$$\begin{pmatrix} i+\alpha & 1 \\ \mu_1 + i\mu_2 & i+\alpha \end{pmatrix}, \tag{48}$$

where α, μ_1, and μ_2 are real parameters and versality is understood with respect to the group of similarity transformations and a real positive scaling. The set of matrices with a resonant Hopf pair is a group orbit [43].

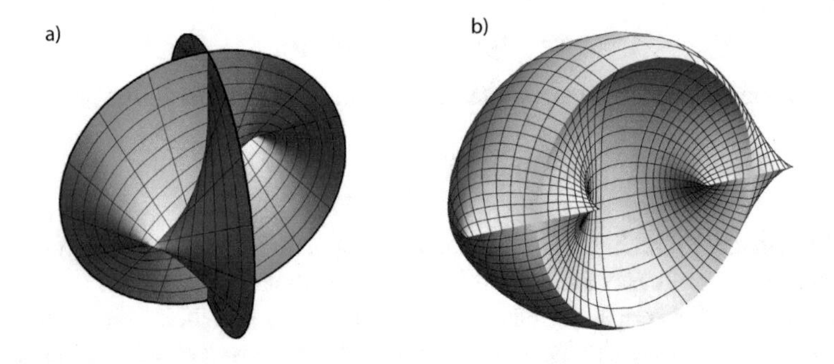

Fig. 13: (a) The Plücker conoid in the unfolding of a semisimple 1 : 1 resonance has a pair of Whitney umbrellas [51]; (b) two pairs of Whitney umbrellas on the boundary of the stability domain of a general 4-degrees-of-freedom dynamical system near a semisimple 1 : 1 resonance [52].

The universal unfolding has a pure imaginary eigenvalue if and only if there exists a real number δ such that $(i(\delta-1)-\alpha)^2-(\mu_1+i\mu_2)=0$. Eliminating δ yields [43]

$$\alpha^2(-4\mu_1+4\alpha^2)=\mu_2^2.$$

Setting $y_2=\alpha$, $y_3=\mu_2$, and $y_1=-4\mu_1+4\alpha^2$ we reduce the equation to the form $y_3^2=y_1y_2^2$, which is nothing else but the normal form (44) for the Whitney umbrella. The double Hopf points of (48) form the half-line $\alpha=\mu_2=0$, $\mu_1=0$. Along the continuation $\mu_1>0$ of this half-line the eigenvalues of (48) are given by $\lambda = i \pm \sqrt{\mu_1}$. We see that the double Hopf points have codimension 2 and the resonant double Hopf points are of codimension 3.

If a family of matrices $\mathbf{A}(\mathbf{p}) = \mathbf{A}(p_1, p_2, p_3, \ldots, p_k)$ has a $1:1$ resonant double Hopf point, the universality of the unfolding (48) means that there exist smooth functions $\alpha(\mathbf{p})$, $\mu_1(\mathbf{p})$, $\mu_1(\mathbf{p})$, such that the Hopf structure of $\mathbf{A}(\mathbf{p})$ near the $1:1$ resonant point is the same as the Hopf structure of the unfolding with α, μ_1, μ_2 replaced by $\alpha(\mathbf{p})$, $\mu_1(\mathbf{p})$, $\mu_2(\mathbf{p})$.

Therefore, the stratified set of Hopf points in the neighborhood of a *non-semisimple* $1:1$ resonance is a Whitney umbrella in $\mathbf{p}$-space too [43]. The functions $\alpha(\mathbf{p})$, $\mu_1(\mathbf{p})$, $\mu_2(\mathbf{p})$ can be found approximately as truncated Taylor series with respect to the components of the vector $\mathbf{p}$ of the parameters [117].

Similar stratification of Hopf points near a *semisimple* $1:1$ resonance involves pairs of Whitney umbrellas forming a Plücker conoid [51, 52], see Figure 13.

Hoveijn and Ruijgrok (1995) were the first who applied these ideas to a practical problem exhibiting the Ziegler paradox. Namely, they considered a problem of widening due to dissipation of the zones of the combination resonance [144] in a system of two parametrically forced coupled oscillators [50]. The system models a rotating disk with oscillating suspension point introduced in [108]. Its linearized equations are

$$\begin{aligned} \ddot{x} + 2\Omega\dot{y} + (1 + \varepsilon\cos\omega_0 t)x + 2\varepsilon\mu\dot{x} &= 0, \\ \ddot{y} + 2\Omega\dot{x} + (1 + \varepsilon\cos\omega_0 t)y + 2\varepsilon\mu\dot{y} &= 0. \end{aligned} \tag{49}$$

It is assumed that for $\varepsilon = 0$ the system (49) has two pairs of imaginary eigenvalues $\pm i\omega_1, \pm i\omega_2$ that depend on the parameter Ω representing the speed of rotation. Of special interest is the case of the sum resonance $\omega_0 = \omega_1 + \omega_2$.

Let parameters δ_1 and δ_2 control the detuning of the frequencies ω_1 and ω_2; then $\delta_+ = \delta_1 + \delta_2$ and $\delta_- = \delta_1 - \delta_2$. The parameter δ_+ is small and represents the detuning of the exact sum resonance: $\omega_0 = \omega_1 + \omega_2 + \delta_+$ where $\delta_+ = 0$. Parameters μ_1 and μ_2 control the detuning of the damping from μ; $\mu_+ = \mu_1 + \mu_2$, $\mu_- = \mu_1 - \mu_2$.

The original nonlinear system that has the linearization (49) at zero detuning can be reduced to the following type of equation [50]

$$\dot{\mathbf{z}} = \mathbf{A}_0\mathbf{z} + \varepsilon\mathbf{f}(\mathbf{z}, \omega_0 t; \mathbf{p}), \quad \mathbf{z} \in \mathbb{R}^4, \tag{50}$$

where $\mathbf{A}_0$ is a 4×4 matrix with the eigenvalues $\pm i\omega_1, \pm i\omega_2$. The vector of parameters $\mathbf{p} = (\delta_+, \delta_-, \mu_+, \mu_-)$ is used to control detuning from resonance and damping.

The vector-valued function $\mathbf{f}$ is 2π-periodic in $\omega_0 t$ and $\mathbf{f}(0, \omega_0 t; \mathbf{p}) = \mathbf{0}$ for all t and $\mathbf{p}$. Since the origin is a stationary point of (50), one may ask how its stability depends on the parameters. Analogous to the case of a single forced oscillator, one can make a planar stability diagram by varying the strength ε and the frequency ω_0 of the forcing while fixing the other parameters. Also in this case one obtains a resonance tongue at $\omega_0 = \omega_1 + \omega_2$. However if damping is varied, the planar stability diagram does not change continuously [108]. For instance, applying zero damping ($\mu = 0$, no damping detuning) we find instability of the trivial solution (equilibrium) if

$$|\delta_+| \leq 1.$$

For $\mu > 0$ the trivial solution is unstable if

$$|\delta_+| \leq \omega_0 \sqrt{\frac{1}{4} - \frac{\mu^2}{\omega_0^2}}.$$

The instability interval depends discontinuously on damping coefficient μ!

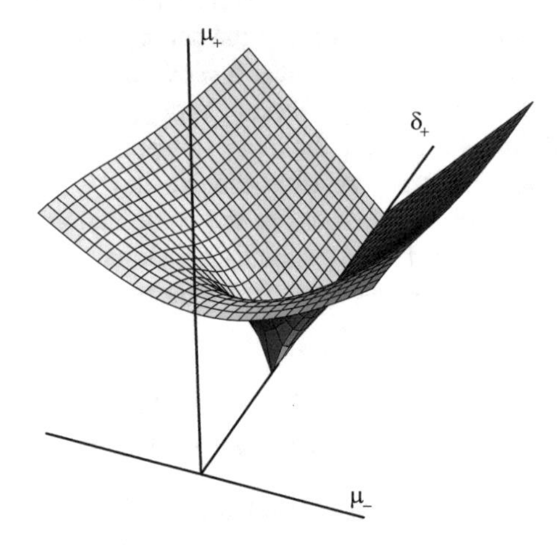

Fig. 14: The critical surface for the damped combination resonance in (μ_+, μ_-, δ_+) space, where $\mu_+ = \mu_1 + \mu_2$, $\mu_- = \mu_1 - \mu_2$, $\delta_+ = \delta_1 + \delta_2$. The parameters δ_1 and δ_2 control the detuning of the frequencies ω_1 and ω_2, the parameters μ_1 and μ_2 the damping of the oscillators. The self-intersection of the surface with the Whitney umbrella singularity is along the δ_+ axis [50].

Hoveijn and Ruijgrok [50] presented a geometrical explanation of this dissipation-induced instability using 'all' the parameters as unfolding parameters first putting the equation (50) into a normal form [7, 110] which is similar to that of the non-semisimple 1 : 1 resonance studied in [137].

In the normalized equation the time dependence appears only in the high order terms. But the autonomous part of this equation contains enough information to determine the stability regions of the origin.

The second step was to test the linear autonomous part $\mathbf{A}(\mathbf{p})$ of the normalized equation for structural stability. This family of matrices is parameterized by the detunings of the frequencies ω_1 and ω_2 and of the damping parameter μ.

Identifying the most degenerate member of this family one can show that $\mathbf{A}(\mathbf{p})$ is its versal unfolding in the sense of Arnold [6, 7]. Put differently, the family $\mathbf{A}(\delta_+, \delta_-, \mu_+, \mu_-)$ is structurally stable, whereas $\mathbf{A}(\delta_+, \delta_-, 0, 0)$ is not. Therefore the stability diagram actually 'lives' in a four dimensional space. In this space, the stability regions of the origin are separated by a critical surface which is the hypersurface where $\mathbf{A}(\mathbf{p})$ has at least one pair of imaginary complex conjugate eigenvalues. This critical surface is diffeomorphic to the Whitney umbrella, see Figure 14.

It is the singularity of the Whitney umbrella that causes the discontinuous behaviour of the planar stability diagram for the combination resonance in the presence of dissipation. The structural stability argument guarantees that the results are 'universally valid' and qualitatively hold for every system in sum resonance. For technical details we refer again to [50].

6 Abscissa minimization, robust stability and heavy damping

Let us return to the work of Bottema [25]. The conditions

$$a_1 > 0, \quad a_3 > 0, \quad a_2 > 2 + \frac{(a_1 - a_3)^2}{a_1 a_3} > 0 \tag{51}$$

are necessary and sufficient for the polynomial

$$p(\lambda) = \lambda^4 + a_1\lambda^3 + a_2\lambda^2 + a_3\lambda + 1 \tag{52}$$

to be Hurwitz. The domain (51) was plotted by Bottema in the (a_1, a_3, a_2)-space, Figure 15a.

A part of the plane $a_1 = a_3$ that lies inside the domain of asymptotic stability constitutes a set of all directions leading from the point (0, 0, 2) to the stability region

$$\{(a_1, a_3, a_2): \ a_1 = a_3, \ a_1 > 0, \ a_2 > 2\}. \tag{53}$$

The *tangent cone* (53) to the domain of asymptotic stability at the Whitney umbrella singularity, which is shown in green in Figure 15a,b, is degenerate in the (a_1, a_3, a_2) – space because it selects a set of measure zero on a unit sphere with the center at the singular point [82, 83].

The singular point $(a_1, a_3, a_2) = (0, 0, 2)$ corresponds to a double complex-conjugate pair of roots $\lambda = \pm i$ of the polynomial (52). The fact that multiple roots of a polynomial are sensitive to perturbation of the coefficients is a phenomenon that was studied already by Isaac Newton, who introduced the so-called Newton polygon to determine the leading terms of the perturbed roots as fractional powers of a perturbation parameter. It follows that, in matrix analysis, eigenvalues are in general not locally Lipschitz at points in matrix space with non-semi-simple eigenvalues [30], and, in the context of dissipatively perturbed Hamiltonian systems, [92]. Thus, it has been well-understood for a long time that perturbation of multiple roots or multiple eigenvalues on or near the stability boundary is likely to lead to instability [59].

Because of the sensitivity of multiple roots and eigenvalues to perturbation, in engineering and control-theoretical applications a natural desire is to "cut the singularities off" by constructing convex inner approximations to the domain of asymptotic stability. Nevertheless, multiple roots per se are not undesirable.

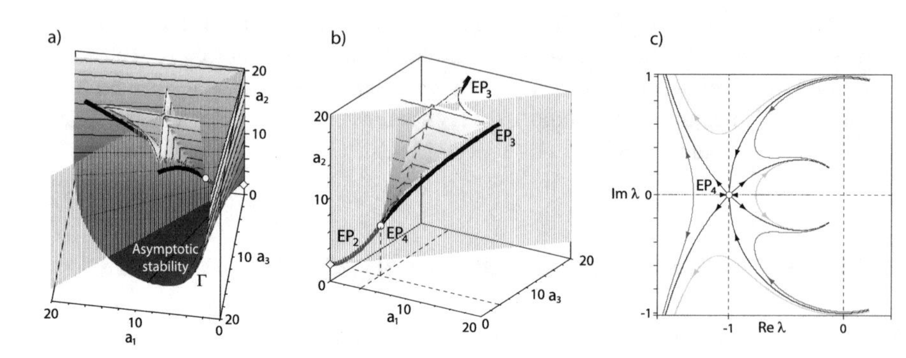

Fig. 15: (a) A singular boundary Γ of the domain of asymptotic stability (51) of the polynomial (52) with the Whitney umbrella singularity at the point $(a_1, a_3, a_2) = (0,0,2)$, marked by the diamond symbol. (b) The tangent cone, the EP-sets, and the discriminant surface with the Swallowtail singularity at EP_4. The domain of heavy damping is inside the 'spire'. (c) Trajectories of roots of the polynomial (52) when a_2 increases from 0 to 15 and: $a_1 = a_3 = 4$ (black); $a_1 = 4$, $a_3 = 3.9$ (red); $a_1 = 3.9$, $a_3 = 4$ (green). The global minimum of the abscissa is attained when all the roots coalesce into the quadruple root $\lambda = -1$ (EP_4) [71].

Indeed, multiple roots also occur deep inside the domain of asymptotic stability. Although it might seem paradoxical at first sight, such configurations are actually obtained by minimizing the so-called *polynomial abscissa* in an effort to make the asymptotic stability of a linear system more *robust*, as we now explain.

Abscissa minimization and multiple roots

The abscissa of a polynomial $p(\lambda)$ is the maximal real part of its roots [18]:

$$a(p) = \max\{\mathrm{Re}\,\lambda : \; p(\lambda) = 0\}. \tag{54}$$

We restrict our attention to monic polynomials with real coefficients and fixed degree n: since these have n free coefficients, this space is isomorphic to $\mathbb{R}^n$. On this space, the abscissa is a continuous but non-smooth, in fact non-Lipschitz, as well as non-convex, function whose variational properties have been extensively studied using non-smooth variational analysis [18, 30].

Now set $n = 4$, consider the set of polynomials $p(\lambda)$ defined in (52), and consider the restricted set of coefficients

$$\left\{(a_1,a_3,a_2):\ a_1=a_3,\ a_2=2+\frac{a_1^2}{4}\right\} \tag{55}$$

On this set the roots are

$$\lambda_1=\lambda_2=-\frac{a_1}{4}-\frac{1}{4}\sqrt{a_1^2-16},\quad \lambda_3=\lambda_4=-\frac{a_1}{4}+\frac{1}{4}\sqrt{a_1^2-16}.$$

When $0\le a_1<4$ ($a_1>4$), the roots $\lambda_{1,2}$ and $\lambda_{3,4}$ are complex (real) with each pair being double, that is with multiplicity two. At $a_1=4$ there is a quadruple real eigenvalue -1. So, we refer to the set (55) as a set of exceptional points [66, 68, 69] (abbreviated as the EP-set).

When $a_1>0$, the EP-set (55) (shown by the red curve in Figure 15a,b) lies within the tangent cone (53) to the domain of asymptotic stability at the Whitney umbrella singularity $(0,0,2)$. The points in the EP-set all define polynomials with two double roots (denoted EP_2) except $(a_1,a_3,a_2)=(4,4,6)$, at which p has a quadruple root and is denoted EP_4; see Figure 15a,b.

Let us consider how the roots move in the complex plane when a_1 and a_3 coincide and are set to specific values and a_2 increases from zero, as shown by black curves in Figure 15c. When $a_1=a_3<4$, the roots that initially have positive real parts and thus correspond to unstable solutions move along the unit circle to the left, cross the imaginary axis at $a_2=2$ and merge with another complex conjugate pair of roots at $a_2=2+\frac{a_1^2}{4}$, i.e., at the EP-set. Further increase in a_2 leads to the splitting of the double eigenvalues, with one conjugate pair of roots moving back toward the imaginary axis. By also considering the case $a_1=a_3>4$, it is clear that when a_1 and a_3 coincide, the choice $a_2=2+\frac{a_1^2}{4}$ minimizes the abscissa, with the polynomial p on the EP-set.

Furthermore, when $a_1=a_3$ is increased toward 4 from below, the coalescence points (EP_2) move around the unit circle to the left. This conjugate pair of coalescence points merges into the quadruple real root $\lambda=-1$ (EP_4) when $a_1=a_3=4$ and hence $a_2=6$, as is visible in Figure 15c. If $a_1=a_3$ is increased above 4 the quadruple point EP_4 splits again into two exceptional points EP_2, one of them inside the unit circle.

Thus, all indications are that the abscissa is minimized by the parameters corresponding to EP_4, with a quadruple root at -1.

In fact, application of the following theorem shows that the abscissa of (52) is globally minimized by the EP_4 parameters.

Theorem 6.1 *([18], Theorems 7 and 14)*

Let $\mathbb{F}$ denote either the real field $\mathbb{R}$ or the complex field $\mathbb{C}$. Let b_0, b_1, ..., $b_n\in\mathbb{F}$ be given (with $b_1,\dots,b_n$ not all zero) and consider the following family of monic polynomials of degree n subject to a single affine constraint on the coefficients:

$$P=\left\{\lambda^n+a_1\lambda^{n-1}+\ldots+a_{n-1}\lambda+a_n:\quad b_0+\sum_{j=1}^{n}b_ja_j=0,\quad a_i\in\mathbb{F}\right\}.$$

Define the optimization problem

$$a^* := \inf_{p \in P} a(p). \tag{56}$$

Let

$$h(\lambda) = b_n \lambda^n + b_{n-1} \binom{n}{n-1} \lambda^{n-1} + \ldots + b_1 \binom{n}{1} \lambda + b_0.$$

First, suppose $\mathbb{F} = \mathbb{R}$. *Then, the optimization problem (56) has the infimal value*

$$a^* = -\max\left\{\zeta \in \mathbb{R}: \ h^{(i)}(\zeta) = 0 \ \ for \ \ some \ \ i \in \{0, \ldots, k-1\}\right\},$$

where $h^{(i)}$ *is the i-th derivative of h and* $k = \max\{j: \ \ b_j \neq 0\}$. *Furthermore, the optimal value* a^* *is attained by a minimizing polynomial* p^* *if and only if* $-a^*$ *is a root of h (as opposed to one of its derivatives), and in this case we can take*

$$p^*(\lambda) = (\lambda - \gamma)^n \in P, \quad \gamma = a^*.$$

Second, suppose $\mathbb{F} = \mathbb{C}$. *Then, the optimization problem (56) always has an optimal solution of the form*

$$p^*(\lambda) = (\lambda - \gamma)^n \in P, \quad \mathrm{Re}\, \gamma = a^*,$$

with $-\gamma$ *given by a root of h (not its derivatives) with largest real part.*

In our case, $\mathbb{F} = \mathbb{R}$, $n = 4$ and the affine constraint on the coefficients of p is simply $a_4 = 1$. So, the polynomial h is given by $h(\lambda) = \lambda^4 - 1$.

Its real root with largest real part is 1, and its derivatives have only the zero root. So, the infimum of the abscissa a over the polynomials (52) is -1, and this is attained by

$$p^*(\lambda) = (\lambda + 1)^4 = \lambda^4 + 4\lambda^3 + 6\lambda^2 + 4\lambda + 1, \tag{57}$$

that is, with the coefficients at the exceptional point EP_4. There is nothing special about $n = 4$ here; if we replace 4 by n we find that the infimum is still -1 and is attained by

$$p^*(\lambda) = (\lambda + 1)^n.$$

Swallowtail singularity as the global minimizer of the abscissa

It is instructive to understand the set in the (a_1, a_3, a_2)-space where the roots of the polynomial (52) are real and negative, but not necessarily simple, which is given by the discriminant surface of the polynomial. A part of it is shown in Figure 15a,b. At the point EP_4 with the coordinates $(4, 4, 6)$ in the (a_1, a_3, a_2)- space the discriminant surface has the *Swallowtail* singularity, which is a generic singularity of bifurcation diagrams in three-parameter families of real matrices [6, 7].

Therefore, the coefficients of the globally minimizing polynomial (57) are exactly at the Swallowtail singularity of the discriminant surface of the polynomial (52).

In the region in side the "spire" formed by the discriminant surface all the roots are simple real and negative. Owing to this property, this region, belonging to the domain of asymptotic stability (see Figure 15a), plays an important role in stability theory. Physical systems with semi-simple real and negative eigenvalues are called *heavily damped*. The solutions of the heavily damped systems do not oscillate and monotonically decrease, which is favorable for applications in robotics and automatic control.

Now we can give the following interpretation of the Bottema stability diagram shown in Figure 15a [71]. The dissipative system with the characteristic polynomial (52) is asymptotically stable inside the domain (51). The boundary of the domain (52) has the Whitney umbrella singular point at $a_1 = 0$, $a_3 = 0$, and $a_2 = 2$.

The domain corresponding to heavily damped systems is confined between three hypersurfaces of the discriminant surface and has a form of a trihedral spire with the Swallowtail singularity at its cusp at $a_1 = 4$, $a_2 = 6$, and $a_3 = 4$. The Whitney umbrella and the Swallowtail singular points are connected by the EP-set given by (55). At the Swallowtail singularity of the boundary of the domain of heavily damped systems, the abscissa of the characteristic polynomial of the damped system attains its global minimum.

Therefore, by minimizing the spectral abscissa one finds points at the boundary of the domain of heavily damped systems. Furthermore, the sharpest singularity at this boundary corresponding to a quadruple real eigenvalue $\lambda = -1$ with the Jordan block of order four is the very point where all the modes of the system with two degrees of freedom are decaying to zero as rapidly as possible when $t \to \infty$.

References

[1] Abbott, B. P., et al.: (LIGO Scientific Collaboration and Virgo Collaboration) GW170817: Observation of gravitational waves from a binary neutron star inspiral. Phys. Rev. Lett. **119**, 161101 (2017)

[2] Alfriend, K. T.: The stability of the triangular Lagrangian points for commensurability of order two. Celest. Mech. **1**(3–4), 351–359 (1970)

[3] Andersson, N.: Gravitational waves from instabilities in relativistic stars. Class. Quantum Grav. **20**, R105–R144 (2003)

[4] Andreichikov, I. P., Yudovich, V. I.: The stability of visco-elastic rods. Izv. Akad. Nauk SSSR. Mekhanika Tverdogo Tela. **9**(2), 78–87 (1974)

[5] Anosov, D. V., Arnold, V. I. (eds.): Dynamical Systems I, Encyclopaedia of Mathematical Sciences, Springer, Berlin, (1988)

[6] Arnold, V. I.: Lectures on bifurcations in versal families. Russ. Math. Surv. **27**, 54–123 (1972)

[7] Arnold, V. I.: Geometrical Methods in the Theory of Ordinary Differential Equations, Springer-Verlag, New York, (1983)

[8] Arnold, V. I. (ed.): Dynamical Systems VIII, Encyclopaedia of Mathematical Sciences, Springer, Berlin, (1993)

[9] Bayly, P. V., Dutcher, S. K.: Steady dynein forces induce flutter instability and propagating waves in mathematical models of flagella. J. R. Soc. Interface **13**, 20160523 (2016)

[10] Berry, M. V., Shukla, P.: Curl force dynamics: symmetries, chaos and constants of motion. New J. Phys. **18**, 063018 (2016)
[11] Banichuk, N. V., Bratus, A. S., Myshkis, A. D.: On destabilizing influence of small dissipative forces to nonconservative systems. Doklady AN SSSR **309**(6), 1325–1327 (1989)
[12] Bialynicki-Birula, I., Kalinski, M., Eberly, J. H.: Lagrange equilibrium points in celestial mechanics and nonspreading wave packets for strongly driven Rydberg electrons. Phys. Rev. Lett. **73**(13), 1777–1780 (1994)
[13] Bialynicki-Birula, I., Charzyǹski, S.: Trapping and guiding bodies by gravitational waves endowed with angular momentum. Phys. Rev. Lett. **121**, 171101 (2018)
[14] Bigoni, D., Noselli, G. Experimental evidence of flutter and divergence instabilities induced by dry friction. J. Mech. Phys. Sol. **59**, 2208–2226 (2011)
[15] Bigoni, D., Kirillov, O. N., Misseroni, D., Noselli, G., Tommasini, M.: Flutter and divergence instability in the Pflüger column: Experimental evidence of the Ziegler destabilization paradox. J. Mech. Phys. Sol. **116**, 99–116 (2018)
[16] Bigoni, D., Misseroni, D., Tommasini, M., Kirillov, O. N., Noselli, G. Detecting singular weak-dissipation limit for flutter onset in reversible systems. Phys. Rev. E **97**(2), 023003 (2018)
[17] Bloch, A. M., Krishnaprasad, P. S., Marsden, J. E., Ratiu, T. S.: Dissipation-induced instabilities. Annales de l'Institut Henri Poincaré **11**(1), 37–90 (1994)
[18] Blondel, V. D., Gürbüzbalaban, M., Megretski, A., Overton, M. L.: Explicit solutions for root optimization of a polynomial family with one affine constraint. IEEE Trans. Autom. Control **57**, 3078–3089 (2012)
[19] Bognár, J.: Indefinite inner product spaces, Springer, New York (1974)
[20] Bolotin, V. V.: Non-conservative Problems of the Theory of Elastic Stability. Fizmatgiz (in Russian), Moscow (1961); Pergamon, Oxford (1963)
[21] Bolotin, V. V., Zhinzher, N. I.: Effects of damping on stability of elastic systems subjected to nonconservative forces. Int. J. Solids Struct. **5**, 965–989 (1969)
[22] Borisov, A. V., Kilin, A. A., Mamaev, I. S.: The Hamiltonian dynamics of self-gravitating liquid and gas ellipsoids. Reg. Chaot. Dyn. **14**(2), 179–217 (2009)
[23] Borisov, A. V., Kilin, A. A., Mamaev, I. S.: A parabolic Chaplygin pendulum and a Paul trap: Nonintegrability, stability, and boundedness. Reg. Chaot. Dyn. **24**(3), 329–352 (2019)
[24] Bottema, O.: On the stability of the equilibrium of a linear mechanical system. Z. Angew. Math. Phys. **6**, 97–104 (1955)
[25] Bottema, O.: The Routh-Hurwitz condition for the biquadratic equation. Indag. Math. **59**, 403–406 (1956)
[26] Bottema, O. Stability of equilibrium of a heavy particle on a rotating surface. Z. angew. Math. Phys. **27**, 663–669 (1976)
[27] Bridges, T. J., Dias, F.: Enhancement of the Benjamin-Feir instability with dissipation. Phys. Fluids **19**, 104104 (2007)
[28] Brouwer, L. E. J.: The motion of a particle on the bottom of a rotating vessel under the influence of the gravitational force, (Dutch). Nieuw Arch. v. Wisk.,

2e reeks **12**, 407–419 (1918) (English transl. in collected works [29], North-Holland Publ. 1976)
[29] Brouwer, L. E. J. : Collected Works vol. 2, Geometry, Analysis, Topology and Mechanics, Freudenthal, H., ed., North-Holland, Amsterdam (1976)
[30] Burke, J. V., Henrion, D., Lewis, A. S., Overton, M. L.: Stabilization via non-smooth, nonconvex optimization. IEEE Trans. Autom. Control **51**(11), 1760–1769 (2006)
[31] Chandrasekhar, S.: Ellipsoidal Figures of Equilibrium. Yale University Press, New Haven (1969)
[32] Chandrasekhar, S.: Solutions of two problems in the theory of gravitational radiation. Phys. Rev. Lett. **24**(11), 611–615 (1970)
[33] Chandrasekhar, S.: On stars, their evolution and their stability. Science **226**(4674), 497–505 (1984)
[34] Cox, G., Levi, M.: Gaussian curvature and gyroscopes. Comm. Pure Appl. Math., **71**, 938–952 (2018)
[35] Crandall, S. H.: The effect of damping on the stability of gyroscopic pendulums. Z. angew. Math. Phys. **46**, S761–S780 (1995)
[36] D'Annibale, F., Ferretti, M.: On the effects of linear damping on the nonlinear Ziegler's column. Nonlinear Dynamics,103, 3149–3164 (2021)
[37] De Canio, G., Lauga, E., Goldstein, R. E.: Spontaneous oscillations of elastic filaments induced by molecular motors. J. R. Soc. Interface **14**, 20170491 (2017)
[38] Dyson, J., Schutz, B. F.: Perturbations and stability of rotating stars-I. Completeness of normal modes. Proc. R. Soc. Lond. A **368**, 389–410 (1979)
[39] Gascheau, G.: Examen d'une classe d'equations differentielles et application á un cas particulier du probleme des trois corps. C. R. Acad. Sci. **16**, 393–394 (1843)
[40] Gladwell, G.: Follower forces – Leipholz early researches in elastic stability. Can. J. Civil Eng. **17**, 277–286 (1990)
[41] Golubitsky, M., Schaeffer, D. G.: Singularities and maps in bifurcation theory. vol. 1, Applied Mathematical Sciences 51, Springer, Berlin (1985)
[42] Golubitsky, M., Schaeffer, D. G., Stewart, I.: Singularities and maps in bifurcation theory. vol. 2, Applied Mathematical Sciences 69, Springer, Berlin (19880
[43] Govaerts, W., Guckenheimer, J., Khibnik, A.: Defining functions for multiple Hopf bifurcations. SIAM J. Numer. Anal. **34**(3), 1269–1288 (1997)
[44] Greenhill, A. G.: On the general motion of a liquid ellipsoid under the gravitation of its own parts. Proc. Cambridge Philos. Soc. **4**, 4–14 (1880)
[45] Greenhill, A. G.: On the strength of shafting when exposed both to torsion and to end thrust. Proc. Inst. Mech. Eng. **34**, 182–225 (1883)
[46] Haller, G.: Gyroscopic stability and its loss in systems with two essential coordinates. Intern. J. of Nonl. Mechs. **27**, 113–127 (1992)
[47] Heisenberg, W.: Über Stabilität und Turbulenz von Flüssigkeitsströmen. Ann. Phys. **379**, 577–627 (1924)

[48] Hoffmann, N., Gaul, L.: Effects of damping on mode-coupling instability in friction-induced oscillations. Z. angew. Math. Mech. **83**, 524–534 (2003)

[49] Holopäinen, E. O.: On the effect of friction in baroclinic waves. Tellus. **13**(3), 363–367 (1961)

[50] Hoveijn, I., Ruijgrok, M.: The stability of parametrically forced coupled oscillators in sum resonance. Z. angew. Math. Phys., **46**, 384–392 (1995)

[51] Hoveijn, I., Kirillov, O. N.: Singularities on the boundary of the stability domain near 1:1-resonance. J. Diff. Eqns, **248**(10), 2585–2607 (2010)

[52] Hoveijn, I., Kirillov, O.: Determining the stability domain of perturbed four-dimensional systems in 1:1 resonance. in "Nonlinear Physical Systems: Spectral Analysis, Stability and Bifurcations" (eds. Kirillov, O. and Pelinovsky, D.), Wiley-ISTE, London, Chapter 8: 155–175 (2014)

[53] Iooss, G., Adelmeyer, M.: Topics in bifurcation theory. World Scientific, Singapore (1992)

[54] Jones, C. A.: Multiple eigenvalues and mode classification in plane Poiseuille flow. Q. J. Mech. appl. Math., **41**(3), 363–382 (1988)

[55] Kirillov, O. N.: How do small velocity-dependent forces (de)stabilize a non-conservative system? DCAMM Report. No. 681. April 2003. 40 pages.

[56] Kirillov, O. N.: How do small velocity-dependent forces (de)stabilize a non-conservative system? Proceedings of the International Conference "Physics and Control". St.-Petersburg. Russia. August 20-22. Vol. 4, 1090–1095 (2003).

[57] Kirillov, O. N.: Destabilization paradox. Doklady Physics. **49**(4), 239–245 (2004)

[58] Kirillov, O. N.: A theory of the destabilization paradox in non-conservative systems. Acta Mechanica. **174**(3-4), 145–166 (2005)

[59] Kirillov, O. N., Seyranian, A. P.: Stabilization and destabilization of a circulatory system by small velocity-dependent forces. J. Sound Vibr. **283**(3-5), 781–800 (2005)

[60] Kirillov, O. N., Seyranian, A. O.: The effect of small internal and external damping on the stability of distributed non-conservative systems. J. Appl. Math. Mech. **69**(4), 529–552 (2005)

[61] Kirillov, O. N.: Destabilization paradox due to breaking the Hamiltonian and reversible symmetry. Int. J. Non-Lin. Mech. **42**(1), 71–87 (2007)

[62] Kirillov, O. N.: Gyroscopic stabilization in the presence of nonconservative forces. Dokl. Math. **76**(2), 780–785 (2007)

[63] Kirillov, O. N.: Campbell diagrams of weakly anisotropic flexible rotors. Proc. of the Royal Society A **465**(2109), 2703–2723 (2009)

[64] Kirillov, O. N., Verhulst, F.: Paradoxes of dissipation-induced instability or who opened Whitney's umbrella? Z. angew. Math. Mech. **90**, 462–488 (2010)

[65] Kirillov, O. N.: Brouwer's problem on a heavy particle in a rotating vessel: wave propagation, ion traps, and rotor dynamics. Phys. Lett. A **375**, 1653–1660 (2011)

[66] Kirillov, O. N.: Nonconservative Stability Problems of Modern Physics. De Gruyter Studies in Mathematical Physics 14, De Gruyter, Berlin, Boston (2013)

[67] Kirillov, O. N.: Stabilizing and destabilizing perturbations of PT-symmetric indefinitely damped systems. Phil. Trans. R. Soc. A **371**, 20120051 (2013)
[68] Kirillov, O. N.: Singular diffusionless limits of double-diffusive instabilities in magnetohydrodynamics. Proc. R. Soc. A **473**(2205), 20170344 (2017)
[69] Kirillov, O. N. Locating the sets of exceptional points in dissipative systems and the self-stability of bicycles. Entropy **20**(7), 502 (2018)
[70] Kirillov, O. N., Levi, M.: A Coriolis force in an inertial frame. Nonlinearity **30**(3), 1109–1119 (2017)
[71] Kirillov, O. N., Overton, M. L.: Robust stability at the swallowtail singularity. Frontiers in Physics **1**, 24 (2013)
[72] Kollàr, R., Miller, P. D.: Graphical Krein signature theory and Evans-Krein functions. SIAM Review **56**(1), 73–123 (2014)
[73] Kopachevskii, N. D., Krein, S. G.: Operator Approach in Linear Problems of Hydrodynamics. Self-adjoint Problems for an Ideal Fluid, Operator Theory: Advances and Applications 1, Birkhauser, Basel (2001)
[74] Krechetnikov, R., Marsden, J. E.: Dissipation-induced instabilities in finite dimensions. Rev. Mod. Phys. **79**, 519–553 (2007)
[75] Krein, M. G.: Topics in differential and integral equations and operator theory, Operator Theory 7, Birkhauser, Basel (1983)
[76] Kuznetsov, Yu. A.: Elements of applied bifurcation theory, Applied Mathematical Sciences 112, Springer, Berlin (2004)
[77] Lamb, H.: On kinetic stability. Proc. R. Soc. London A **80**(537), 168–177 (1908)
[78] Lancaster, P.: Stability of linear gyroscopic systems: a review. Lin. Alg. Appl. **439**, 686–706 (2013)
[79] Langford, W. F.: Hopf meets Hamilton under Whitney's umbrella. In: IUTAM symposium on nonlinear stochastic dynamics. Proceedings of the IUTAM symposium, Monticello, IL, USA, Augsut 26-30, 2002, Solid Mech. Appl. 110, S. N. Namachchivaya, et al., eds., pp. 157–165, Kluwer, Dordrecht (2003)
[80] Lebovitz, N. R.: Binary fission via inviscid trajectories. Geoph. Astroph. Fluid. Dyn. **38**(1), 15–24 (1987)
[81] Leipholz, H.: Stability theory: an introduction to the stability of dynamic systems and rigid bodies. 2nd ed., Teubner, Stuttgart (1987)
[82] Levantovskii, L. V.: The boundary of a set of stable matrices. Uspekhi Mat. Nauk **35**, no. 2(212), 213–214 (1980)
[83] Levantovskii, L. V.: Singularities of the boundary of a region of stability, (Russian). Funktsional. Anal. i Prilozhen. **16**, no. 1, 44–48, 96 (1982)
[84] Levi, M.: Geometrical aspects of rapid vibrations and rotations. Phil. Trans. R. Soc. A **377**, 20190014 (2019).
[85] Lindblom, L., Detweiler, S. L.: On the secular instabilities of the Maclaurin spheroids. Astrophys. J. **211**, 565–567 (1977)
[86] Liouville, J.: Memoire sur les figures ellipsoidales a trois axes inegaux, qui peuvent convenir a l'equilibre d'une masse liquide homogene, douee d'un mouvement de rotation. Journal de mathematiques pures et appliquees. 1re serie, tome **16**, 241–254 (1851)

[87] Luongo, A., Ferretti, M., D'Annibale, F.: Paradoxes in dynamic stability of mechanical systems: investigating the causes and detecting the nonlinear behaviors. Springer Plus. **5**, 60 (2016)
[88] Lyapunov, A. M.: The general problem of the stability of motion (translated into English by A. T. Fuller). Int. J. Control. **55**, 531–773 (1992)
[89] MacKay, R. S.: Stability of equilibria of Hamiltonian systems. In: Nonlinear Phenomena and Chaos, (ed. S. Sarkar), Adam Hilger, Bristol, pp. 254–270 (1986)
[90] MacKay, R. S. Movement of eigenvalues of Hamiltonian equilibria under non-Hamiltonian perturbation. Phys. Lett. A **155**, 266–268 (1991)
[91] Maclaurin, C. A. A Treatise of Fluxions: In Two Books. 1. Vol. 1. Ruddimans (1742)
[92] Maddocks, J., Overton, M. L.: Stability theory for dissipatively perturbed Hamiltonian systems. Comm. Pure and Applied Math. **48**, 583–610 (1995)
[93] McHugh, K. A., Dowell, E. H.: Nonlinear response of an inextensible, cantilevered beam subjected to a nonconservative follower force. J. Comput. Nonlinear Dynam. **14**(3), 031004 (2019)
[94] McIntyre, M. E.: Diffusive destabilisation of the baroclinic circular vortex. Geophys. Astrophys. Fluid Dyn. **1**(1-2), 19–57 (1970)
[95] Meyer, C. O.: De aequilibrii formis ellipsoidicis. J. Reine Angew. Math. **24**, 44–59 (1842)
[96] Montgomery, M.: Hartmann, Lundquist, and Reynolds: the role of dimensionless numbers in nonlinear magnetofluid behavior. Plasma Phys. Control. Fusion **35**, B105–B113 (1993)
[97] Nezlin, M. V.: Negative-energy waves and the anomalous Doppler effect. Soviet Physics Uspekhi. **19**, 946–954 (1976)
[98] Nicolai, E. L.: On the stability of the rectilinear form of equilibrium of a bar in compression and torsion. Izvestia Leningradskogo Politechnicheskogo Instituta. **31**, 201–231 (1928)
[99] O'Reilly, O. M., Malhotra, N. K.,Namachchivaya, N. S.: Some aspects of destabilization in reversible dynamical systems with application to follower forces. Nonlin. Dyn. **10**, 63–87 (1996)
[100] Perego, A. M., Turitsyn, S. K., Staliunas, K.: Gain through losses in nonlinear optics, Light: Science and Applications **7**, 43 (2018)
[101] Phillips, D., Simpson, S., Hanna, S.: Chapter 3 - Optomechanical microtools and shape-induced forces, in: Light Robotics: Structure-Mediated Nanobiophotonics, Glückstad, J., Palima, D. eds., Elsevier, Amsterdam, pp. 65–98 (2017)
[102] Pigolotti, L., Mannini, C., Bartoli, G.: Destabilizing effect of damping on the post-critical flutter oscillations of flat plates. Meccanica **52**(13), 3149–3164 (2017)
[103] Poincaré, H.: Leçons sur les hypothèses cosmogoniques (ed. Henri Vergne), Librairie Scientifique A. Hermann et fils, Paris (1913)
[104] Pontryagin, L.: Hermitian operators in a space with indefinite metric. Izv. Akad. Nauk. SSSR Ser. Mat. **8**, 243–280 (1944)

[105] Riemann, B.: Ein Beitrag zu den Untersuchungen über die Bewegung eines gleichartigen flüssigen ellipsoides. Abh. d. Königl. Gesell. der Wiss. zu Göttingen. **9**, 3–36 (1861)
[106] Roberts, G. E.: Linear stability of the elliptic Lagrangian triangle solutions in the three-body problem. J. Diff. Eqns **182**, 191–218 (2002)
[107] Roberts, R. H., Stewartson, K.: On the stability of a Maclaurin spheroid of small viscosity. Astrophys. J. **137**, 777–790 (1963)
[108] Ruijgrok, M., Tondl, A., Verhulst, F.: Resonance in a Rigid Rotor with Elastic Support. Z. angew. Math. Mech. **73**, 255–263 (1993)
[109] Samantaray, A. K., Bhattacharyya, R., Mukherjee, A.: On the stability of Crandall gyropendulum. Phys. Lett. A **372**, 238–243 (2008)
[110] Sanders, J. A., Verhulst, F., Murdock, J.: Averaging methods in nonlinear dynamical systems. Applied Math. Sciences 59, Springer, Berlin (2007)
[111] Saw, S. S., Wood, W. G.: The stability of a damped elastic system with a follower force. J. Mech. Eng. Sci. **17**(3), 163–176 (1975)
[112] Schindler, J., Li, A., Zheng, M. C., Ellis, F. M., Kottos, T.: Experimental study of active LRC circuits with PT symmetries. Phys. Rev. A **84**, 040101(R) (2011)
[113] Schutz, B. F.: Perturbations and stability of rotating stars-II. Properties of the eigenvectors and a variational principle. Mon. Not. R. astr. Soc. **190**, 7–20 (1980)
[114] Schutz, B. F.: Perturbations and stability of rotating stars-III. Perturbation theory for eigenvalues. Mon. Not. R. astr. Soc. **190**, 21–31 (1980)
[115] Seyranian, A. P.: On stabilization of non-conservative systems by dissipative forces and uncertainty of critical load. Doklady Akademii Nauk. **348**, 323–326 (1996)
[116] Seyranian, A. P., Kirillov, O. N., Mailybaev, A. A.: Coupling of eigenvalues of complex matrices at diabolic and exceptional points. J. Phys. A: Math. Gen. **38**(8), 1723–1740 (2005)
[117] Seyranian, A. P., Mailybaev, A. A.: Multiparameter stability theory with mechanical applications. World Scientific, Singapore (2003)
[118] Shapiro, V. E. The gyro force of high-frequency fields lost by the concept of effective potential. Phys. Lett. A **238**, 147–152 (1998)
[119] Sinou, J.-J., Jezequel, L.: Mode coupling instability in friction-induced vibrations and its dependency on system parameters including damping. Eur. J. Mech. A. **26**, 106–122 (2007)
[120] Smith, D. M.: The motion of a rotor carried by a flexible shaft in flexible bearings. Proc. Roy. Soc. London A **142**, 92–118 (1933)
[121] Sobolev, S. On motion of a symmetric top with a cavity filled with fluid. in: G. V. Demidenko and V. L. Vaskevich (eds.), Selected Works of S. L. Sobolev, pp. 333–382, Springer (2006)
[122] Stewartson, K.: On the stability of a spinning top containing liquid. J. Fluid Mech. **5**, 577–592 (1959)
[123] Sturrock, P. A.: Kinematics of growing waves. Phys. Rev. **112**, 1488–1503 (1958)

[124] Sturrock, P. A.: In what sense do slow waves carry negative energy. J. Appl. Phys. **31**, 2052–2056 (1960)

[125] Sugiyama, Y., Langthjem, M., Katayama, K.: Dynamic Stability of Columns under Nonconservative Forces: Theory and Experiment. Solid Mechanics and its Applications 255, Springer, Berlin (2019)

[126] Sukhov, S., Dogariu, A.: Non-conservative optical forces. Rep. Prog. Phys. **80**, 112001 (2017)

[127] Thom, R.: Structural Stability and Morphogenesis, W. A. Benjamin (1972)

[128] Thomson, W., Tait, P. G.: Treatise on Natural Philosophy, Vol. I, Part I, New Edition, Cambridge Univ. Press, Cambridge, pp. 387–391 (1879)

[129] Thompson, J. M. T.: Instabilities of elastic and spinning systems: concepts and phenomena. Int. J. Bif. Chaos. **27**(09), 1730029 (2017)

[130] Thompson, J. M. T., Virgin, L.: Instabilities of nonconservative fluid-loaded systems. Int. J. Bif. Chaos., in press (2019)

[131] Thomson, W.: On an experimental illustration of minimum energy. Nature **23**, 69–70 (1880)

[132] Thorpe, S. A., Smyth, W. D., Li, L.: The efect of small viscosity and diffusivity on the marginal stability of stably stratified shear flows. J. Fluid. Mech. **731**, 461–476 (2013)

[133] Tommasini, M., Kirillov, O. N., Misseroni, D., Bigoni, D.: The destabilizing effect of external damping: Singular flutter boundary for the Pflüger column with vanishing external dissipation. J. Mech. Phys. Sol. **91**, 204–215 (2016)

[134] Verhulst, F.: Parametric and Autoparametric Resonance. Acta Appl. Math. **70**(1-3), 231–264 (2002)

[135] Verhulst, F.: Perturbation analysis of parametric resonance. Encyclopedia of Complexity and Systems Science, Springer, Berlin (2009)

[136] Verhulst, F.: Brouwer's rotating vessel I: stabilization. Z. angew. Math. Phys. **63**(4), 727–736 (2012)

[137] van Gils, S. A., Krupa, M., Langford, W. F.: Hopf bifurcation with non-semisimple 1:1 resonance. Nonlinearity **3**, 825–850 (1990)

[138] von Laue, M.: Die Fortpflanzung der Strahlung in dispergierenden und absorbierenden Medien. Ann. Physik **18**(4), 523–566 (1905)

[139] Whitney, H.: The general type of singularity of a set of $2n-1$ smooth functions of n variables. Duke Math. J. **10**, 161–172 (1943)

[140] Whitney, H.: The singularities of a smooth n-manifold in $(2n-1)$-space. Ann. of Math. **45**(2), 247–293 (1944)

[141] Williamson, J.: On the algebraic problem concerning the normal forms of linear dynamical systems. American Journal of Mathematics. **58**, 141–163 (1936)

[142] Williamson, J.: On the normal forms of linear canonical transformations in dynamics. American Journal of Mathematics. **59**, 599–617 (1937)

[143] Willcocks, B. T., Esler, J. G.: Nonlinear baroclinic equilibration in the presence of Ekman friction. J. Phys. Oceanogr. **42**, 225–242 (2012)

[144] Yakubovich, V. A., Starzhinskii, V. M.: Linear differential equations with periodic coefficients. 2 vols., John Wiley, New York (1975)

[145] Yang, C., Khesin, B.: Averaging, symplectic reduction, and central extensions. Nonlinearity 33, 1342–1365 (2020).
[146] Yurkin, M. Y.: The finite dimension property of small oscillations of a top with a cavity filled with an ideal fluid. Funct. Anal. Appl. **31**, 40–51 (1997)
[147] Ziegler, H.: Die Stabilitätskriterien der Elastomechanik. Ing.-Arch. **20**, 49–56 (1952)
[148] Ziegler, H.: Linear elastic stability: A critical analysis of methods. Z. Angew. Math. Phys, **4**, 89–121 (1953)
[149] Zhang, R., Qin, H., Davidson, R. C., Liu, J., Xiao, J.: On the structure of the two-stream instability-complex G-Hamiltonian structure and Krein collisions between positive- and negative-action modes. Phys. Plasmas **23**, 072111 (2016)
[150] Zhang, R., Qin, H., Xiao, J.: PT-symmetry entails pseudo-Hermiticity regardless of diagonalizability. J. Math. Phys. 61, 012101 (2020)

Part II
Data Analysis and Finance

Industrial problems today are faced with huge amounts of data. One main task is to discover or dig for useful information, which may unveil structural properties of the system, which have to be represented in models of the system, which nowadays have to combine physical reasoning with data-driven approaches. Although data analysis usually is regarded as an emerging discipline within computer science, new mathematics has been essential in its progress. The contributions in this part on topologicial data analysis, functional analysis and quantization define examples for the latter, with applications in finance and energy markets.

The first chapter in this part on data analysis and finance written by Jean-Daniel Boissonnat and Frédéric Chazal introduces into the field of topological data analysis, which aims at inferring and exploiting interesting topological and geometric structures hidden in given data. Characterizing and exploiting such structures is a challenging problem that asks for new mathematics and that is motivated by a real need from industrial applications.

The second chapter on functional data analysis is written by a group of Spanish scientists, Manuel Febrero–Bande, Wenceslao González–Manteiga and Manuel Oviedo de la Fuente. Starting point is the observation that variables used in energy market applications such as the price and the energy demand are based on quite complex information due to the huge amount and/or high frequency of the data. Functional data analysis is based on functional descriptions evaluated in fixed intervals and thus able to exploit the continuous nature of the information in a better way than . classical methods such as time series models. This approach has not only been successfully to price and energy demand prediction, but also to other fields inside industrial environment: quality control, environmental monitoring and predictive maintenance among others industrial applications.

The last contribution in this part by Jörg Kienitz, Thomas McWalter, Ralph Rudd and Eckard Platen defines a joint work over three continents, Europe, Africa and Australia. The authors nicely show that a classical mathematical tool successfully developed for and applied to one specific field of applications can be generalized to and applied in a rather different field of application by creatively combining different tools and ideas in a new manner: vector quantization (VQ), a method originally devised for lossy signal compression, if applied to probability distributions, defines the starting point of a new technique, recursive marginal quantization, which allows for fast and reliable (for example, positivity preserving) numerical approximation of stochastic differential equations in computational finance.

Topological Data Analysis

Jean-Daniel Boissonnat and Frédéric Chazal and Bertrand Michel

Abstract It has been observed since a long time that data are often carrying interesting topological and geometric structures. Characterizing such structures and providing efficient tools to infer and exploit them is a challenging problem that asks for new mathematics and that is motivated by a real need from applications. This paper is an introduction to Topological Data Analysis (TDA), a new field that emerged during the last two decades with the objective of understanding and exploiting the topological structure of modern and complex data. The paper surveys some important mathematical and algorithmic developments in TDA as well as software solutions that are currently used to address various applied and industrial problems.

1 Introduction

The recent years have seen all domains of science, economy and even everyday life overwhelmed with massive amounts of data. Bringing scientists, industrials and citizens to the most relevant, often unexpected, features and giving them the tools to discover and extract the best knowledge out of their data are fundamental challenges for our modern society.

During the last decades, the wide availability of measurement devices and simulation tools has not only led to an explosion in the amount of available data, but also in a spectacular increase in their complexity, making their analysis more and more challenging: data are often represented as points in a high dimensional space, or as

J.-D. Boissonnat
Université Côte d'Azur Inria e-mail: `jean-daniel.boissonnat@inria.fr`

F. Chazal
Inria Saclay e-mail: `frederic.chazal@inria.fr`

B. Michel
Ecole Centrale de Nantes e-mail: `Bertrand.Michel@ec-nantes.fr`

M. Günther, W. Schilders (eds.), *Novel Mathematics Inspired by Industrial Challenges*, Mathematics in Industry 38, https://doi.org/10.1007/978-3-030-96173-2_9

complex objects like a 2D or 3D image, a meshed shape, a multivariate time series, a graph...

This challenge has led to the development of a wide variety of new mathematical theories and tools among which topology and geometry have recently shown to be particularly relevant. Indeed, a closer look at data often shows that they carry geometric and topological patterns and structures that are immensely helpful in analyzing the systems and phenomena from which they have been generated and that can be used for further machine learning tasks - see Figure 1 for a concrete illustrative example. From this observation, made by practitioners in many academic and industrial domains, emerged the need to develop new mathematical and effective tools to capture and exploit topological information from data. This gave rise to a new research domain known as Topological Data Analysis (TDA). Although one can trace back geometric approaches in data analysis quite far in the past, TDA emerged from various works in applied (algebraic) topology and computational geometry during the first decade of the century. It really started as a field with the pioneering works of [52] and [79] in persistent homology and was popularized in a landmark paper in 2009 [21]. It is interesting to mention that persistent homology was already introduced and developed earlier, in some restricted setting, in the 90's by [55, 74] under the name of size theory and, in a pure mathematics context, by [4] under the name of framed Morse complex.

TDA aims at designing and providing relevant and efficient methods to infer, analyze and exploit the complex topological and geometric structures underlying data that are often represented as point clouds in Euclidean or more general metric spaces. During the last few years, considerable efforts have been made to settle the mathematical and algorithmic foundations of TDA and to provide robust, efficient and easy to use software such as the Gudhi library (C++ and Python) [64] and its R software interface [53]. Although this new mathematical field is still young and evolves rapidly, TDA already provides a set of mature and efficient tools that are complementary to other tools in data science. Adding them to the toolbox of the data scientist turned out to be remarkably useful in several applications and industrial problems.

About this paper

This paper is not a survey on TDA and does not aim at presenting an exhaustive overview of the field and its numerous applications. Its goal is to explain and illustrate, through a few selected topics reflecting the experience of the authors in the field, and without technical details, how the need to understand and exploit the topological structure of data has led to new mathematical ideas. The resulting developments are playing an essential role to bring TDA from a set of ad-hoc or heuristic methods to a well-founded field. They also contribute to bring experimental algorithms and prototype codes to efficient software for industrial applications.

To avoid the introduction of too technical notions and digressions, and make the paper as easy-to-read as possible for non experts, a brief glossary recalling some

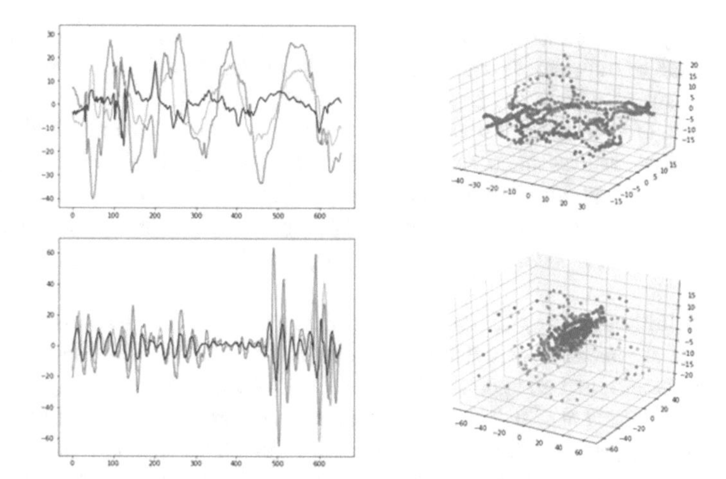

Fig. 1: The left plots represent the 3D acceleration of the arm of a subject making specific disordered movements, measured by an inertial device developed by Sysnav, a French SME, during a short interval of time. The 3 colored curves correspond to the 3 coordinates of the acceleration vector measured in the device coordinate system. The right point clouds are plots of the trajectory of the acceleration vector in $\mathbb{R}^3$. It shows that the "chaotic" behavior of the considered multivariate time-series carries interesting topological patterns that are typical of the movements. The aim of TDA is to provide relevant and efficient tools to characterize qualitatively and quantitatively such patterns in order to exploit them -in combination with other features - for further analysis of the movement.

classical but important definitions has been included in the last section of the paper[1]. References to specific aspects of TDA are given all along the paper. Additionally, the reader can find several surveys [47, 77] and textbooks [12, 51, 56, 68], covering different aspects of the field.

2 The need of new mathematical and algorithmic tools

Formalizing the notion of topological and geometric structure of data is a tedious question. Inferring relevant topological and geometric information from data raises difficult problems requiring new mathematical tools. The reasons for these difficulties are many but are closely related to the three fundamental following facts.

- First of all, data are discrete, i.e. finite sets of observations, while topological and geometric quantities are usually associated to continuous shapes. It is thus

[1] The first occurences, in the paper, of the notions defined in the glossary are put in italic in the text

necessary to introduce intermediate geometric models that both faithfully approximate or summarize the data and carry information about the underlying shapes around which they have been sampled. The highly non linear nature of the space of such models makes the approximation theory for topological and geometric invariants much more difficult than the classical approximation theories for functions. This problem has been addressed from different perspectives among which the distance-based approaches, presented in the next section, have given rise to new mathematical developments. This is still an active research area.

- Second, although they appear to be concentrated around geometric shapes, real data are usually corrupted by noise and outliers. It is also often observed that the topological and geometric features underlying data are strongly dependent on the scale at which they are considered. Quantifying and distinguishing topological/geometric noise from topological/geometric signal to infer relevant scale-dependent or multi-scale information is a subtle problem that does not benefit from the standard signal processing and statistical tools. It requires the development of new tools and approaches. Persistent homology, presented in Section 3.2, emerged at the beginning of the century as a new approach to address such problems. Since then it has grown as a new mathematical theory with impressive developments going from fundamental mathematics to algorithms and concrete applications.
- Third, even when the data are concentrated around low dimensional shapes, the possibly high dimensionality of the spaces in which they are embedded raises severe algorithmic and practical issues. Classical data structures and algorithms from computational topology and geometry quickly become inefficient in practice when the dimensionality of the data increases. Although these questions, at the crossing of computer science and mathematics, are not discussed in details in this paper, they are of fundamental importance to provide efficient and implemented software tools. They are subject to intense research activities.

3 The emergence of geometric inference and persistent homology

Many approaches have been proposed, to address the difficulties mentioned above. However, since these difficulties are often driven by the constraints and needs of specific applications, the solutions are mostly ad-hoc and are not easy to generalize or exploit in other settings. During the last two decades many efforts have been made to address them in general mathematical frameworks that have led to new mathematical developments and significant progress on the practical side.

The general problem underlying TDA can be roughly summarized in the following ill-posed question: given a finite set of points $\mathbb{X}_n = \{x_0, \cdots, x_n\}$ in $\mathbb{R}^d$, or in a more general metric space, is it possible to reliably and efficiently estimate topological and geometric properties reflecting the global structure of $\mathbb{X}_n$?

This question can be more precisely formalized when the points of $\mathbb{X}_n$ are assumed to be sampled around a given compact subset K. In that case, it boils down to the estimation of topological or geometric invariants of K. However, assuming the existence of a well-defined shape underlying the data to which the estimated topological invariants could be compared, is in many cases a too restrictive assumption. Then, the estimation of geometric and topological invariants is relevant only if one can establish that the invariants remain stable under perturbations of the input data.

3.1 Distance-based geometric inference

An intuitive way to associate a continuous geometric structure to a discrete data set is to consider union of balls centered on the data points. Given a compact subset K of $\mathbb{R}^d$, and a non negative real number r, the union of balls of radius r centered on K, $K^r = \cup_{x \in K} B(x,r) = d_K^{-1}([0,r])$, called the r-offset of K, is the r-sublevel set of the distance function $d_K : \mathbb{R}^d \to \mathbb{R}$ defined by $d_K(x) = \inf_{y \in K} \|x-y\|$. The idea underlying distance-based approaches is, instead of directly comparing $\mathbb{X}_n$ and K, to compare the topology of the foliations defined by the level sets of d_K and $d_{\mathbb{X}_n}$.

This is made possible thanks to a fundamental property of the squared distance function: for any compact set K, d_K is semi-concave, i.e. $x \to \|x^2\| - d_K^2(x)$ is convex. Distance functions inherit interesting differential properties from semi-concavity that allow to relate the topology of the offsets of compact sets that are close to each other with respect to the *Hausdorff distance* $d_H(.,.)$ between compact sets. For example, when K is a smooth and compact submanifold of $\mathbb{R}^d$, this leads to a basic method to reliably estimate the topology (homotopy type) and the homology groups of K from well-chosen offsets of $\mathbb{X}_n$ under explicit mild conditions on $d_H(\mathbb{X}_n, K)$ [45, 67]. As geometric structures underlying data are not always as regular as smooth manifolds, the result has been extended to a larger class of non smooth compact sets K and led to stronger results on the inference of the topology of the offsets of K [32]. This approach, based on the study of distance functions, also led to results on the estimation of other geometric and differential quantities of K such as normal cones [31], curvature measures [33, 34] and boundary measures [35]. Some of these estimators were adapted to find applications for feature detection - such as sharp edges or corners - on 3D shapes [65].

Covers and nerves to compute the topology of union of balls

From an algorithmic perspective, the advantage of estimating topological features of data from union of balls is that, thanks to the so-called *Nerve Theorem* in algebraic topology, their topology is fully described by an *abstract simplicial complex* (a purely combinatorial structure) encoding the intersection patterns of the balls. More precisely, given a union of balls in Euclidean space, $\bigcup_{i=1}^n B(x_i, r_i)$, connecting the data points x_i, x_j by edges whenever the two corresponding balls $B(x_i, r_i), B(x_j, r_j)$

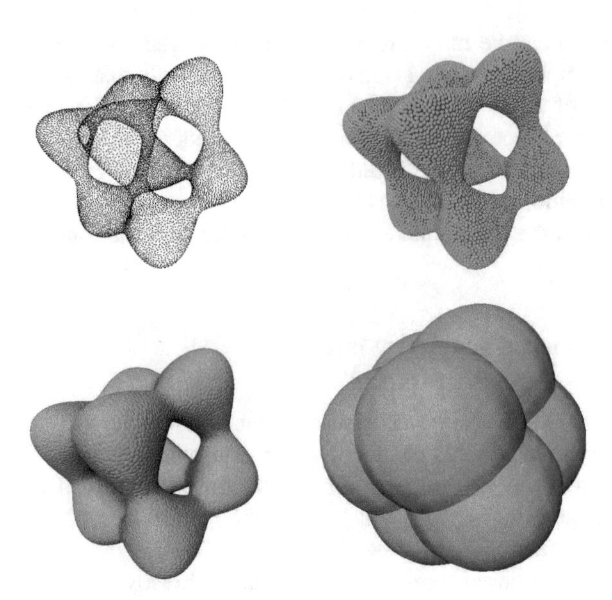

Fig. 2: The level sets of the distance function to a point cloud sampled on a smooth surface (upper left picture), the tangle cube defined by the implicit equation $x^4 - 5x^2 + y^4 - 5y^2 + z^4 - 5z^2 + 11.8 = 0$, for different values. The offset on the upper left picture has the same topology as the sampled surface.

intersect gives rise two a graph with the same connectivity as the union of balls. To go beyond connectivity, one can also connect $(k+1)$-uple of points $x_{i_0}, \cdots x_{i_k}$ whenever $\bigcap_{j=0}^{k} B(x_{i_j}, r_{i_j}) \neq \emptyset$ - see Figure 3. The resulting object, called the *nerve* of the union of balls is a *simplicial complex* and allows to identify higher dimensional topological features such as cycles, voids and their higher dimensional counterparts. It follows from the Nerve Theorem that the homotopy type of $\bigcup_{i=1}^{n} B(x_i, r_i)$ is determined by its nerve. As a consequence, many topological invariants such as the *homology groups*, the *Betti numbers* and the Euler characteristic of a union of ball can be efficiently computed from its nerve.

Indeed, simplical complexes play a fundamental role in TDA where they are widely used to infer topological features from data. On one hand, they are classical well-studied topological objects that are well adapted to model geometric structures from data. One the other hand, they are combinatorial objects that are building blocks of most TDA algorithms. As a consequence, they are perfectly well-suited to bridge the gap between continuous spaces and discrete data structures that can be processed by computer programs.

The previous approach faces two issues. First, computing a union of balls in high dimensions is very difficult and takes time that is exponential in the ambient dimension even if the object of interest is low dimensional. Second, only the homotopy type of the geometric shape can be recovered, not a triangulation homeomorphic to

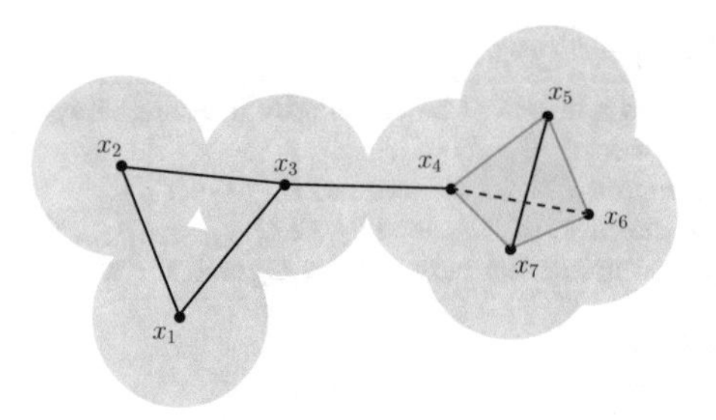

Fig. 3: The nerve of a union of balls in $\mathbb{R}^2$. Notice that, although the balls are included in $\mathbb{R}^2$, their nerve contains the tetrahedron $[x_4, x_5, x_6, x_7]$ and cannot be embedded in $\mathbb{R}^2$: this is an abstract simplicial complex.

the shape. These restrictions can be removed when the shape is a manifold and the amount of noise in the data is limited. In particular, the tangential Delaunay complex [9, 13] can construct a triangulation of a manifold from a dense sample in time only linear in the dimension of the ambient space. This result has been further studied in a statistical context, proving that the reconstruction is minimax optimal [1].

Another use of covers and nerves: the Mapper algorithm

Nerves are more generally defined for any family of subsets covering a data set. They provide, in many cases, an interesting way to summarize, visualize and explore the topological structure of data. This natural idea was first proposed for TDA in [72], giving rise to the so-called Mapper algorithm 1. Despite its simplicity, this algorithm has been successfully used in a large variety of problems as an exploratory tool for clustering and identify important features of data - see, for example, [63, 78]. It has also given birth to a successfull company, Ayasdi (`https://www.ayasdi.com/`) that have made it one of its core technology. Despite its successes, the Mapper algorithm remains an exploratory tool that is very sensitive to various parameters requiring a tedious involvement of the user to be correctly tuned. Overcoming this problem to make Mapper a fully automated tool is an active mathematical research topic. Based on preliminary results on the stability of Mapper proposed in [27], advances towards a statistically well-founded version of Mapper have been obtained recently in [26], but new mathematical ideas are necessary to go beyond these preliminary results and make significant progress.

Distance-based inference with noisy data

Real world data are often corrupted by noise and the distance-based approach may fail completely in the presence of outliers. Indeed, adding even a single outlier to the point cloud can dramatically change the distance function and make the above methods fail.

Algorithm 1 The Mapper algorithm

Input: A data set $\mathbb{X}$ with a metric or a dissimilarity measure between data points, a function $f : \mathbb{X} \to \mathbb{R}$ (or $\mathbb{R}^d$), and a cover $\mathscr{U}$ of $f(\mathbb{X})$.
for each $U \in \mathscr{U}$, decompose $f^{-1}(U)$ into clusters $C_{U,1}, \cdots, C_{U,k_U}$.
Compute the nerve of the cover of X defined by the $C_{U,1}, \cdots, C_{U,k_U}$, $U \in \mathscr{U}$
Output: a simplicial complex, the nerve (often a graph for well-chosen covers $\to$ easy to visualize):
- a vertex $v_{U,i}$ for each cluster $C_{U,i}$,
- an edge between $v_{U,i}$ and $v_{U',j}$ iff $C_{U,i} \cap C_{U',j} \neq \emptyset$

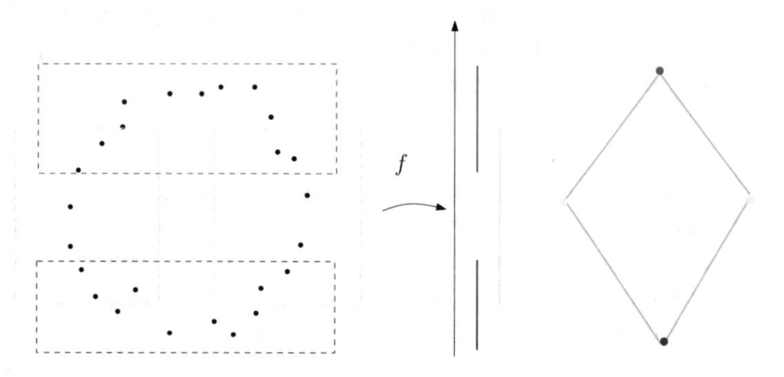

Fig. 4: The mapper algorithm applied to a point cloud sampled around a circle and the height function.

To circumvent this problem, we adopted a measure theoretic approach and introduced an alternative distance-like function, the distance-to-a-measure (DTM) [36]. DTM is robust to noise and its sublevel sets still carry relevant topological information.

Given a probability distribution P in $\mathbb{R}^d$ and a real parameter $0 \leq u \leq 1$, the notion of distance to the support of P may be generalized as the function

$$\delta_{P,u} : x \in \mathbb{R}^d \mapsto \inf\{t > 0; P(B(x,t)) \geq u\},$$

where $B(x,t)$ is the closed Euclidean ball of center x and radius t. Intuitively, $\delta_{P,u}(x)$ is the smallest ball centered at x that contains a fraction u of the total mass of the probability distribution P. To avoid issues due to discontinuities of the map $P \to \delta_{P,u}$, the distance-to-measure function (DTM) with parameter $m \in [0,1]$ is then defined by averaging the functions $\delta_{P,u}$ for $u \in [0,m]$:

$$d_{P,m}(x) : x \in \mathbb{R}^d \mapsto \left(\frac{1}{m} \int_0^m \delta_{P,u}^2(x) du \right)^{1/2}. \tag{1}$$

A nice property of the DTM proved in [36] is its stability with respect to perturbations of P in the Wasserstein metric. More precisely, the map $P \to d_{P,m,r}$ is $m^{-\frac{1}{r}}$-Lipschitz, i.e. if P and $\widetilde{P}$ are two probability distributions on $\mathbb{R}^d$, then

$$\|d_{P,m} - d_{\widetilde{P},m}\|_\infty \leq m^{-\frac{1}{2}} W_2(P,\widetilde{P}) \tag{2}$$

where W_2 is the Wasserstein distance [2] for the Euclidean metric on $\mathbb{R}^d$, with exponent 2. This property implies that the DTM associated to close distributions in the Wasserstein metric have close sublevel sets. Moreover, the function $d^2_{P,m}$ is semiconcave ensuring strong regularity properties on the geometry of its sublevel sets. Using these observations, [36] show that, under general assumptions, if $\widetilde{P}$ is a probability distribution approximating P, then the sublevel sets of $d_{\widetilde{P},m,2}$ provide a topologically correct approximation of the support of P.

In practice, the measure P is usually only known through a finite set of observations $\mathbb{X}_n = \{x_1,\ldots,x_n\}$ sampled from P, the DTM, for $m = k/n$, can be approximated using the empirical measure P_n instead of P which is defined by the simple following formula,

$$d^r_{P_n,k/n,r}(x) := \frac{1}{k} \sum_{j=1}^{k} \|x - \mathbb{X}_n\|^r_{(j)},$$

where $\|x - \mathbb{X}_n\|_{(j)}$ denotes the distance between x and its j-th neighbor in $\{x_1,\ldots,x_n\}$. This quantity can be easily computed in practice since it only requires the distances between x and the sample points, making it well-adapted for algorithmic and practical use. Moreover, its convergence properties and, more generally approximations of the DTM, have been carefully studied [20, 40, 46, 57, 69]. The practical use of the DTM raises the question of the choice of the parameter m that has been partly addressed in [40, 46] but still remains largely open. New mathematical ideas will probably be necessary to provide a complete and practically rigorous answer to this problem. The study and results on DTM have led to further works and applications in various directions such as topological data analysis [19], GPS traces analysis [28], density estimation [6], hypothesis testing [17] or clustering [44]. The use of the DTM in the setting of an industrial research project, with the Japanese company Fujitsu, whose goal was to address an anomaly detection problem from inertial sensor data in bridge and building monitoring [60] has also recently led to new mathematical developments on the inference of relevant topological features from data with noise and outliers [3].

3.2 Persistent homology

Distance-based inference methods presented in the previous section offer a set of efficient methods to estimate topological features (homology, homotopy type) of data, or even reconstruct an approximating shape, at a given scale r, by considering union of balls of radius r or r-sublevel sets of the DTM. Their validity relies on specific sampling assumptions and on the regularity of some fixed shape underlying the data. However, in many real cases, such assumptions are difficult to assert, or

[2] See [76] for a definition of the Wasserstein distance between probability distributions.

are not even satisfied. Moreover, the underlying structure of the data may strongly depend on the scale r at which it is considered as illustrated on Figure 5.

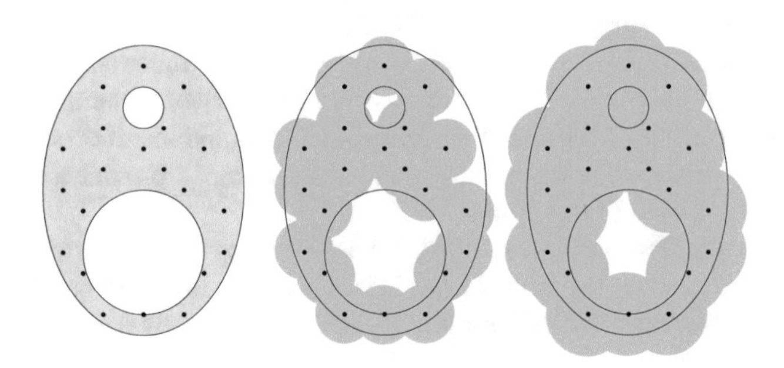

Fig. 5: A sampled 2D domain in $\mathbb{R}^2$ with a small and big holes. Depending on the choice of the radius r, the union of balls centered on the data points may have the topology of a disk with one or two holes. In such a case, selecting the appropriate radius from the data to infer the correct topology is an ill-posed problem.

In some cases, it also happens that none of the offsets of the data have the correct topology as illustrated on Figure 6.

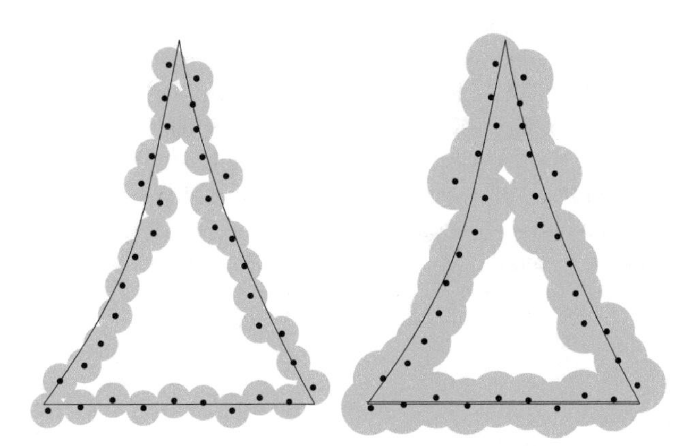

Fig. 6: When the sampling and regularity conditions provided by distance-based inference results are not satisfied, it may happen that there is no correct choice of radius to infer the topology of the underlying shape. Here, none of the union of balls centered on the data point succeeds to recover the topology of the sampled simple curve.

To overcome these problems and give a well-defined and formal meaning to multiscale topological structure of a wide range of data, it was thus necessary to introduce new mathematical ideas and tools. This is where persistent homology, a central concept in TDA, comes into play! Preliminary ideas related to persistent homology can be traced back quite far in the past in pure mathematics [4, 66]. The theory emerged in its present form at the begining of the century [52, 52, 75] as a tool to compute, study and efficiently encode multiscale topological features of data. Persistent homology provides a framework and efficient algorithms to encode the evolution of the *homology* of families of nested topological spaces or simplicial complexes indexed by a set of real numbers, called *filtrations*.

In TDA, two different kinds of filtrations are classically considered. When data come as a point cloud in metric space (not necessarily Euclidean), one can construct various filtered abstract simplicial complexes whose vertex set is the set of data points. An example of such a filtration is the νCech complex filtration, which consists of the nerves of a growing family of balls centered on the data points (see the previous section). In that case, the persistent homology of such a filtration reflects the global topological structure of the data at different scales corresponding to the radii of the considered balls. When data come as a collection of already complex objects such as images, 3D shapes or graphs, functions defined on each of them may be used to highlight some of their features. The persistent homology of the sublevel or upperlevel set filtrations of these functions then provide topological information. This new type of information can be fruitfully used to compare and classify the data elements, and can be combined with other non topological features to enhance the results.

Given a filtration $\mathrm{Filt} = (F_r)_{r \in T}$ of a simplicial complex or a topological space, the homology of F_r changes as r increases: new connected components can appear, existing components can merge, loops and cavities can appear or be filled, etc... Persistent homology tracks these changes, identifies the appearing features and associates a life time to them. The resulting information is encoded as a set of intervals called a *barcode* or, equivalently, as a multiset of points, called a *persistence diagram* in $\mathbb{R}^2$ where the coordinates of each point are the starting and end points of the corresponding intervals, as illustrated on a simple example on Figure 7. The length of the persistence intervals - or equivalently the vertical distance of the persistence point to the diagonal in $\mathbb{R}^2$ - reflects the life span of the topological features along the filtration. Intuitively, the longer an interval is, the more relevant is the corresponding topological feature.

A simple but fundamental observation, at the root of important mathematical developments in the theory of persistent homology, is that persistence diagrams can be defined in a purely algebraic way. Given a filtration $\mathrm{Filt} = (F_r)_{r \in T}$, a non negative integer k, and considering the homology groups $H_k(F_r)$, we obtain a sequence of vector spaces where the inclusions $F_r \subset F_{r'}$, $r \leq r'$ induce linear maps between $H_k(F_r)$ and $H_k(F_{r'})$. Such a sequence of vector spaces together with the linear maps connecting them is called a *persistence module*.

More generally, a persistence module $\mathbb{V}$ over a subset T of the real numbers $\mathbb{R}$ is an indexed family of vector spaces over a field $\mathbb{K}$, $(V_r \mid r \in T)$ and a doubly-indexed family of linear maps $(v_s^r : V_r \to V_s \mid r \leq s)$ which satisfy the composition law $v_t^s \circ v_s^r = v_t^r$ whenever $r \leq s \leq t$, and where v_r^r is the identity map on V_r. In favorable cases, a persistence module can be uniquely decomposed, up to a re-ordering of the

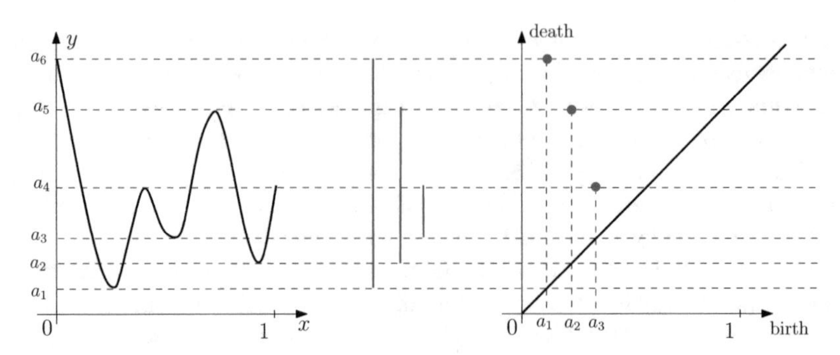

Fig. 7: The persistence barcode and the persistence diagram of the sublevel set filtration $(F_r = f^{-1}((-\infty, r]))_{r\in\mathbb{R}}$ of a function $f : [0,1] \to \mathbb{R}$. All the sublevel sets of f are either empty or a union of interval, so the only non trivial topological information they carry is their 0-dimensional homology, i.e. their number of connected components. At $r = a_1$, persistent homology registers the birth of a first connected component in F_{a_1} and start a first interval. Then, F_r remains connected until r reaches the value a_2 where a second connected component appears and a second interval is created. Similarly, when r reaches a_3, a new connected component appears and a new interval is created. When r reaches a_4, the two connected components created at a_1 and a_3 merges together to give a single larger component. At this step, persistent homology follows the rule that this is the most recently appeared component in the filtration that dies: the interval started at a_3 is thus ended at a_4. When r reaches a_5, the component born at a_2 dies and the persistent interval (a_2, a_5) is created. The interval created at a_1 remains until the end of the filtration giving rise to the persistent interval (a_1, a_6). The obtained set of intervals encoding the span life of the different homological features encountered along the filtration is called the *persistence barcode* of f. Each interval (a, a') can be represented by the point of coordinates (a, a') in $\mathbb{R}^2$ plane. The resulting set of points is called the *persistence diagram* of f.

terms, in a direct sum of irreducible modules of the form $\mathbb{I}_{(b,d)} = (I_r \mid r \in T)$ where $I_r = \mathbb{K}$ if $r \in (b,d)$ and $I_r = 0$ otherwise, and where all the linear maps from $\mathbb{K}$ to $\mathbb{K}$ are the identity. The persistence barcode/diagram of $\mathbb{V}$ is then defined as the set of intervals (b,d) in this decomposition.

The study of general algebraic persistence modules opened a new research direction in mathematics that is knowing various and important developments. In [29, 48], it is proven that the notion of persistence diagram can be defined for a large class of abstract persistence modules, that are not necessarily decomposable. This generalization, which may appear of purely theoretical interest, reveals extremely useful to study the convergence and statistical properties of persistence diagrams and related topological features under various models - see for example [39, 42]. It is interesting to notice that these mathematical developments were initially motivated by an applied problem of topological inference [43]. This problem involvesromov persistence modules that were not the homology groups of a filtration and to which the already existing theory did not applied.

A fundamental property of persistent homology in TDA is that persistence diagrams are stable, ensuring that the topological features of data obtained from persistence diagrams are robust to various types of perturbations. In the case of filtrations defined by the sublevel sets of functions, this property can be stated in the following way.

Theorem 3.1 (Stability of persistence diagrams for functions [49]) *Let $f, g : X \to \mathbb{R}$ be two real-valued continuous functions defined on a compact triangulable topological space X. Then for any integer k,*

$$d_B(\mathrm{dgm}_k(f), \mathrm{dgm}_k(g)) \leq \|f - g\|_\infty = \sup_{x \in M} |f(x) - g(x)|$$

where $\mathrm{dgm}_k(f)$ *(resp.* $\mathrm{dgm}_k(g)$*) is the persistence diagram of the persistence module* $(H_k(f^{-1}(-\infty, r]))|r \in \mathbb{R})$ *(resp.* $(H_k(g^{-1}(-\infty, r]))|r \in \mathbb{R})$*) and* d_B *the* bottleneck distance *between diagrams.*

The above result appeared to be a particular case of a much more general algebraic stability result in persistence theory. This later revealed to be a powerful tool to derive stability results in various settings [48]. In particular, it led to establish the stability of the persistence diagrams of classical filtrations built on top of data sets for perturbations with respect to the Gromov-Hausdorff metric [37].

Persistent homology for machine learning

Thanks to the stability results and to natural invariance properties, persistence diagrams can be used as robust and relevant multiscale topological signatures of data. Moreover, in many cases, persistence diagrams can be easily interpreted by the user since persistence intervals represent the life span of some topological features. This makes them particularly attractive to be used as discriminative features for classification and other machine learning tasks. See [30, 62] for some examples in classification, or [50] for an industrial application to arrhythmia classification that led to a patent application.

However, persistence diagrams cannot be directly used for such tasks: indeed persistence diagrams come as unordered finite sets of points in the plane while most of classical data analysis and machine learning algorithms require vectors as input. This linearization problem has been the subject of an intense research activity during the last few years, motivated in part by concrete applied and industrial problems. Some approaches intend to design representations that preserve some of the stability properties of persistence diagrams, like for example persistence landscapes [18] and images [2], while some others follow more heuristic approaches driven by specific needs, like for example [73]. The construction of kernels on the space of diagrams have also been proposed in order to use persistence diagrams in combination with kernel-based methods such as SVM. See, e.g. [25, 70]. More recent works have also proposed data-driven methods to automatically learn an adapted representation of persistence diagrams for a given task [59]. Despite various applied and industrial successes, these works remain very dependent on a specific considered data or problem and usually require a deep understanding of persistence theory to be applied. This prevents non expert users and engineers to really benefit from the resulting tools. To facilitate the work of non expert users to chose and correctly use the appropriate TDA tools for their problems, there is a need to better understand and analyze, from a mathematical perspective, the existing linearization methods and to provide global frameworks in which they can be compared. This problem is at the core of an industrial research partnership between the authors and the Japanese company Fujitsu, that goes from the most mathematical aspects to concrete industrial problems. It already led to several new tools to overcome this issue [24, 71].

The algorithmic and software challenges of TDA

TDA tools and persistent homology computations raise challenging algorithmic problems due to the size of the combinatorial objects (simplicial complexes and filtrations) involved in computations. Many efforts and progress have been made in computational topology to address these problems and improve the whole pipeline of TDA. On one end of the pipeline, new efficient data structures have been proposed to represent simplicial complexes and filtrations compactly [11, 14, 16]. New algorithms have also been designed to approximate filtrations in a controlled way [15]. Algorithms for computing persistent homology or cohomology have also made remarkable progress [5, 8, 10]. Lastly, efficient algorithms are now available to process persistent diagrams and, in particular, to compute distances between persistent diagrams and to compare and cluster such diagrams [61].

On the practical side, reliable and efficient state-of-the-art software have been developed to support industrial applications of TDA. These software also help mathematical research by providing tools to experiment and explore theoretical ideas and new research directions. Building on the mathematical and algorithmic progress of TDA during the last decade, several individuals and teams working on computational topology have developed their own software - e.g. DIONYSUS [3], PHAT [4], RIPSER [5] - implementing specific methods and functionalities. Taking a more global approach, funded by the ERC project GUDHI, a high quality open source software platform called GUDHI[6] has been developed to provide a unified framework for the central data structures and algorithms in TDA. The library is written in C++ with a Python interface that makes the functionalities of the library easily accessible to data scientists [64]. An interface with R has also been developed [53]. The development of the GUDHI library is a long-term project, supervised by an editorial board, and submitted

[3] http://www.mrzv.org/software/dionysus/

[4] https://bitbucket.org/phat-code/phat

[5] https://github.com/Ripser/ripser

[6] http://gudhi.gforge.inria.fr/

to a rigorous review process. The library is already used by industrial partners of the authors (Sysnav, Fujitsu, IFPEN) and serves as a basis for joint research and concrete developments.

4 New research directions

New research directions in TDA are motivated by the increasing variety of applied and industrial problems where TDA is relevant. Most of them require innovative mathematical ideas and approaches. Providing a global picture of these new research directions is beyond the scope of this paper. We only list a few of them that are of particular importance, both from the mathematical and applied side.

- *Statistical aspects of geometric inference and* TDA*:* the study of inference and estimation problems in TDA from a statistical perspective has started to attract some attention since a few years ago and is now an active research theme. It gave rise to a significant literature that has led down the mathematical foundations of Statistical Topological Data Analysis. Beyond its mathematical interest, this research direction also intend to provide effective new approaches to address concrete problems. As an example, one can mention the study of bootstraping and subsampling strategies for persistent homology. They led to practical tools able to infer relevant topological features from data sets that are too large to be handled by classical TDA algorithms [39, 41]. Another example is the design of confidence intervals and statistical tests [7, 17, 54].
- *Persistent homology in an algebraic framework:* persistent homology, considered in an algebraic and category theory perspective, is knowing impressive developments and generalizations in pure mathematics. They raise deep mathematical questions, reveal unexpected connections with other areas of mathematics like, e.g. symplectic geometry, but they also find their roots in very concrete motivations. As an example, one can mention the study of multidimensional persistence [23] that, instead of considering filtrations indexed by real numbers, consider filtrations with multidimensional indices in $\mathbb{R}^k$, $k > 1$, or zig-zag persistence that consider non increasing sequences of spaces [22].
- TDA *and machine learning:* as already mentioned in the previous section, combining TDA with other machine learning approaches and algorithms is a research direction motivated by applied and industrial problems that currently widely contributes to stimulate TDA. Many applied and experimental works and results demonstrate the interest and the potential of this research direction and there is a real need, to go further, to address it from a mathematical perspective. For example, understanding the structure and the properties of all the existing representations of persistent homology - persistence landscapes [18], persistence images [2], Betti curves [73],... - in a general framework is an important question for practical purposes. Despite recent works in this direction [38], this kind of question remains rather unexplored.

- Software tools: Software tools have made tremendous progress in the recent past. It is expected that this will continue in near future both in terms of new functionalities associated to new theoretical advances and in terms of efficiency. As the field of TDA expand and is becoming an essential tool in modern data analysis, feedback from applied fields and industry will push further experimental research and software development.

5 Conclusion

Giving sense to the notion of topological and geometric structures, that are concepts usually associated to continuous spaces, to discrete data is a natural but important general challenge. Although it has been largely explored for a long time in very particular cases[7], there was a real need to propose new mathematical and algorithmic approaches to address the problem in the more general settings provided by data science and its numerous applied problems. During the last decade, TDA has largely contributed to serve this need by developing a new mathematical research area at the crossing of mathematics, statistics and computer science. An important specificity of TDA is that if it has known applied and industrial successes, this is mainly because it has established its roots in pure and applied mathematics. This enabled to address the challenges at the right level of generality and to develop concepts and models that turned out to be relevant and useful in a large variety of settings. It is also important to underline that part of TDA, still motivated by concrete problems, is also actively expanding towards more theoretical domains resulting in many publications in pure mathematics journals and books.

A brief glossary

- **Hausdorff distance.** Given a compact subset $K \subset \mathbb{R}^d$, the distance function from K, $d_K : \mathbb{R}^d \to [0,+\infty)$, is defined by $d_K(x) = \inf_{y \in K} d(x,y)$. The Hausdorff distance between two compact subsets $K, K' \subset \mathbb{R}^d$ is defined by $d_H(K,K') = \|d_K - d_{K'}\|_\infty = \sup_{x \in \mathbb{R}^d} |d_K(x) - d_{K'}(x)|$.
- **Homotopy type.** Given two topological spaces X and Y, two maps $f_0, f_1 : X \to Y$ are *homotopic* if there exists a continuous map $H : [0,1] \times X \to Y$ such that for all $x \in X$, $H(0,x) = f_0(x)$ and $H(1,x) = f_1(x)$. The two spaces X and Y are said to be *homotopy equivalent*, or to *have the same homotopy type* if there exist two continuous maps $f : X \to Y$ and $g : Y \to X$ such that $g \circ f$ is homotopic to the identity map in X and $f \circ g$ is homotopic to the identity map in Y. Spaces with the same homotopy type have isomorphic homology groups.

---

[7] for example in CAD industry, the construction of numerical geometric models (meshes) of real world shapes from scans or other measurements has been widely studied

- **Geometric and abstract simplicial complexes.** Simplicial complexes can be seen as higher dimensional generalization of graphs. They are mathematical objects that are both topological and combinatorial, a property making them particularly useful for TDA. Given a set $\mathbb{X} = \{x_0, \cdots, x_k\} \subset \mathbb{R}^d$ of $k+1$ affinely independent points, the *k-dimensional simplex* $\sigma = [x_0, \cdots x_k]$ spanned by $\mathbb{X}$ is the convex hull of $\mathbb{X}$. The points of $\mathbb{X}$ are called the *vertices* of σ and the simplices spanned by the subsets of $\mathbb{X}$ are called the *faces* of σ. A *geometric simplicial complex* K in $\mathbb{R}^d$ is a collection of simplices such that:
  $i)$ any face of a simplex of K is a simplex of K,
  $ii)$ the intersection of any two simplices of K is either empty or a common face of both.
  The union of the simplices of K is a subset of $\mathbb{R}^d$ called the underlying space of K that inherits from the topology of $\mathbb{R}^d$. So, K can also be seen as a topological space through its underlying space. Notice that once its vertices are known, K is fully characterized by the combinatorial description of a collection of simplices satisfying some incidence rules.
  Given a set V, an *abstract simplicial complex* with vertex set V is a set $\widetilde{K}$ of finite subsets of V such that the elements of V belongs to $\widetilde{K}$ and for any $\sigma \in \widetilde{K}$ any subset of σ belongs to $\widetilde{K}$. The elements of $\widetilde{K}$ are called the faces or the simplices of $\widetilde{K}$. The combinatorial description of any geometric simplicial K obviously gives rise to an abstract simplicial complex $\widetilde{K}$. The converse is also true: one can always associate to an abstract simplicial complex $\widetilde{K}$, a topological space $|\widetilde{K}|$ such that if K is a geometric complex whose combinatorial description is the same as $\widetilde{K}$, then the underlying space of K is homeomorphic to $|\widetilde{K}|$. Such a K is called a *geometric realization* of $\widetilde{K}$. As a consequence, one can consider simplicial complexes at the same time as combinatorial objects that are well-suited for effective computations and as topological spaces from which topological properties can be inferred.
- **Covers and nerves.** A *cover* $\mathscr{U} = (U_i)_{i \in I}$ of a topological space X is a family of sets U_i such that $X = \cup_{i \in I} U_i$. The *nerve of* $\mathscr{U}$ is the abstract simplicial complex $C(\mathscr{U})$ whose vertices are the U_i's and such that

$$\sigma = [U_{i_0}, \cdots, U_{i_k}] \in C(\mathscr{U}) \text{ if and only if } \bigcap_{j=0}^{k} U_{i_j} \neq \emptyset.$$

The nerve of a union of balls of given radius r centered on a set of points in Euclidean space, or in a more general metric space, is also known as the $\boldsymbol{\nu}$*Cech complex*. The nested family of $\boldsymbol{\nu}$Cech complexes, for $r \geq 0$ is called the $\boldsymbol{\nu}$Cech filtration of the set of points.

Theorem 5.1 (Nerve theorem) *Let $\mathscr{U} = (U_i)_{i \in I}$ be a cover of a topological space X by open sets such that the intersection of any subcollection of the U_i's is either empty or contractible, i.e having the homotopy type of a point. Then, X and the nerve $C(\mathscr{U})$ are homotopy equivalent.*

- **Homology and Betti numbers.** Homology (with coefficient in a given field) is a classical object from algebraic topology that associates to any topological space X, a family of vector spaces, the so-called homology groups $H_k(X)$, $k = 0, 1, \ldots$, each of them encoding k-dimensional topological features of X - see [58] for a formal definition. The kth Betti number of X, denoted $\beta_k(X)$, is the rank of $H_k(X)$. It corresponds to the number of "independent" k-dimensional features of X: for example, $\beta_0(X)$ is the number of connected components of M, $\beta_1(X)$ is the number of independent cycles or tunnels, $\beta_2(X)$ the number of cavities, etc... A fundamental property of homology is that any continuous function $f : X \to Y$ induces a linear map $f_* : H_k(X) \to H_k(Y)$ between homology groups that encodes the way the topological features of X are mapped to the topological features of Y by f. This linear map is an isomorphism when f is an homeomorphism or an homotopy equivalence.
- **Filtrations.** Given a simplicial complex C and a finite or infinite subset $A \subset \mathbb{R}$, a *filtration of* C is a family $(C_\alpha)_{\alpha \in A}$ of subcomplexes of C such that for any $\alpha \leq \alpha'$, $C_\alpha \subseteq C_{\alpha'}$ and $C = \cup_{\alpha \in A} C_\alpha$. Given a topological space X and a function $f : X \to \mathbb{R}$, the *sub level set filtration of* f is the nested family of sublevel sets of f: $(f^{-1}((-\infty, \alpha]))_{\alpha \in \mathbb{R}}$.
- **Bottleneck distance.** Given two persistence diagrams, D and D', the *bottleneck distance* $d_B(D, D')$ is defined as the infimum of $\delta \geq 0$ for which there exists a matching between the diagrams, such that two points can only be matched if their distance is less than δ and all points at distance more than δ from the diagonal must be matched.

References

[1] Aamari, E., Levrard, C.: Stability and Minimax Optimality of Tangential Delaunay Complexes for Manifold Reconstruction. Discrete and Computational Geometry (2018). URL `https://hal.archives-ouvertes.fr/hal-01245479`

[2] Adams, H., Emerson, T., Kirby, M., Neville, R., Peterson, C., Shipman, P., Chepushtanova, S., Hanson, E., Motta, F., Ziegelmeier, L.: Persistence images: a stable vector representation of persistent homology. Journal of Machine Learning Research **18**(8), 1–35 (2017)

[3] Anai, H., Chazal, F., Glisse, M., Ike, Y., Inakoshi, H., Tinarrage, R., Umeda, Y.: Dtm-based filtrations. arXiv preprint arXiv:1811.04757. To appear in the Proc. of the Abel Symposium 2018. (2019)

[4] Barannikov, S.A.: The framed Morse complex and its invariants. Adv. Soviet Math. **21**, 93–115 (1994)

[5] Bauer, U., Kerber, M., Reininghaus, J., Wagner, H.: PHAT – persistent homology algorithms toolbox. Journal of Symbolic Computation **78**, 76–90 (2017)

[6] Biau, G., Chazal, F., Cohen-Steiner, D., Devroye, L., Rodriguez, C.: A weighted k-nearest neighbor density estimate for geometric inference. Electronic Journal

of Statistics **5**, 204–237 (2011)

[7] Blumberg, A., Gal, I., Mandell, M., Pancia, M.: Robust statistics, hypothesis testing, and confidence intervals for persistent homology on metric measure spaces. Foundations of Computational Mathematics **14**(4), 745–789 (2014). DOI 10.1007/s10208-014-9201-4. URL http://dx.doi.org/10.1007/s10208-014-9201-4

[8] Boissonnat, J., Dey, T.K., Maria, C.: The compressed annotation matrix: An efficient data structure for computing persistent cohomology. Algorithmica **73**(3), 607–619 (2015). DOI 10.1007/s00453-015-9999-4. URL https://doi.org/10.1007/s00453-015-9999-4

[9] Boissonnat, J., Dyer, R., Ghosh, A., Oudot, S.Y.: Only distances are required to reconstruct submanifolds. Comput. Geom. **66**, 32–67 (2017). DOI 10.1016/j.comgeo.2017.08.001. URL https://doi.org/10.1016/j.comgeo.2017.08.001

[10] Boissonnat, J., Maria, C.: Computing persistent homology with various coefficient fields in a single pass. In: Algorithms - ESA 2014 - 22th Annual European Symposium, Wroclaw, Poland, September 8-10, 2014. Proceedings, pp. 185–196 (2014). DOI 10.1007/978-3-662-44777-2_16. URL https://doi.org/10.1007/978-3-662-44777-2_16

[11] Boissonnat, J., Maria, C.: The simplex tree: An efficient data structure for general simplicial complexes. Algorithmica **70**(3), 406–427 (2014). DOI 10.1007/s00453-014-9887-3. URL https://doi.org/10.1007/s00453-014-9887-3

[12] Boissonnat, J.D., Chazal, F., Yvinec, M.: Geometric and Topological Inference, vol. 57. Cambridge University Press (2018)

[13] Boissonnat, J.D., Ghosh, A.: Manifold reconstruction using tangential Delaunay complexes. Discrete and computational Geometry (November) (2103)

[14] Boissonnat, J.D., Karthik, C.: An Efficient Representation for Filtrations of Simplicial Complexes. ACM Transactions on Algorithms **14** (2018). URL https://hal.inria.fr/hal-01883836

[15] Boissonnat, J.D., Pritam, S., Pareek, D.: Strong Collapse for Persistence. In: ESA 2018 - 26th Annual European Symposium on Algorithms, pp. 67:1–67:13. Helsinki, Finland (2018). DOI 10.4230/LIPIcs. URL https://hal.inria.fr/hal-01886165

[16] Boissonnat, J.D., Srikanta, K.C., Tavenas, S.: Building Efficient and Compact Data Structures for Simplicial Complexes. Algorithmica (2016). DOI 10.1007/s00453-016-0207-y. URL https://hal.inria.fr/hal-01364648

[17] Brécheteau, C.: The DTM-signature for a geometric comparison of metric-measure spaces from samples (2017)

[18] Bubenik, P.: Statistical topological data analysis using persistence landscapes. Journal of Machine Learning Research **16**, 77–102 (2015)

[19] Buchet, M., Chazal, F., Dey, T.K., Fan, F., Oudot, S.Y., Wang, Y.: Topological analysis of scalar fields with outliers. In: Proc. Sympos. on Computational Geometry (2015)
[20] Buchet, M., Chazal, F., Oudot, S., Sheehy, D.R.: Efficient and robust persistent homology for measures. In: Proceedings of the 26th ACM-SIAM symposium on Discrete algorithms. SIAM. SIAM (2015)
[21] Carlsson, G.: Topology and data. AMS Bulletin **46**(2), 255–308 (2009)
[22] Carlsson, G., De Silva, V.: Zigzag persistence. Foundations of computational mathematics **10**(4), 367–405 (2010)
[23] Carlsson, G., Zomorodian, A.: The theory of multidimensional persistence. Discrete & Computational Geometry **42**(1), 71–93 (2009)
[24] Carrière, M., Chazal, F., Ike, Y., Lacombe, T., Royer, M., Umeda, Y.: A general neural network architecture for persistence diagrams and graph classification. arXiv preprint arXiv:1904.09378 (2019)
[25] Carriere, M., Cuturi, M., Oudot, S.: Sliced wasserstein kernel for persistence diagrams (2017). To appear in ICML-17
[26] Carrière, M., Michel, B., Oudot, S.: Statistical analysis and parameter selection for mapper (2017)
[27] Carrière, M., Oudot, S.: Structure and stability of the 1-dimensional mapper. arXiv preprint arXiv:1511.05823 (2015)
[28] Chazal, F., Chen, D., Guibas, L., Jiang, X., Sommer, C.: Data-driven trajectory smoothing. In: Proc. ACM SIGSPATIAL GIS (2011)
[29] Chazal, F., Cohen-Steiner, D., Glisse, M., Guibas, L., Oudot, S.: Proximity of persistence modules and their diagrams. In: SCG, pp. 237–246 (2009)
[30] Chazal, F., Cohen-Steiner, D., Guibas, L.J., Mémoli, F., Oudot, S.Y.: Gromov-hausdorff stable signatures for shapes using persistence. Computer Graphics Forum (proc. SGP 2009) pp. 1393–1403 (2009)
[31] Chazal, F., Cohen-Steiner, D., Lieutier, A.: Normal cone approximation and offset shape isotopy. Computational Geometry **42**(6), 566–581 (2009)
[32] Chazal, F., Cohen-Steiner, D., Lieutier, A.: A sampling theory for compact sets in euclidean space. Discrete & Computational Geometry **41**(3), 461–479 (2009)
[33] Chazal, F., Cohen-Steiner, D., Lieutier, A., Mérigot, Q., Thibert, B.: Inference of curvature using tubular neighborhoods. In: Modern Approaches to Discrete Curvature, pp. 133–158. Springer (2017)
[34] Chazal, F., Cohen-Steiner, D., Lieutier, A., Thibert, B.: Stability of Curvature Measures. Computer Graphics Forum (proc. SGP 2009) pp. 1485–1496 (2008)
[35] Chazal, F., Cohen-Steiner, D., Mérigot, Q.: Boundary measures for geometric inference. Found. Comp. Math. **10**, 221–240 (2010)
[36] Chazal, F., Cohen-Steiner, D., Mérigot, Q.: Geometric inference for probability measures. Foundations of Computational Mathematics **11**(6), 733–751 (2011)
[37] Chazal, F., De Silva, V., Oudot, S.: Persistence stability for geometric complexes. Geometriae Dedicata **173**(1), 193–214 (2014)
[38] Chazal, F., Divol, V.: The density of expected persistence diagrams and its kernel based estimation. In: 34th International Symposium on Computational

Geometry (SoCG 2018) - LIPIcs-Leibniz International Proceedings in Informatics, vol. 99. Schloss Dagstuhl-Leibniz-Zentrum fuer Informatik (2018)
[39] Chazal, F., Fasy, B., Lecci, F., Michel, B., Rinaldo, A., Wasserman, L.: Subsampling methods for persistent homology. In: D. Blei, F. Bach (eds.) Proceedings of the 32nd International Conference on Machine Learning (ICML-15), pp. 2143–2151. JMLR Workshop and Conference Proceedings (2015). URL http://jmlr.org/proceedings/papers/v37/chazal15.pdf
[40] Chazal, F., Fasy, B.T., Lecci, F., Michel, B., Rinaldo, A., Wasserman, L.: Robust topological inference: Distance to a measure and kernel distance. Journal of Machine Learning Research (2014)
[41] Chazal, F., Fasy, B.T., Lecci, F., Rinaldo, A., Wasserman, L.: Stochastic convergence of persistence landscapes and silhouettes. Journal of Computational Geometry **6**(2), 140–161 (2015)
[42] Chazal, F., Glisse, M., Labruère, C., Michel, B.: Convergence rates for persistence diagram estimation in topological data analysis. Journal of Machine Learning Research **16**, 3603–3635 (2015). URL http://jmlr.org/papers/v16/chazal15a.html
[43] Chazal, F., Guibas, L., Oudot, S., Skraba, P.: Scalar field analysis over point cloud data. Discrete & Computational Geometry **46**(4), 743–775 (2011). DOI 10.1007/s00454-011-9360-x. URL http://dx.doi.org/10.1007/s00454-011-9360-x
[44] Chazal, F., Guibas, L.J., Oudot, S.Y., Skraba, P.: Persistence-based clustering in riemannian manifolds. Journal of the ACM (JACM) **60**(6), 41 (2013)
[45] Chazal, F., Lieutier, A.: Smooth manifold reconstruction from noisy and non uniform approximation with guarantees. Computational Geometry Theory and Applications **40**, 156–170 (2008)
[46] Chazal, F., Massart, P., Michel, B.: Rates of convergence for robust geometric inference. Electron. J. Statist **10**, 2243–2286 (2016)
[47] Chazal, F., Michel, B.: An introduction to topological data analysis: fundamental and practical aspects for data scientists. arXiv preprint arXiv:1710.04019 (2017)
[48] Chazal, F., de Silva, V., Glisse, M., Oudot, S.: The structure and stability of persistence modules. SpringerBriefs in Mathematics. Springer (2016)
[49] Cohen-Steiner, D., Edelsbrunner, H., Harer, J.: Stability of persistence diagrams. Discrete & Computational Geometry **37**(1), 103–120 (2007)
[50] Dindin, M., Umeda, Y., Chazal, F.: Topological data analysis for arrhythmia detection through modular neural networks. arXiv preprint arXiv:1906.05795 (2019)
[51] Edelsbrunner, H., Harer, J.: Computational Topology: An Introduction. AMS (2010)
[52] Edelsbrunner, H., Letscher, D., Zomorodian, A.: Topological persistence and simplification. Discrete Comput. Geom. **28**, 511–533 (2002)
[53] Fasy, B.T., Kim, J., Lecci, F., Maria, C.: Introduction to the R package TDA. arXiv preprint arXiv:1411.1830 (2014)

[54] Fasy, B.T., Lecci, F., Rinaldo, A., Wasserman, L., Balakrishnan, S., Singh, A.: Confidence sets for persistence diagrams. The Annals of Statistics **42**(6), 2301–2339 (2014)
[55] Frosini, P.: Measuring shapes by size functions. In: Intelligent Robots and Computer Vision X: Algorithms and Techniques, vol. 1607, pp. 122–133. International Society for Optics and Photonics (1992)
[56] Ghrist, R.W.: Elementary applied topology, vol. 1. CreateSpace Independent Publishing Platform
[57] Guibas, L., Morozov, D., Mérigot, Q.: Witnessed k-distance. Discrete Comput. Geom. **49**, 22–45 (2013)
[58] Hatcher, A.: Algebraic Topology. Cambridge Univ. Press (2001)
[59] Hofer, C., Kwitt, R., Niethammer, M., Uhl, A.: Deep learning with topological signatures. In: Advances in Neural Information Processing Systems, pp. 1634–1644 (2017)
[60] Laboratories, F.: Estimating the degradation state of old bridges-fijutsu supports ever-increasing bridge inspection tasks with AI technology. Fujitsu Journal (2018). URL `https://journal.jp.fujitsu.com/en/2018/03/01/01/`
[61] Lacombe, T., Cuturi, M., Oudot, S.: Large Scale computation of Means and Clusters for Persistence Diagrams using Optimal Transport. In: NIPS. Montreal, Canada (2018). URL `https://hal.inria.fr/hal-01966674`
[62] Li, C., Ovsjanikov, M., Chazal, F.: Persistence-based structural recognition. In: Computer Vision and Pattern Recognition (CVPR), 2014 IEEE Conference on, pp. 2003–2010 (2014). DOI 10.1109/CVPR.2014.257
[63] Lum, P., Singh, G., Lehman, A., Ishkanov, T., Vejdemo-Johansson, M., Alagappan, M., Carlsson, J., Carlsson, G.: Extracting insights from the shape of complex data using topology. Scientific reports **3** (2013)
[64] Maria, C., Boissonnat, J.D., Glisse, M., Yvinec, M.: The GUDHI library: Simplicial complexes and persistent homology. In: International Congress on Mathematical Software, pp. 167–174. Springer (2014)
[65] Mérigot, Q., Ovsjanikov, M., Guibas, L.: Robust Voronoi-based Curvature and Feature Estimation. In: Proc. SIAM/ACM Joint Conference on Geometric and Physical Modeling, pp. 1–12 (2009)
[66] Morse, M.: Rank and span in functional topology. Annals of Mathematics **41**(2), 419–454 (1940)
[67] Niyogi, P., Smale, S., Weinberger, S.: Finding the homology of submanifolds with high confidence from random samples. Discrete & Computational Geometry **39**(1-3), 419–441 (2008)
[68] Oudot, S.Y.: Persistence Theory: From Quiver Representations to Data Analysis, *AMS Mathematical Surveys and Monographs*, vol. 209. American Mathematical Society (2015)
[69] Phillips, J.M., Wang, B., Zheng, Y.: Geometric inference on kernel density estimates. arXiv preprint 1307.7760 (2014)

[70] Reininghaus, J., Huber, S., Bauer, U., Kwitt, R.: A stable multi-scale kernel for topological machine learning. In: Proceedings of the IEEE Conference on Computer Vision and Pattern Recognition, pp. 4741–4748 (2015)
[71] Royer, M., Chazal, F., Ike, Y., Umeda, Y.: Atol: Automatic topologically-oriented learning. arXiv preprint arXiv:1909.13472 (2019)
[72] Singh, G., Mémoli, F., Carlsson, G.E.: Topological methods for the analysis of high dimensional data sets and 3d object recognition. In: SPBG, pp. 91–100. Citeseer (2007)
[73] Umeda, Y.: Time series classification via topological data analysis. Transactions of the Japanese Society for Artificial Intelligence **32**(3), D–G72_1 (2017)
[74] Verri, A., Uras, C., Frosini, P., Ferri, M.: On the use of size functions for shape analysis. Biological cybernetics **70**(2), 99–107 (1993)
[75] Verri, A., Uras, C., Frosini, P., Ferri, M.: On the use of size functions for shape analysis. Biological Cybernetics **70**(2), 99–107 (1993). DOI 10.1007/BF00200823. URL `http://dx.doi.org/10.1007/BF00200823`
[76] Villani, C.: Topics in Optimal Transportation. American Mathematical Society (2003)
[77] Wasserman, L.: Topological data analysis. Annual Review of Statistics and Its Application **5**, 501–532 (2018)
[78] Yao, Y., Sun, J., Huang, X., Bowman, G.R., Singh, G., Lesnick, M., Guibas, L.J., Pande, V.S., Carlsson, G.: Topological methods for exploring low-density states in biomolecular folding pathways. The Journal of chemical physics **130**(14), 144115 (2009)
[79] Zomorodian, A., Carlsson, G.: Computing persistent homology. Discrete Comput. Geom. **33**(2), 249–274 (2005)

Prediction Models with Functional Data for Variables Related with Energy Production

Manuel Febrero–Bande, Wenceslao González–Manteiga and Manuel Oviedo de la Fuente

Abstract In this chapter, different dynamic regression models designed for the prediction of variables related with energy production, mainly, with variables associated with the price and the energy demand are presented. This type of information is quite complex due to the huge amount and/or high frequency of the data and the new paradigm in industry needs new statistical methods that can be able to treat this class of problems associated with the complexity of the data in an accurate and efficient way. This leads to the emerging context of Object–Oriented Statistics in which Functional Data Analysis is a representative branch. In the first part of the chapter, different models for the prediction of the future values of the response are reviewed. These models have in common that make use of Functional Data Analysis (FDA) i.e. the covariates involved in the regression model are functions evaluated in a fixed interval $[a,b]$. For instance, for predicting the price at certain hour of the next day, the model uses the continuous trajectory along the present day (interval $[0,24]$) of the price or the energy demand. FDA has emerged as an alternative in the industry to classical models (like Time Series models) due to the increasing capacity of the computational resources to store and record information in denser grids. The advantage of FDA over classical models is that it is able to exploit this continuous nature of the information in a better way. The second part of the chapter is devoted to the problem of variable selection. In industrial environments it is common to have a huge set of covariates related with the main goal. A variable selection method

Manuel Febrero-Bande
Department of Statistics, Mathematical Analysis and Optimization, Universidade de Santiago de Compostela, e-mail: manuel.febrero@usc.es

Wenceslao González-Manteiga
Department of Statistics, Mathematical Analysis and Optimization, Universidade de Santiago de Compostela, e-mail: wenceslao.gonzalez@usc.es

Manuel Oviedo de la Fuente
Technological Institute for Industrial Mathematics and Department of Statistics, Mathematical Analysis and Optimization, Universidade de Santiago de Compostela e-mail: manuel.oviedo@usc.es

M. Günther, W. Schilders (eds.), *Novel Mathematics Inspired by Industrial Challenges*,
Mathematics in Industry 38, https://doi.org/10.1007/978-3-030-96173-2_10

designed for mixed covariates of different nature: scalar, multivariate, directional, functional, ... is proposed. The proposal begins with a simple null model and sequentially selects a new variable to be incorporated into the final prediction model. The algorithm have showed quite promising results when applied to real data set and, in particular, here it is applied to price and energy demand prediction. All the previous developments can be applied to other fields inside industrial environment. For instance, Functional Data Analysis is been successfully applying to Quality Control, Environmental Monitoring and Predictive Maintenance among others industrial applications.

1 Introduction

Energy and, in particular, Electricity is one of the resources that most contribute to the development of a country and contribute to the comfort of the citizens of a society. So, the access to the energy is a big issue in all developed countries and this access is closely related with an efficient setting of the electricity price. An important characteristic of the electricity is that it cannot be easily stored for large quantities (like national scales) and so, in every moment as much must be produced as is required. The electricity price, in most of European countries, is the final result of a negotiation among providers, distributors and customers through an organized market that ensures certain principles in the negotiation (the overall process must be open, competitive, transparent and readily accessible). The wholesale electricity market for Spain and Portugal is managed by OMIE (see `http://www.omie.es/`) that must maintain a close coordination with the agents of the market to match supply and demand curves. The whole process of setting the electricity price must be quick and feasible to meet the requirements of the agents. And, of course, the main information for the bids/sells of the agents is the one related with accurate predictions of the electrical loads and the recent evolution of price (and all possible covariates of them). Indeed, electricity is considered a clean energy because it is growing the amount that can be generated using renewable means (wind, flowing water) even though an important part of the demand is also generated using fossil fuels or nuclear reactions. The availability of renewal sources reduces the energy price because this type of generation has certain bonuses in the national regulation. This means that the price may be affected by meteorological covariates that can also affect the total load. For instance, the worst scenario is a calm day (no wind) in a hot summer because the lack of wind does not allow to generate from wind mills and, on the other hand, the load can be notably increased by the use of air-conditioned devices. So, one important objective in the electricity industry is the short term load (and price) forecasting. The short term, traditionally, makes reference for periods between one hour and seven days and more recently, for the 24 hours of the next day. This is an important task for system operators, regulatory agencies and producers.

This is the starting point for our motivating example. Figure 1 shows the daily profile of Price and Electricity Market Demand (both measured hourly) at the Iberian

Energy Market for December, 2014 with a color code that considers two groups: M-F (Monday to Friday) and WEnd (Saturday and Sunday). We are interested in predicting the price or the demand at a certain hour of the next day using the available information. We dispose of many different sources to do this. These sources may include other variables related with energy market or generation, meteorological information, calendar effects or any transformation/filter of the preceding. The nature of the covariates can be different but one important thing here is that the trajectory of the price (or the demand) can be measured every hour or every ten minutes.

Figures 2 and 3 show a small sample of selected covariates. Figure 2 includes some of the energy generation variates (by type) included in the energy generation pool. The total demand can be decomposed as a sum of the different energy generations covering the demand. Figure 3 is a small sample of calendar and meteorological information. Note that the number of variables that may be included in a possible prediction model is rather high (over 150).

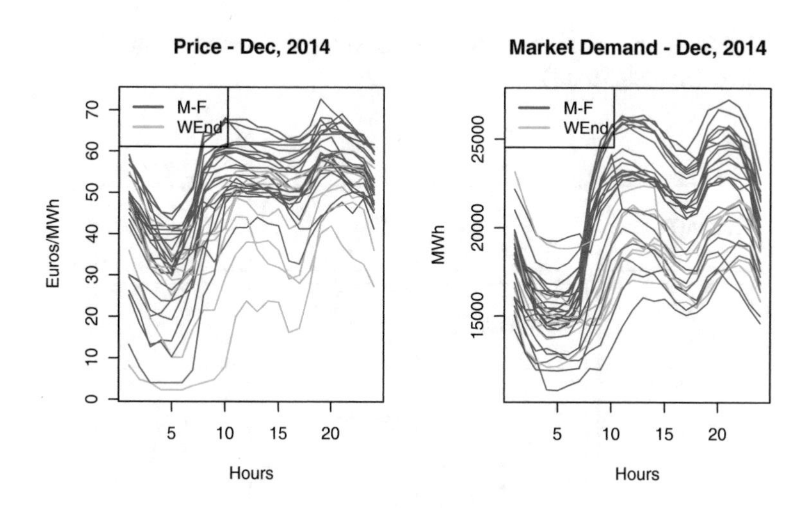

Fig. 1: Price and Energy Market Demand curves negotiated in the Iberian Energy Market. Period: December, 2014 (Source omie.es).

The curves in Figure 2 are the components of the energy pool by generation type, which may be useful for predicting price or demand. Due to the Iberian Energy Market regulations, the odds of becoming part of the final energy pool consumed are unalike among all the types of energy (due to its price or availability). For instance, hydroelectric energy is only offered to the market when the price is high in the presence of founded expectations on refilling the reservoir (using the weather forecasts). During those days, the type Fuel/Gas had little chance to become part of the pool because the demand was completely satisfied with other (cheaper) energy sources.

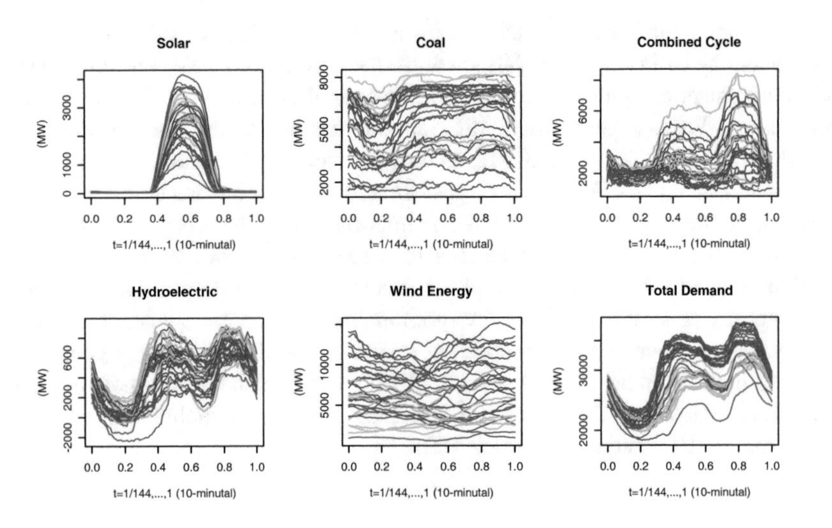

Fig. 2: Daily profile curves of amount of generated energy by type. Period: December, 2014 (Source demanda.ree.es).

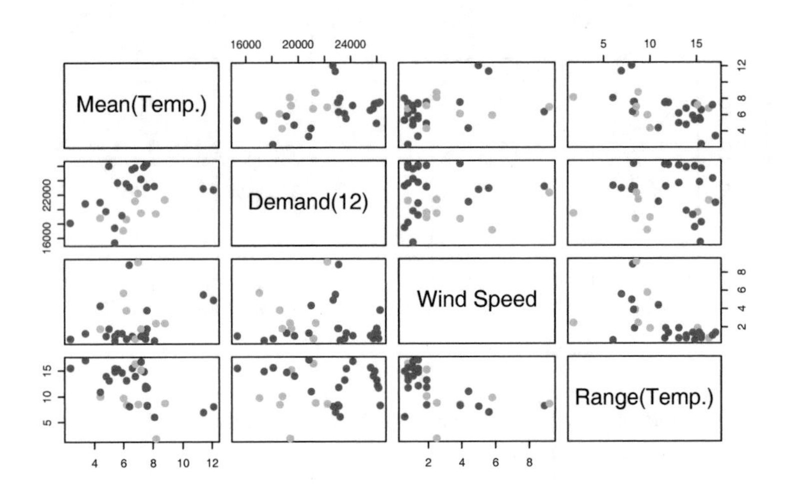

Fig. 3: Examples of available scalar covariates for the energy prediction problem. December, 2014

The scalar covariates for our prediction problem can be linked to calendar effects (like month or day-of-week), meteorological information (like temperature or wind speed) and transformations from the functional variables (such as the market demand at midday $X(t)\mathbb{1}_{\{t=12h\}}$). This means that many new variables can be created from the original ones (for instance, by using derivatives or considering certain subintervals), many of which may certainly be closely related.

Until, about ten years ago, the models designed for the prediction were based on structures using ARIMA time series analysis (see, for instance, [6]). For example, the specification model for a daily time series C_t of electricity consumption on the day t could be the next additive model:

$$\ln C_t = p_t + s_t + CSD_t + CWEA_t + U_t \tag{1}$$

where p_t denotes the trend, s_t the seasonality, CSD_t the contribution of special days, $CWEA_t$ the contribution of meteorological variables and U_t the stationary innovation random term.

Considering in model 1: a) $X_t = p_t + s_t + U_t$ following an ARIMA model $\Phi(L)\Delta(L)X_t = \Theta(L)a_t$, where $\Phi(L)$ and $\Theta(L)$ are polynomials in the lag operator L with all of their roots outside the unit circle, $\Delta(L)$ a polynomial with unit roots and a_t a white noise process, b) CSD_t the time series associated to the special days using dummy variables and c) $CWEA_t$ a time series basically based on the temperature effects, in [8] the model:

$$\ln C_t = \sum_{i=1}^{m} \alpha_i(L)SD_{i,t} + \sum_{j=1}^{n} \beta_j(L)CWEA_{j,t} + \frac{\Theta(L)\Theta_7(L^7)\Theta_{365}(L^{365})}{\Phi(L)\Phi_7(L^7)\Phi_{365}(L^{365})\Delta\Delta_7} a_t \tag{2}$$

was constructed for the forecasting of daily and hourly data for 2006 in *Red Eléctrica de España* (REE), the main government agency for electricity distribution. Here, $SD_{1,t}, ..., SD_{m,t}$ are m dummy variables that define the different classes of special days; $CWEA_{1,t}$, ..., $CWEA_{n,t}$ represent n transformations of the observed meteorological variables and $\alpha_i(L)$, $\beta_j(L)$, $i = 1,\dots,m$ and $j = 1,\dots,n$ are lag polynomials. For more details see the aforementioned paper as an excellent work of reference in the literature making use of Box–Jenkins models.

The above example is only one reference in the collection of statistical models (using dynamic regression, transfer functions, general time series, exponential smoothing, among others). The book by [47] and the reviews [39] and [22] include in one additional way different procedures also based on neural networks, support vector machines, etc.

2 Functional Data Models

In the recent years, the use of functional data has been considered in the model designed for forecasting in the time series related with the electricity production.

The main reasons of this change in the methodology are the big collection of high frequency data (included in a "Big Data" framework) and the new regression models that appeared in the statistical literature since the seminal books for functional data of [34], [18] and more recently [23], [24] and [37]. Functional Data Models are specially suitable for the Industry 4.0 framework where the information comes from multiple automatic sensors that provide signals in almost continuous time that must be related with production variables. So, instead of taking decisions with a vector of variables, the Functional Data framework propose to consider the vector of curves, i.e. the evolution of that signals in the near past to explain the status of the system or to provide future predictions. In this section, we will present several dynamic regression models designed with functional covariates related with energy production.

Under the assumption that the time series of interest (for example, electricity load or price) can be considered high frequency data, with observations every minute or ten minutes, the models given with ARIMA structures (models 1 and 2 above) may not be useful. In this scenario, the data to be used can be got of one stochastic process $\{\xi_t\}_{t\in\mathbb{R}}$, where t is the unit of time of observations. For instance, thinking in hours as unit and considering that such process was observed on the interval of time $[a,b)$ with $b=a+N\tau$, one can be interested in predicting $Y_{N+t}=\xi_{a+N\tau+t}$, $t\in\{1,2,\ldots,\tau=24\}$ using the historical matrix of information given by the functional data $\mathscr{X}=\{X_i=\xi_{a+i\tau+t}, t\in[0,\tau)=[0,24)\}_{i=1}^{N-1}$. Here $\tau=24$ represent the 24 hours of one day, N is the number of functional data observations corresponding to the evolution of the process ξ_t along of N days and the objective is to predict the process in the 24 hours of the $N+1$ day.

In this context the regression models with functional covariable and scalar response can be very useful to explain the dynamics of the process. In the following, having in mind the nature of the prediction problem in time series related with energy production, we will describe some important models introduced in the last years.

1. Functional Linear Regression (FLR) with scalar response and one functional covariate.

$$Y_i=\alpha+\int_a^b \beta(t)X_i(t)dt+\varepsilon_i=\alpha+\langle\beta,X_i\rangle+\varepsilon_i,\ i=1,\ldots,N \quad (3)$$

   where X_i is the trajectory of the process in the ith day, Y_i the value of the process in certain hour of the $(i+1)$day, α the intercept, β the functional parameter of interest to be estimated and ε_i the error of mean 0. β and X_i are belonging to the L^2 Hilbert space in $[a,b)=[0,\tau)$ with $\langle\cdot,\cdot\rangle$ as inner product.
2. Functional Linear Regression with J functional covariates

$$Y_i=\alpha+\sum_{j=1}^{J}\int_{a_j}^{b_j}\beta_j(t)X_i^j(t)dt+\varepsilon_i=\alpha+\sum_{j=1}^{J}\left\langle\beta_j,X_i^j\right\rangle+\varepsilon_i,\ i=1,\ldots,N \quad (4)$$

   Here $\{X^j\}_{j=1}^{J}$ can be the covariates of the evolution of the process in some previous days or linked to the trajectories of other processes (along different intervals) as, for instance, the curve of temperature along certain hours of the precedent days. The functional linear model (models 3 and 4) has been widely studied in the literature and the functional coefficients can be estimated in a different ways, as can been seen in the books aforementioned. The notation $\langle \cdot, \cdot \rangle$ suggests that the extension to incorporate scalar covariates in this model can be done in the same way as in the classical multivariate case.

3. Functional Additive Regression (FAR). The model given by 4 involve some functional predictors but in an inflexible way because it assumes linear relationship between the predictors and the response. Just, as in the context of the standard regression with scalar covariates, more accurate fits can be got by modelling a nonlinear relationship. In this case, a more flexible regression model is given by:

$$Y_i = \alpha + \sum_{j=1}^{J} m_j(X_i^j) + \varepsilon_i, i = 1, \ldots, N \tag{5}$$

When $J = 1$ we have the nonparametric regression model with functional covariate. The estimation for m is given using different procedures based on smoothing techniques: kernel smoothing, splines, etc., as can be seen in the book of [18] and related papers.
For the general case $J \geq 1$, this model represents the generalization of the additive models with scalar covariates and the estimation of the nonparametric components was developed in different ways:

a. [19] uses one cyclic conditional algorithm.
b. [31] uses the functional principal component scores of the functional covariates X_j
c. [13] and [17] develop the special case where $m_j(X^j) = g_j(\langle \beta_j, X^j \rangle)$ being $g_j, j = 1, \ldots, J$ nonparametric link functions.
d. [15] computes the estimation for the nonparametric components based on a generalization of the backfitting algorithm developed in [20].

4. Functional Semi–Linear Models (FSLM). Very often, there exist exogenous scalar variables with the additional information given by the functional covariates. For example, when the objective is the prediction of the electricity demand, the temperature can have a nonlinear effect. In this case, an appropriate model is the Functional partially linear model proposed by [25]:

$$Y_i = \int_a^b \beta(s) X_i(s) ds + g(T_i) + \varepsilon_i, i = 1, \ldots, N \tag{6}$$

where new T_i is a vector of scalar variables and g a vectorial function or, in general, we could have another functional covariate T_i with a nonparametric influence in the answer variable Y_i.
Other possibility is to generalize the nonparametric model with functional covariates incorporating in the regression function a linear component with J exogenous scalar variables $\mathbf{Z} = \{Z^j\}_{j=1}^{J}$:

$$Y_i = \mathbf{Z}_i^\top \beta + m(X_i) + \varepsilon_i, i = 1, \ldots, N \tag{7}$$

This model is also called Functional Semi–Linear Regression introduced and was studied in [2, 3]. The difference among (6) and (7) is the nature of the variable that plays the nonparametric role in the formula: Scalar in the first case and functional in the second one.

A recent paper oriented to Spanish data but with functional methodology is the paper by [45]. Other important papers have illustrated different models with data in different European countries. For example, [9] used an hybrid approach with time series with scalar and functional data in the French electricity loads between 1996 and 2009. See also [5]. In [26], some models are provided to predict spot prices in the German Power market. In [4] models are given to describe the electricity consumption in Sardinia.

So, in summary, energy forecasting covers a wide range of prediction problems related with energy production, where the more recent models have the challenge of obtaining business value from the big amount of information we have actually at hand. One hot point reference of the actual interest in these prediction tasks is the IEEE Working Group of Energy Forecasting (WGEF) who organize the global Energy Forecasting Competitions.

2.1 Application to Iberian Market Energy

To check the weakness and strengths of the precedent models, we have applied some of them to our motivating example. In our case we have information about $N = 2459$ days from August, 2008 to December, 2014. In this set of information, we have daily profiles of price, demand and types of energy that can be considered functional variates jointly with scalar or categorical variables derived from meteorological stations or related with calendar effects. Our aim is to consider lagged versions of information to predict the price or the demand on a certain hour H of a certain day t. In all the examples of this section, we will use the period 2008–2013 as the training sample ($N = 2094$) and the data from 2014 for testing purposes ($n = 365$). For all the applications, we have used the package `fda.usc` ([14]) where most of the functional models described in the previous subsection are included. In our first example, we will try to forecast the price at hour $H = 12$ of day t (with the notation $\mathrm{Pr}_t(H)$) using the daily profile of price one week ($\mathbf{Pr}(t-7)$) and one day before ($\mathbf{Pr}(t-1)$) plus several scalar covariates: Temperature average (T_m), day-of-week (DoW) and wind speed (WS).

The diagnosis of the functional regression model is similar but adapting the classical tools to the new nature of the variates. For instance, the summary of this model (shown after this paragraph) contains the coefficients for each variable showing several rows for categorical and functional variables. In the first case, the rows begining with `DoW` are the classical trick of converting each group of the variate into an indicator whereas for functional variables, the coefficients are showing the representation in a basis chosen by the user. In this case, the two functional covariates are represented using its respective basis of four first principal components.

```
....
>
> Coefficients:
>                  Estimate Std. Error t value Pr(>|t|)
> (Intercept)    53.731447   0.765886  70.156  < 2e-16 ***
> DoW2           -2.268208   0.769031  -2.949 0.003219 **
> DoW3           -2.816168   0.733271  -3.841 0.000126 ***
> DoW4           -2.879316   0.735521  -3.915 9.34e-05 ***
> DoW5           -2.224218   0.728649  -3.053 0.002298 **
> DoW6           -6.435733   0.842593  -7.638 3.34e-14 ***
> DoW7          -10.214512   0.793929 -12.866  < 2e-16 ***
> WSpeed         -0.849461   0.085454  -9.941  < 2e-16 ***
> Tmed            0.158747   0.026840   5.915 3.88e-09 ***
> Pr1.PC1        -0.172459   0.003748 -46.009  < 2e-16 ***
> Pr1.PC2         0.061018   0.008797   6.936 5.37e-12 ***
> Pr1.PC3         0.120083   0.014589   8.231 3.24e-16 ***
> Pr1.PC4        -0.106724   0.021462  -4.973 7.14e-07 ***
> Pr7.PC1        -0.037124   0.003713  -9.997  < 2e-16 ***
> Pr7.PC2        -0.009968   0.008983  -1.110 0.267265
> Pr7.PC3        -0.031959   0.014571  -2.193 0.028398 *
> Pr7.PC4        -0.062523   0.021434  -2.917 0.003572 **
....
```

In the output, the categorical variables are easy to interpret taking into account that Mondays are set as the reference (1=Monday for *DoW*) and the effect of the others are computed against it. The estimates for variate *DoW* tell us that the price on any other day of the week is cheaper than in Monday (see the negative values for `DoW2`,..., `DoW7`). The negative value of WS is simply indicating that the pool of energy in windy days has an increased percentage from windmills which is cheaper than other sources. And finally, the positive value of T_m points out that the demand of energy (and so, the price) is higher in hot days by the use of air conditioned equipments. For functional variates the interpretation is more difficult but possible. We have to reconstruct the β from the coefficients associated to the functional variates to try to interpret this output. In the Figure 4 the estimation of β is provided for two functional regression models using the functional covariates $\mathbf{Pr}(t-1)$ and $\mathbf{Pr}(t-7)$ with and without the scalar covariates (T_m, *DoW* and WS).

To interpret the β parameters associated to functional covariates in Figure 4, we must take as reference the zero line with the mean of the functional covariate in the following sense: If we have a trajectory that it is over the mean in certain interval and the β parameter is also over the zero line then that part of the trajectory contributes with positive values to the response. On the contrary, if the trajectory is over the mean and the β parameter is below the zero line, that contributes with negative values to the response. In this case, and paying attention to **Pr**(t-1), those previous days with price over the mean predict higher prices for $\mathrm{Pr}_t(12)$ (note that except for a small interval, β is always over the zero line). Also the maximums of the $|\beta|$ can be interpreted as the most influencers points on the response. The β parameter associated with

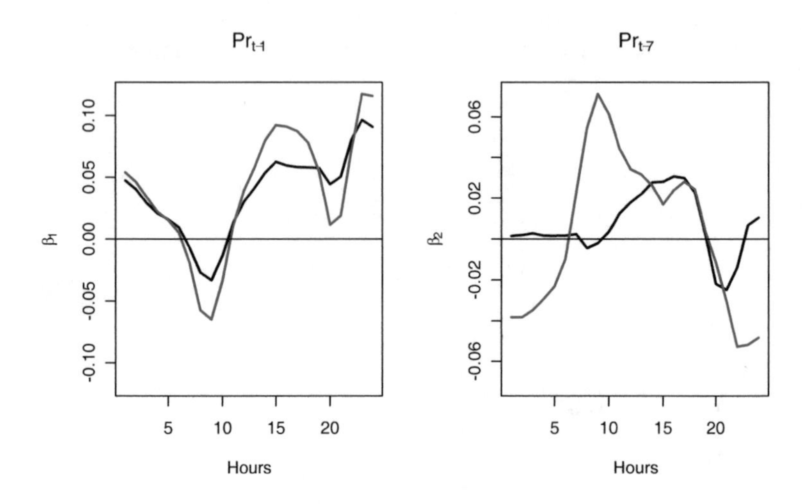

Fig. 4: Estimation of β parameters for functional variates under model 4 using $\mathbf{Pr}(t-1)$, $\mathbf{Pr}(t-7)$ with and without T_m, DoW and WS (resp. red and black) for predicting $\text{Pr}_t(12)$

$\mathbf{Pr}(t-7)$ changes dramatically when the scalar covariates are included in the model on the contrary with the β associated with $\mathbf{Pr}(t-1)$. This means that the effect of $\mathbf{Pr}(t-1)$ over the response is more stable and persistent than the effect of the price one week ago or that the information from $\mathbf{Pr}(t-7)$ is shared with the scalar covariates. A way to assess the relevance of each covariate is to compute the "partial" determination coefficient derived from its contribution, i.e. $\rho\left(Y,\left\langle\widehat{\beta}_j,X^j\right\rangle\right)^2$ being ρ the classical Pearson's linear correlation. In order of importance, the covariates in the model are: $\mathbf{Pr}(t-1)$ (0.724), $\mathbf{Pr}(t-7)$ (0.519), DoW (0.053), WS (0.048) and T_m (0.041). The rest of diagnostics are the typical ones for linear model to check the structural hypothesis (linear form and homoskedasticity) using the residuals and the fitted values (see Figure 5).

A more flexible model can be applied to this information using additive models (eq. 5) with the same set of covariates as in the previous example: $\mathbf{Pr}_{t-1}, \mathbf{Pr}_{t-7}, DoW$, WS and T_m. Except DoW which is categorical, the rest of covariates add a smooth contribution to the model that in the case of functional covariates corresponds to estimate smooth functions of the scores of the principal components decomposition (see [31]).

```
....

> Parametric coefficients:
>                Estimate Std. Error t value Pr(>|t|)
> (Intercept)   54.1257     0.6084  88.970  < 2e-16 ***
> DoW2          -3.3834     0.8106  -4.174 3.12e-05 ***
> DoW3          -3.7383     0.7759  -4.818 1.56e-06 ***
```

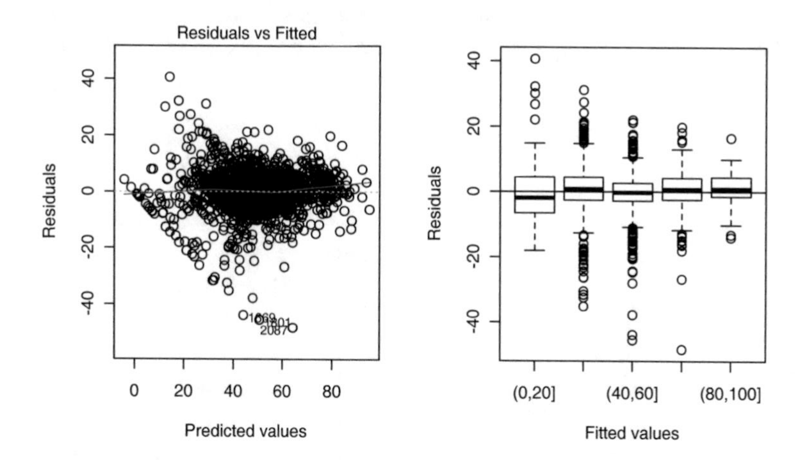

Fig. 5: Plot of Predicted values vs Residuals (left) and Boxplots of residuals respect Fitted values

```
> DoW4           -3.9136     0.7816  -5.007 5.99e-07 ***
> DoW5           -3.3083     0.7693  -4.301 1.78e-05 ***
> DoW6           -7.6950     0.8921  -8.626  < 2e-16 ***
> DoW7          -10.5524     0.8158 -12.934  < 2e-16 ***
> Approximate significance of smooth terms:
>                edf Ref.df       F  p-value
> s(WSpeed)    5.573  6.609  19.741  < 2e-16 ***
> s(Tmed)      5.785  6.948   4.910 1.80e-05 ***
> s(Pr1.PC1) 6.213  7.378 262.038  < 2e-16 ***
> s(Pr1.PC2) 1.000  1.000  41.133 1.75e-10 ***
> s(Pr1.PC3) 8.295  8.860   9.993 1.71e-14 ***
> s(Pr1.PC4) 2.602  3.379   5.506 0.000584 ***
> s(Pr7.PC1) 8.687  8.966  15.482  < 2e-16 ***
> s(Pr7.PC2) 3.201  4.120   1.709 0.145709
> s(Pr7.PC3) 1.692  2.158   2.305 0.104870
> s(Pr7.PC4) 1.000  1.000   5.707 0.016983 *
> ---
> Signif. codes:
> 0 '***' 0.001 '**' 0.01 '*' 0.05 '.' 0.1 ' ' 1
....
```

Now, the interpretation of this model is clearly more difficult than the linear one. In Figures 6–8 it can be seen the contributions in the additive model of the scalar covariates (WS, T_m) and the contribution of the first four PC's of $\mathbf{Pr}_{t-1}$ and $\mathbf{Pr}_{t-7}$, respectively. The functions associated with functional covariates are anchored in the interpretation of the principal components and so, to obtain a clear insight we have to interpret two chained things: the principal component and the function which is applied to every PC. There are obvious insights. For instance, the scale of each contribution (or the difference respect the zero constant) gives an idea about the importance of each contribution. In this case, the effect (almost linear) of the first PC of $\mathbf{Pr}_{t-1}$ is clearly the big one whereas the effect of the second, third and fourth PC of $\mathbf{Pr}_{t-7}$ are almost flat.

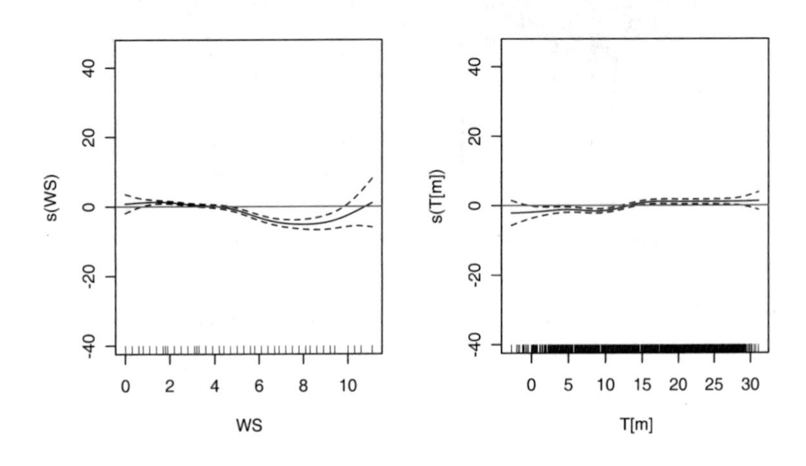

Fig. 6: Smooth contributions of the scalar covariates

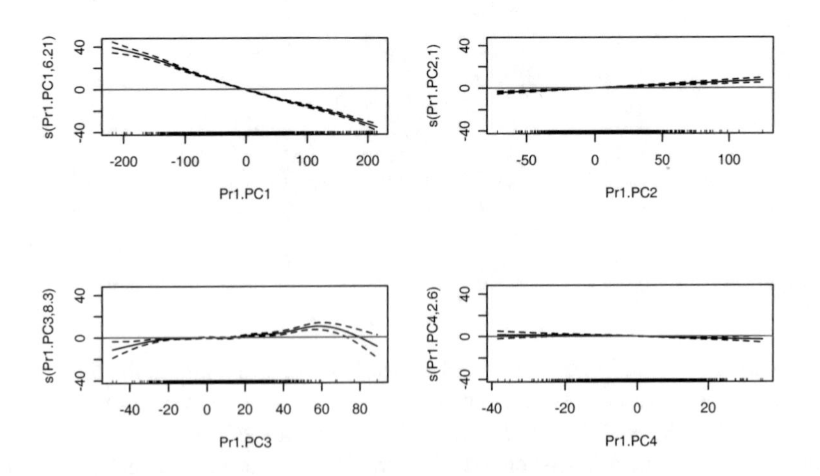

Fig. 7: Smooth contributions of the first four Principal Components of $\mathbf{Pr}_{t-1}$

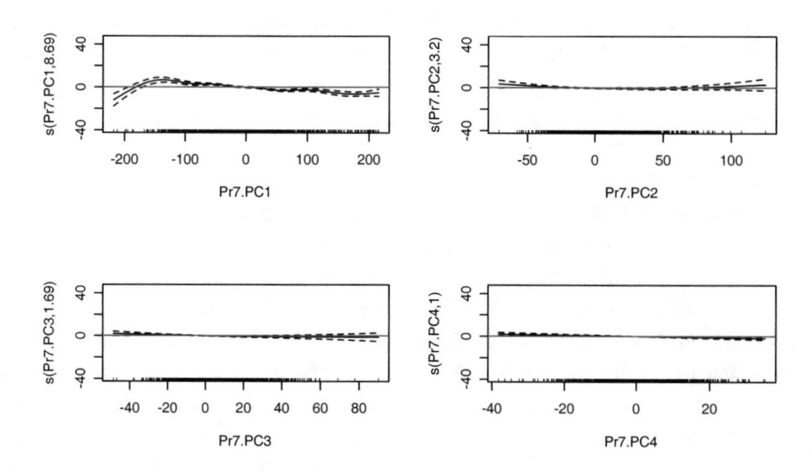

Fig. 8: Smooth contributions of the first four Principal Components of $\mathbf{Pr}_{t-7}$

Of course, the last model is an extension of the linear one and obtains better results. The explained deviance is 83.13% better than 81.26% of the linear one and with a better AIC (13725.946 vs 13877.246).

3 Variable selection

The variable selection problem in a general regression model tries to find the subset of covariates that best predicts or explains a response. In the classical approach, the covariates and the response are scalar (or multivariate) and the model established among them is linear. In a general framework, this problem is even more important because now, in the Big Data era, huge amounts of information are available but the information is contained on variables of different nature: functional, scalar, directional, categorical, etc. Our purpose in the rest of this work is to provide an automatic procedure for selecting regression models with a subset of the available covariates that, as in the case of our motivating example, are of different nature (mainly functional, scalar and categorical) and its number is quite huge (more than 400 possible covariates).

The stepwise regression, the most widely-used model selection technique throughout the 80's and the 90's, is rooted in the classical papers by [1], [30], [36] and [38]. The main idea is to use some diagnostic tools, directly derived from the linear model, to valuate the contribution of a new covariate to the model and decide whether it should be included in the model. The final subset is usually constructed using two main strategies: the forward selection that begins with a simple null model and tests, at each step, the inclusion of a new covariate in the model; and the backward selection that starts with the full model including all candidate variables and removes the most insignificant one at each step. It is also possible to mix both strategies, testing at each step which variables can be included or excluded in the optimal regression subset of covariates. In any case, the stepwise regression is anchored in the diagnostics of the linear model; it is therefore blind to detecting contributions other than the linear one among the covariates and the response.

The work by [44] proposing the LASSO estimator opens a new direction on variable selection procedures. The main innovation of the LASSO estimator is that it includes a L_1-type constraint for the coefficient vector β to force some parameters (components of β) to equal zero and, thereby, obtains the optimal subset of covariates such as those with non-zero coefficients. The effect of the constraint also helps in the optimization step and satisfactorily deals with the sparsity phenomenon. See [50] for a revision of the oracle properties of LASSO. Interesting examples may be found in the literature following the same line but using several penalties, constraints or using different structures in the regression such as: LARS ([11]), SCAD ([12]), COSSO ([27]), additive models ([48]) and extensions to partial linear, additive or semiparametric models like PLM ([10]), APLM ([28]) or GAPLM ([46]). All these works have two common characteristics: each paper is based on a specific model and all the covariates must be included in the model at the same time. The latter leads to highly demanding computing algorithms that sometimes are difficult to implement particularly for high-dimensional or functional data problems. See, for example, [21] for a review of some of the aforementioned methods.

Another type of strategy is a pure feature selection where the covariate is selected without a model. This is the approach employed in mRMR (minimum Redundancy Maximum Relevance, [32]). To enter into the model, a new candidate covariate must have a great relevancy with the response while maintaining a lower redundancy with the covariates already selected in the model. The main advantage of this approach is that it is an incremental rule; once a variate has been selected, it cannot be deselected in a later step. On the other hand, the measures for redundancy and relevancy must be chosen in function of the regression model applied to ensure good predictive results. The FLASH method proposed in [33] is a modification of the LASSO technique that sequentially includes a new variate changing the penalty at each step. This greedy increasing strategy is also employed by Boosting (see, for example, [7] or [19] in a functional data context). Boosting is not a purely feature selection method but rather a predictive procedure that selects at each step the best covariate/model respect to the unexplained part of the response. The final prediction is constructed as a combination of the different steps. All the previous solutions are not completely satisfactory in a functional data framework, specially when the number of possible covariates can be arbitrarily large.

Our aim is to select significant covariates for a general additive regression model with scalar response:

$$Y_i = \alpha + \sum_{j=1}^{J} f_j\left(X_i^{(j)}\right) + \varepsilon_i, \quad i = 1, \ldots, N$$

where the covariates are chosen from the set $S = \{X^1, X^2, \cdots, X^k, \cdots\}$ of different potential covariates (functional, vectorial, ...). The notation $X^{(j)}$ refers to the j-th covariate selected for the model. The number of variates can be extraordinarily large, so we intent to construct the regression model sequentially, i.e. from the trivial model up to the one that includes all the useful information provided by the covariates in the set S. We have chosen this type of model because it is available for any kind of covariates and it is a good compromise between flexibility and simplicity.

3.1 The selection algorithm

Our proposal that will be detailed in the next paragraphs combines some of the above strategies. As the pure feature selection methods we will select each candidate without thinking in a particular model/contribution involving the covariate. Like boosting we will correlate the residuals of the current model with the possible candidates and finally, the contribution of the candidate will be assessed checking the improvement in the model respect to the possible contribution of each candidate. Typically, we must decide the contribution of the new variable jointly with the selected ones in precedent steps. The new model is compared with the previous one using an appropriate test to decide whether to definitely incorporate the new variate into the model. If the new model is not better than the previous one, the candidate is discarded and the current model remains invariant. If the new model is better then the new model becomes the current model. Finally, using the residuals of the current model, new candidates are tried following the same cycle. This procedure is repeated until no more variables enter the model.

As aforementioned, the key idea for the selection of a feature is to find a tool that can be homogeneously applied to variates of different nature. This work is based on the exhaustive use of the distance correlation, $\mathscr{R}$, proposed by [43], and recently reviewed in [42]. The distance correlation fulfills the following two conditions:

i) $\mathscr{R}(X,Y)$ is defined for X and Y random vector variables in arbitrary finite dimension spaces.
ii) $\mathscr{R}(X,Y) = 0$ characterizes independence of X and Y.

Both conditions indicate that $\mathscr{R}$ is able to detect relationships other than linear among variates of arbitrary dimensions. Indeed, the work by [29] extends the properties for metric spaces of a certain type (*strong negative type*). In particular, Hilbert spaces or any space that can be embedded in a Hilbert space are of this type. Along this work, only Hilbert spaces are used (Euclidean spaces for scalar variables and L_2 for functional ones). Also, a test for independence using distance correlation can be derived following [43] and [40] that allows us to contrast when a variable could be a reasonable candidate to enter the model.

The idea of using the distance correlation for variable selection is not completely new. In [49], a definition of partial distance correlation (PDC) among X and Y given Z was introduced based on computing the distance correlation among the residuals of two models: Y respect to Z and X respect to Z. The PDC is the key of that paper for a variable selection method. But, PDC has two important drawbacks: First, PDC is constructed under linear relationship assumptions among variables and, in the original paper, it is only applied to classical linear models with scalar variates (the authors called it "linear PDC"). Second, there is no implementation of the measure. In [41] a different proposal for partial distance correlation is provided which is implemented in the function `pdcor` of R–package `energy` ([35]). This last definition seems more general and it is based on the linear projection of the distance matrices of X and Y over Z (and so, using implicitly again a linear assumption). Specifically, Z could be a mix of functional, scalar or multivariate variables where a distance using all of them must be computed. Even restricting ourselves to the scalar case, those variables should have a similar scale. On the contrary, our proposal can be applied to any mix of variables given that in every step, the distance correlation is computed among the residuals of the current model with each candidate. Taking into account that the residuals have the same nature as the response variable, the distance correlation can always be computed at each step, supposing that we are able to compute $\mathscr{R}(X,Y)$ at first step.

The computation of $\mathscr{R}(X,Y)$ for a given sample (empirical distance correlation) is quite straightforward because it only depends on the distances among data. See again [43]. Specifically, supposing we are interested in computing the distance correlation among X^j and Y, let $a_{kl} = d(X_k^j, X_l^j)$, the distance in the Hilbert space of X^j among cases k and l, and $A_{kl} = a_{kl} - \overline{a}_{k\cdot} - \overline{a}_{\cdot l} + \overline{a}_{\cdot\cdot}$ and respectively, $b_{kl} = d(Y_k, Y_l)$, and $B_{kl} = b_{kl} - \overline{b}_{k\cdot} - \overline{b}_{\cdot l} + \overline{b}_{\cdot\cdot}$.

The (empirical) distance correlation is simply computed as:

$$\mathscr{R}(X^j,Y) = \frac{\sum_k \sum_l A_{kl} B_{kl}}{\sqrt{\sum_k \sum_l A_{kl}^2}\sqrt{\sum_k \sum_l B_{kl}^2}}$$

Our proposal can be outlined as follows (see [16]):

1. Let Y the response and $S = \left\{X^1, \ldots, X^J\right\}$ the set of all variables that can be included in the model.
2. Set $\widehat{Y} = \overline{Y}$, and let $M^{(0)} = \emptyset$ the initial set of the variates included in the model. Set $i = 0$.
3. Compute the residuals of the current model: $\widehat{\varepsilon} = Y - \widehat{Y}$.
4. Choose $X^j \in S \neq \emptyset$ such that: 1) maximizes $\mathscr{R}\left(\widehat{\varepsilon}, X^k\right), \forall k \in S$ and 2) the null hypothesis for the test of independence among $\left\{X^j\right\}$ and $\widehat{\varepsilon}$ is rejected. IF NOT, END.
5. Update the sets M and S: $M^{(i+1)} = M^{(i)} \cup \left\{X^j\right\}$, and $S = S \setminus \left\{X^j\right\}$.
6. Compute the new model for Y using $M^{(i+1)}$ choosing the best contribution of the new covariate. Typically, there will be several possibilities for constructing correct models with the variates in $M^{(i+1)}$ fixing the contributions of the variates in $M^{(i)}$ and adding the new one.
7. Analyze the contribution of X^j in the new model respect to the current:
   IF this contribution is not relevant (typically comparing with the previous model)

THEN $M^{(i+1)} = M^{(i+1)} \setminus \{X^j\}$ and the current model remains unalterable
ELSE the new model becomes the current model and provides new predictions $(\widehat{Y})$. The simplest approach employed in this work is to consider an additive model: $\widehat{Y} = \overline{Y} + \sum_{m \in M} \widehat{f_m}\left(X^{(m)}\right)$ where at each step $\widehat{f_m}$ could be linear or smooth.

8. Update the number of iterations: $i = i+1$ and go to 3
9. END. The current model is the final model with the variates included in $M^{(i)}$. S is either the $\emptyset$ or contains those variables that do not pass the filter in step 4.

Steps 1–3 establish the null model as the initial model for beginning the procedure. Step 4 picks up the best variable from the set of available ones S. A new candidate is selected when maximizes the distance correlation among available ones and the test of independence between $\widehat{\varepsilon}$ and the candidate rejects the null hypothesis. For a better reading of the results, every time a distance correlation is showed, the value is filtered by the test of independence, i.e. all tables show $\mathscr{R}\left(X^k, \widehat{\varepsilon}\right) \mathbb{1}_{\{H_1\}}$ that only takes values distinct from zero for those covariates that rejects the null hypothesis of independence.

Step 5 updates the sets of variates. The selected variable is included as a possible candidate in set M and it is removed from set S. This is a clear forward strategy. Each candidate has only a chance to become part of the model.

Step 6 tries the possible models that can be conformed with the variables included in the updated M. Here, the previous contributions of the variables already in the model remain fixed and a list of models is used to check what could be the contribution of the new candidate. For instance, it may happen that my list only contains linear models or models with a limited number of nonlinear terms. In the latter case, if that limit is reached in previous steps, the new candidate can only be added with a linear contribution. This means that the list of possible models may not suffice to explain the relationship shown by the distance correlation. As a diagnostic tool, in latter iterations the test of independence using $\mathscr{R}\left(X^k, \widehat{\varepsilon}\right)$ can be computed for those variates in M to analyze if the contribution of X^k in the model is collecting all the possible information. If the test is rejected for any of the variables of M, it is an evidence that the contribution of the variable is not completely exploiting its information. Please note that this diagnostic tool is never done in step 5 that only takes a forward direction. This is done to avoid an infinite loop. Of course, a final arrangement is leaved to the user when the procedure ends. For instance, a certain variate entering the model at an earlier stage, may become insignificant some iterations later, when other variates add their contributions to the model. Of course, a final diagnostic of the model is leaved to the user to make some final arrangements. In this work, we have only applied additive models because this type of model is available for (nearly) any kind of variates and is quite flexible taking into account that each contribution may be linear or nonlinear (smooth).

Step 7 analyzes the contribution of the new candidate in the model with respect to the previous one. Typically, depending on the model, this can be done with a Generalized Likelihood Ratio Test or using a simple measure like AIC or BIC. If the new model is better than the previous one, the new residuals are computed and the algorithm backs to step 3 to recompute the distance correlation among these residuals and the variables remaining in set S. The procedure ends when no more variables can be added to the model because S is empty or all variables in S accepts the independence null hypothesis. Note that the rejection of H_0 does not guarantee that the contribution of the new candidate is relevant. Here, it is important to have a list of models powerful enough in order to exploit all the information of the candidate.

3.2 Numerical results

In the general framework, there is no previous works that can be considered direct competitors of our proposal and so there are no previous simulation scenarios to compare with using distance correlation measures. As commented before, [49] proposes an algorithm valid for scalar variates under linear assumptions about the model. In that paper, a simulation study using only scalar variables is provided that we have compared with our proposal in [16]. The main conclusion is that in linear scenarios our proposal performs similarly as [49] (or slightly better) and it is clearly superior (in terms of predictive performance and identification of the variables) in nonlinear ones even in the case of strong dependence among covariates.

For assessing the situation when the variables are of different nature, we include a simulation study (originally included in [16]) to check the performance of the algorithm in a mixed scenario with functional and scalar variables. In this simulation study, five functional and five scalar variables were simulated and the response was constructed as a function of the two first of each class. The functional variables: $\{\mathscr{X}_1,\ldots,\mathscr{X}_5\}$ were generated following independent Ornstein-Uhlenbeck processes in $[0,1]$, and the scalar variables $\{Z_1,\ldots,Z_5\}$ following, respectively, $U[0,1]$, $N[0,1]$, $N[0,1]$, $U[0,1]$ and $N[0,1]$ independently of each other. The response is constructed as follows:

$$Y = 10 + a_1 \langle \mathscr{X}_1, \beta_1 \rangle + a_2 \left\| \mathscr{X}_2^2 \right\|_2 + 3a_3 Z_1 + a_4 Z_2^2 + \varepsilon$$

with $\beta_1 = 2t + \sin 4\pi t + 0.1, t \in [0,1]$ and $\varepsilon \sim N(0, .25^2)$. The elements $\{a_1, a_2, a_3, a_4\}$ were introduced to emphasize/mask each part of the model. We estimated the model through a Functional Additive Model ([31]) using the first four principal components for functional covariates and a standard additive model for the scalar ones. Samples with $N = 200$ were generated and the process was repeated $B = 500$ times to count the proportion that a particular covariate enters the model. Table 1 shows the excellent results for different combinations of $\{a_1, a_2, a_3, a_4\}$. In all cases, $\mathscr{X}_1$, Z_1 and Z_2 are selected with very high frequencies in the 500 repetitions. $\mathscr{X}_2$ is also selected most of the times unless $a_2 = \frac{1}{4}$ that corresponds to a small contribution of that variate to the model. This can be explained because the $\mathrm{Var}\left(0.25\left\|\mathscr{X}_2^2\right\|_2\right) \approx 0.02$ that it is about $\frac{1}{3}$ of the residual variance.

We repeated the simulation study but forcing that some irrelevant covariates ($\mathscr{X}_3$, $\mathscr{X}_4$, $\mathscr{X}_5$, Z_3, Z_4, Z_5) may have a strong relationship with any of the important ones

Table 1: % of times that the variate was included in the model for $B = 500$ replications.

$\{a_1,a_2,a_3,a_4\}$	$\mathscr{X}_1$	$\mathscr{X}_2$	$\mathscr{X}_3$	$\mathscr{X}_4$	$\mathscr{X}_5$	z_1	z_2	z_3	z_4	z_5
$\{1,1,1,1\}$	1	0.996	0	0	0	1	1	0	0	0.002
$\{\frac{1}{2},\frac{1}{2},\frac{1}{2},\frac{1}{2}\}$	1	0.828	0	0	0	1	1	0	0.004	0
$\{\frac{1}{4},\frac{1}{4},\frac{1}{4},\frac{1}{4}\}$	0.962	0.078	0	0	0	1	0.998	0	0.002	0.002
$\{\frac{1}{4},\frac{1}{4},1,1\}$	0.960	0.076	0	0	0	1	1	0	0	0.002
$\{1,1,\frac{1}{4},\frac{1}{4}\}$	1	0.956	0	0.002	0	0.966	0.918	0	0	0

($\mathscr{X}_1$, $\mathscr{X}_2$, Z_1, Z_2) to check how the procedure approaches covariates with strong collinearities.

To this end, we computed new covariates $\mathscr{X}_3^*, \mathscr{X}_4^*, Z_3^*$ and Z_4^* in the following way: $\mathscr{X}_3^* = 0.95\mathscr{X}_1 + 0.05\mathscr{X}_3$, $\mathscr{X}_4^* = 0.95\mathscr{X}_2 + 0.05\mathscr{X}_4, Z_3^* = 0.95Z_1 + 0.05Z_3$ and $Z_4^* = 0.95Z_2 + 0.05Z_4$. This ensures that $\mathscr{X}_3^*$ (respectively, $\mathscr{X}_4^*$, Z_3^*, Z_4^*) has nearly the same information as $\mathscr{X}_1$ (respectively, $\mathscr{X}_2, Z_1, Z_2$). Table 2 shows the proportion of times that every variate is included in the model for this new scenario. Now the proportion of times shown in Table 1 for any of the relevant variables is shared among that variable and its (very close) copy. The important message is that when a variate is included in the model, its copy is banned for consequent steps and the model can be safely estimated without redundant information. Typically now, the sum of proportion of times of each covariate and its (almost) copy in Table 2 is the proportion of times that can be seen in Table 1 for the covariate.

Table 2: % of times that the variate was included in the model for $B = 500$ replications under strong collinearities among covariates.

$\{a_1,a_2,a_3,a_4\}$	$\mathscr{X}_1$	$\mathscr{X}_2$	$\mathscr{X}_3^*$	$\mathscr{X}_4^*$	$\mathscr{X}_5$	Z_1	Z_2	Z_3^*	Z_4^*	Z_5
$\{1,1,1,1\}$	0.560	0.670	0.440	0.328	0	0.736	0.532	0.264	0.468	0
$\{\frac{1}{2},\frac{1}{2},\frac{1}{2},\frac{1}{2}\}$	0.560	0.528	0.440	0.294	0	0.740	0.560	0.260	0.440	0
$\{\frac{1}{4},\frac{1}{4},\frac{1}{4},\frac{1}{4}\}$	0.500	0.058	0.476	0.046	0.002	0.678	0.548	0.322	0.452	0
$\{\frac{1}{4},\frac{1}{4},1,1\}$	0.510	0.066	0.448	0.026	0	0.832	0.560	0.168	0.440	0
$\{1,1,\frac{1}{4},\frac{1}{4}\}$	0.610	0.636	0.390	0.316	0	0.666	0.504	0.308	0.400	0

Table 3: $\mathscr{R}(X,\widehat{\varepsilon})\mathbb{1}_{\{\mathscr{H}_1\}}$ for one run. At each step, a variable enters the model when maximizes the row.

	$\mathscr{X}_1$	$\mathscr{X}_2$	$\mathscr{X}_3$	$\mathscr{X}_4$	$\mathscr{X}_5$	Z_1	Z_2	Z_3	Z_4	Z_5
1	0.040	0.029	0.000	0.000	0.000	0.177	0.108	0.000	0.000	0.000
2	0.122	0.049	0.012	0.000	0.000	0.000	0.141	0.000	0.000	0.000
3	0.220	0.062	0.000	0.000	0.000	0.000	0.000	0.000	0.000	0.000
4	0.000	0.103	0.014	0.000	0.000	0.000	0.000	0.000	0.000	0.000
5	0.000	0.000	0.000	0.000	0.000	0.000	0.000	0.000	0.000	0.000

Table 3 shows the distance correlation with the residuals for a single run at each step of the procedure under independence among covariates. The first column of this table is the iteration step. This allows us to analyze the order in which the covariates enter the model: Z_1 (0.177), Z_2 (0.141), $\mathscr{X}_1$ (0.22), $\mathscr{X}_2$ (0.103). In all cases, $\mathscr{R}(X,\widehat{\varepsilon})$ increases with the iterations until that covariate is chosen. This is the expected behavior in a case in which all the covariates are independent of each other. So the relationship with the residual becomes stronger as previous effects of other covariates are removed. Also, every time a variate is chosen, the distance correlation of that covariate with residuals in future steps is zero although zero means here that the null hypothesis of independence among covariate and residuals is accepted. This suggests that the model was able to incorporate all the important information of that covariate. In this case, the algorithm ends in iteration five when no more covariates are added to the model because none of the remaining ($\mathscr{X}_3, \mathscr{X}_4, \mathscr{X}_5$, Z_3, Z_4, Z_5) rejects the null hypothesis of independence with respect to the residuals.

4 Real data application

We applied our proposal to the forecast of the Market Demand Energy in the Iberian Energy Market and its Price using the available information from different sources. Indeed, it is possible to create new variables as transformations/functions of the original ones. For instance, all the information included as a functional variable was also included as scalar covariates (each discretization point was a new one). We employed the FSAM/FAM model in all cases given its flexibility and availability not only for multivariate variables, but also for functional ones (see, for instance, [31] or [15]).

4.1 Energy Market Demand

We have information about $N = 2459$ days from August, 2008 to December, 2014 of the Market Demand Energy at day t and hour H ($\mathrm{En}_t(H)$) in the IntraDay Electricity Iberian Market. Our aim is to forecast this variable with the available information up to day $t-1$ obtained from the sources that measure energy or price. The meteorological information is considered concurrent in time because the forecasting of these variates can be done accurately. In addition, we consider the calendar effects for every day. The following list summarizes all the variates involved:

- Energy Market Variables (source:www.omie.es): Daily profiles of Energy (**En**) and Price (**Pr**) at $t-1$ and $t-7$.
- Total Load and Generated energy type (source:demanda.ree.es): Daily profiles of Load (every ten minutes) (**Lo**), Nuclear (**Nu**), Fuel/Gas (**Fu**), Coal (**Ca**), Combined Cycle (**Cc**), Solar (**So**), Wind Energy (**WE**), Hydroelectric (**Hy**), Cogeneration (**Co**), Rest (**Re**).

- Meteorological information at Madrid-Barajas airport (source:aemet.es): Temperatures (T_{Max}, T_{Min}, T_{Med}, $T_A = T_{Max} - T_{Min}$), Wind Speed (WS), Solar Radiation (SR), Precipitation, Pressure, ...
- Every discretization value of the functional variates ($t-1$ and $t-7$).
- Categorical: Year(YY), Month(MM), Day-of-Week: $DoW = \{\mathbb{1}_{\{Mon\}}, \ldots, \mathbb{1}_{\{Sun\}}\}$.

Specifically, we are interested on estimating the model 5 where $Y_i = \text{En}_t(H)$, the number J and the covariates $X_i^{(j)}$ must be selected from the above list but evaluated before day t (except for meteorological and calendar variables), m_j is the contribution of each variable and ε_i is the error. The sample size is the number of days: $N = 2459$.

Table 4 summarizes the results of this application for four specific hours along the day where the variates are listed in order of inclusion into the model and the functional variables are marked in bold. In all those models, the deviance is mostly explained by the effect of the three first variables selected. As an example, the evolution in deviance of the model for $\text{En}_t(18)$ was 59.9%, 71.9%, 80.0%, 80.7% and 81.7%.

Table 4: Summary of models for energy with its selected variables (in order of entering)

Resp.	Covariates (in order)	Dev. expl.	σ_ε
$\text{En}_t(6)$	$\text{En}_{t-1}(6)$, $\mathbf{En}_{t-7}$, WS, $\mathbf{Pr}_{t-1}$, $\mathbf{Pr}_{t-7}$, $\text{En}_{t-1}(24)$, $\mathbf{Re}_{t-1}$	84.4%	1207.6
$\text{En}_t(12)$	$\mathbf{En}_{t-7}$, $\text{En}_{t-1}(16)$, DoW, $\mathbf{Cc}_{t-1}$, $\text{En}_{t-1}(23)$	84.2%	1540.3
$\text{En}_t(18)$	$\mathbf{En}_{t-7}$, $\text{En}_{t-1}(17)$, DoW, WS, $\mathbf{Pr}_{t-1}$,	81.7%	1662.6
$\text{En}_t(24)$	$\text{En}_{t-1}(24)$, $\mathbf{Lo}_{t-7}$, DoW, $\mathbf{Cc}_{t-1}$	86.0%	1221.0

In these models, the information of the first contributor tends to be the same as that of a week before rather than that of the previous day except for the consumption in the early morning hours (06:00, 24:00). This suggests a strong weekly pattern in the market demand profile complemented with the appearance of the indicators for Day-of-Week as a contributor in the third place. The different behaviour of the early morning hours models comes from the fact that the energy demand at that time has no main dependence on the activity of a given day of the week. Therefore, it is enough to consider the value of the day before. The inclusion into the models of variates like $\mathbf{Pr}_{t-1}$ or $\mathbf{Cc}_{t-1}$ although shocking at first glance, has a simple explanation. These two variates $\mathbf{Pr}_{t-1}$ and $\mathbf{Cc}_{t-1}$ are closely related to each other because, following the rules of Iberian Energy Market, the final price is fixed as the maximum of all types of energy required to cover all the demands and, the combined cycle is one of the expensive sources of energy. The price is therefore higher when the proportion of energy produced by combined cycle power stations is also high. These variates are probably included in the models to improve the prediction on the days characterized as having an energy demand that cannot be solved by using renewable sources (or other cheaper ones).

4.2 Energy Price

The second application example corresponds to the negotiated price at day t and hour H ($\mathrm{Pr}_t(H)$) in the Iberian Energy Market. The set of possible covariates is the same as in the previous example and the results are shown in Table 5.

Table 5: Summary of models for price with the selected variables (in order of entering)

Resp.	Covariates (in order)	Dev. expl.	c
$\mathrm{Pr}_t(6)$	$\mathbf{Pr}_{t-1}$, T_A, T_{Max}, T_{Med}, SR, $\mathrm{Pr}_{t-1}(24)$, $\mathbf{Cc}_{t-1}$	79.7%	6.
$\mathrm{Pr}_t(12)$	$\mathrm{Pr}_{t-1}(23)$, T_{Max}, $\mathbf{Lo}_{t-7}$, $\mathrm{Pr}_{t-1}(9)$, WS, $\mathbf{Cc}_{t-1}$, $\mathbf{Pr}_{t-7}$, $\mathbf{Cc}_{t-7}$	83.6%	6.
$\mathrm{Pr}_t(18)$	$\mathrm{Pr}_{t-1}(18)$, DoW, WS, $\mathbf{Pr}_{t-7}$, T_{Med}, $\mathrm{Pr}_{t-1}(16)$, $\mathbf{Cc}_{t-1}$, $\mathbf{Cc}_{t-7}$, $\mathbf{Pr}_{t-1}$, $\mathbf{WE}_{t-1}$	82.7%	6.
$\mathrm{Pr}_t(24)$	$\mathrm{Pr}_{t-1}(24)$, $\mathbf{Pr}_{t-7}$, $\mathbf{Cc}_{t-1}$, $\mathbf{Cc}_{t-7}$, YY, T_A, T_{Med}	74.1%	7.

The main contributor in all price models is the price of the preceding day, which indicates a strong persistence of this variable. Also, all models include the combined cycle generation (in positions more or less advanced). Surprisingly, now, the variables related with demand do not enter into models. This is contrary to what is expected by the classical economic theory. The meteorological variates supply this gap. For instance, the model for $\mathrm{En}_t(18)$ includes the mean of wind speed and the mean of temperature as covariates. These two variates can jointly explain the high price of energy on summer days with high temperatures but no wind, when the energy system must provide high amounts of energy (for feeding the air conditioning equipments) with low probability of using wind sources. The inclusion of other meteorological variates into the models follows similar rules although too much influence from these variables is not expected because the meteorological information comes from a particular site (Barajas airport) and it would be hard to explain in detail the price or the consumption of energy for a whole country using only the information of one site. Surely, meteorological information is useful for predicting demand in small regions but its contribution for price can only be explained in terms of general tendencies. The calendar effects only appear in two models. The Day-of-Week effect appears in a prominent position in the model for $\mathrm{En}_t(18)$. Checking the coefficients obtained in the model, the difference in price between Mondays and Sundays (the maximum difference) is about twelve euros and the minimum is around eight. The effect of the years only appears in the model for $\mathrm{En}_t(24)$ showing a clearly increasing pattern along the years, which were particularly intense in the period 2009-2011.

An important remark concerns the practical issues of the implementation of the distance correlation when the sample size grows. The main difficulty is unrelated to the ease of the implementation rather with the memory consumption of the method. The best strategy for implementing this procedure making an extensive use of the distance correlation with respect to the same covariates, is to compute and store the distance matrix for every covariate in advance. However, this strategy is impossible when the sample size or the number of covariates grows. The overall consumption of memory when storing the distance matrices for p covariates with N elements each is $(p+1)(N(N-1)/2-N)$. In this example, we have $p \geq 500$ covariates (many of which correspond to the discretization points of functional ones) with a sample size $N = 2459$. This amount typically exceeds the available resources of a desktop computer (even for the powerful ones). To overcome this difficulty, the obvious response is to use a HPC facility where the computation of distance correlation for each pair can be distributed along the available nodes (storing the distance matrices in the shared storage device). When it is impossible to access to a HPC facility, one may still compute the distance correlation for an arbitrary sample size by simply dividing all the computations in blocks using submatrices of dimension $L \times L$. As an example, Table 6 shows the maximum consumed memory and the execution time (in seconds) when the test of independence based on distance correlation for different sample sizes and two different dimensions of the submatrices is executed in a Intel Core i7-3770 with 32GB of RAM. The last row of the table shows that with $N = 25000$ the computer is unable to reserve enough memory. The submatrix trick can manage that sample size maintaining a limit for the memory consumption, but it does so with the cost of larger execution times. The execution time using $L = 500$ seems slightly better than with $L = 1000$. However we cannot conclude a general rule. For instance, for $N = 10000$ the execution times with $L = 100, 250$ and 2000 were, respectively, 52.10, 39.84 and 64.93 seconds.

Table 6: Maximum memory consumption and computation times (s.) for computing the test of independence using distance correlation with different computational strategies involving submatrices.

N	Memory consumption			Time(sec.)		
	Direct	$L = 500$	$L = 1000$	Direct	$L = 500$	$L = 1000$
1000	7.63MB	7.63MB	7.63MB	0.32	0.94	0.48
2500	47.68MB	7.63MB	30.52MB	1.58	4.40	5.40
5000	190.7MB	7.63MB	30.52MB	6.57	13.08	15.47
10000	762.9MB	7.63MB	30.52MB	26.83	45.91	53.00
25000	CRASH	7.63MB	30.52MB	-	290.13	309.91

5 Conclusions

In this chapter, we have revised the development of functional regression models with scalar response which are extremely useful for predicting variables related with energy due to the high frequency at which these variables are measured. So, the functional data analysis has become in the last years the perfect framework for treating this kind of problems. Energy market is a complex structure that evolves depending on many sources of information, some of them internal to the market and depending on the producers and distributors and some other externals like socioeconomic or meteorological information. FDA has developed a lot of models in the last years capable of treating such variety of information that must be filtered and selected to become useful. In the second part of the chapter, we have introduced an algorithm that automatically selects the variates for a regression model. The procedure operates in a forward way adding a variable to the model at each iteration. The key of the whole procedure is the extensive use of the distance correlation that presents two important advantages: the choice of the variate is made without considering a particular model and it is possible to compute this quantity for variates of different nature (functional, multivariate, circular, ...) as it is only computed from distances. The simplicity of the latter is also its main drawback because the number of operations (and memory consumption) is of a quadratic order respect to the sample size. But fortunately, it is possible to compute the distance correlation when the sample size is huge while the consumption of resources is maintained under certain limitations. Our proposal is presented is a very general way although in the applications in this work, we have restricted ourselves to additive models that offer a balanced compromise between predictive ability and simplicity. The obtained results are quite promising in scenarios where no competitors are available because no other procedure can deals with variates of different nature in a homogeneous way. As a final comment, the procedure was applied to a real problem related with the Iberian Energy Market (Price and Demand) where the number of possible covariates is really big. The algorithm was able to find synthetic regression models offering interesting insights about the relationship among the response and the covariates. The final selected models mix functional, scalar and categorical information.

Acknowledgements The authors acknowledge financial support from Ministerio de Economía y Competitividad grant MTM2016-76969-P and European Regional Development Fund (ERDF).

References

[1] Akaike, H.: Maximum likelihood identification of Gaussian autoregressive moving average models. Biometrika **60**(2), 255–265 (1973)
[2] Aneiros-Pérez, G., Vieu, P.: Semi-functional partial linear regression. Stat Probabil Lett **76**(11), 1102–1110 (2006)
[3] Aneiros-Pérez, G., Vieu, P.: Nonparametric time series prediction: A semi-functional partial linear modeling. J Multivariate Anal **99**(5), 834–857 (2008)
[4] Antoch, J., Prchal, L., Rosaria De Rosa, M., Sarda, P.: Electricity consumption prediction with functional linear regression using spline estimators. J Appl Stat **37**(12), 2027–2041 (2010)
[5] Antoniadis, A., Brossat, X., Cugliari, J., Poggi, J.M.: A prediction interval for a function-valued forecast model: Application to load forecasting. Int J Forecasting **32**(3), 939–947 (2016)
[6] Box, G.E., Jenkins, G.M., Reinsel, G.C., Ljung, G.M.: Time series analysis: forecasting and control. John Wiley & Sons (2015)

[7] Bühlmann, P., Yu, B.: Boosting with the L_2 loss: regression and classification. J Am Stat Assoc **98**(462), 324–339 (2003)
[8] Cancelo, J.R., Espasa, A., Grafe, R.: Forecasting the electricity load from one day to one week ahead for the spanish system operator. Int J Forecasting **24**(4), 588–602 (2008)
[9] Cho, H., Goude, Y., Brossat, X., Yao, Q.: Modeling and forecasting daily electricity load via curve linear regression. In: Lecture Notes in Statistics 217: Modeling and Stochastic Learning for Forecasting in High Dimensions, pp. 35–54. Springer (2015)
[10] Du, P., Cheng, G., Liang, H.: Semiparametric regression models with additive nonparametric components and high dimensional parametric components. Comput Stat Data Anal **56**(6), 2006–2017 (2012)
[11] Efron, B., Hastie, T., Johnstone, I., Tibshirani, R.: Least angle regression. Technical report, Department of Statistics (2002)
[12] Fan, J., Li, R.: Variable selection via nonconcave penalized likelihood and its oracle properties. J Am Stat Assoc **96**(456), 1348–1360 (2001)
[13] Fan, Y., James, G.M., Radchenko, P.: Functional additive regression. Ann Stat **43**(5), 2296–2325 (2015)
[14] Febrero-Bande, M., Oviedo de la Fuente, M.: fda.usc: Functional Data Analysis. Utilities for Statistical Computing. (2012). URL `https://CRAN.R-project.org/package=fda.usc`. R package version 1.4.0
[15] Febrero-Bande, M., González-Manteiga, W.: Generalized additive models for functional data. TEST **22**(2), 278–292 (2013)
[16] Febrero-Bande, M., González-Manteiga, W., Oviedo de la Fuente, M.: Variable selection in functional additive regression models. Comput Stat (2018). DOI 10.1007/s00180-018-0844-5
[17] Ferraty, F., Goia, A., Salinelli, E., Vieu, P.: Functional projection pursuit regression. TEST **22**(2), 293–320 (2013)
[18] Ferraty, F., Vieu, P.: Nonparametric functional data analysis: theory and practice. Springer (2006)
[19] Ferraty, F., Vieu, P.: Additive prediction and boosting for functional data. Comput Stat Data Anal **53**(4), 1400–1413 (2009)
[20] Hastie, T., Tibshirani, R.: Generalized additive models: some applications. J Am Stat Assoc **82**(398), 371–386 (1987)
[21] Hastie, T., Tibshirani, R., Wainwright, M.: Statistical learning with sparsity: the LASSO and generalizations. CRC Press (2015)
[22] Hong, T., Fan, S.: Probabilistic electric load forecasting: A tutorial review. Int J Forecasting **32**(3), 914–938 (2016)
[23] Horváth, L., Kokoszka, P.: Inference for functional data with applications, vol. 200. Springer Science & Business Media (2012)
[24] Hsing, T., Eubank, R.: Theoretical foundations of functional data analysis, with an introduction to linear operators. John Wiley & Sons (2015)
[25] Lian, H.: Functional partial linear model. J Nonparametr Stat **23**(1), 115–128 (2011)

[26] Liebl, D.: Modeling and forecasting electricity spot prices: A functional data perspective. Ann Appl Stat **7**(3), 1562–1592 (2013)
[27] Lin, Y., Zhang, H.H.: Component selection and smoothing in smoothing spline analysis of variance models. Ann Stat **34**(5), 2272–2297 (2006)
[28] Liu, X., Wang, L., Liang, H.: Estimation and variable selection for semiparametric additive partial linear models. Stat Sinica **21**(3), 1225–1248 (2011)
[29] Lyons, R.: Distance covariance in metric spaces. Ann Probab **41**(5), 3284–3305 (2013)
[30] Mallows, C.L.: Some comments on C_p. Technometrics **15**(4), 661–675 (1973)
[31] Müller, H.G., Yao, F.: Functional additive models. J Am Stat Assoc **103**(484), 1534–1544 (2008)
[32] Peng, H., Long, F., Ding, C.: Feature selection based on mutual information criteria of max-dependency, max-relevance, and min-redundancy. IEEE Transactions on pattern analysis and machine intelligence **27**(8), 1226–1238 (2005)
[33] Radchenko, P., James, G.M.: Improved variable selection with Forward-Lasso adaptive shrinkage. Ann Appl Stat **5**(1), 427–448 (2011)
[34] Ramsay, J., Silverman, B.: Functional Data Analysis. Springer (2005)
[35] Rizzo, M.L., Székely, G.J.: energy: E-statistics. R package version 1.6.2 (2014)
[36] Schwarz, G.: Estimating the dimension of a model. Ann Stat **6**(2), 461–464 (1978)
[37] Srivastava, A., Klassen, E.P.: Functional and shape data analysis. Springer (2016)
[38] Stone, M.: Comments on model selection criteria of Akaike and Schwarz. J Roy Stat Soc B Met **41**(2), 276–278 (1979)
[39] Suganthi, L., Samuel, A.A.: Energy models for demand forecasting–A review. Renewable Sustainable Energy Rev. **16**(2), 1223–1240 (2012)
[40] Székely, G.J., Rizzo, M.L.: The distance correlation t-test of independence in high dimension. J Multivariate Anal **117**, 193–213 (2013)
[41] Székely, G.J., Rizzo, M.L.: Partial distance correlation with methods for dissimilarities. Ann Stat **42**(6), 2382–2412 (2014)
[42] Székely, G.J., Rizzo, M.L.: The Energy of Data. Annu Rev Stat Appl **4**(1), 447–479 (2017)
[43] Székely, G.J., Rizzo, M.L., Bakirov, N.K.: Measuring and testing dependence by correlation of distances. Ann Stat **35**(6), 2769–2794 (2007)
[44] Tibshirani, R.: Regression shrinkage and selection via the LASSO. J Roy Stat Soc B Met **58**(1), 267–288 (1996)
[45] Vilar, J., Aneiros, G., Raña, P.: Prediction intervals for electricity demand and price using functional data. Int J Elec Power **96**, 457–472 (2018)
[46] Wang, L., Liu, X., Liang, H., Carroll, R.J.: Estimation and variable selection for generalized additive partial linear models. Ann Stat **39**(4), 1827–1851 (2011)
[47] Weron, R.: Modeling and forecasting electricity loads and prices: A statistical approach. John Wiley & Sons (2007)
[48] Xue, L.: Consistent variable selection in additive models. Stat Sinica **19**(3), 1281–1296 (2009)

[49] Yenigün, C.D., Rizzo, M.L.: Variable selection in regression using maximal correlation and distance correlation. J Stat Comput Sim **85**(8), 1692–1705 (2015)

[50] Zou, H.: The adaptive LASSO and its oracle properties. J Am Stat Assoc **101**(476), 1418–1429 (2006)

Quantization Methods for Stochastic Differential Equations

J. Kienitz, T.A. McWalter, R. Rudd, and E. Platen

Abstract In this paper we provide an introduction to quantization with applications in quantitative finance. We start with a review of vector quantization (VQ), a method originally devised for lossy signal compression. The basics of VQ are presented and applied to probability distributions. In order to solve stochastic differential equations (SDEs), a recursive algorithm based on the Euler-Maruyama approximation was devised by Pagès and Sagna [48]. The approach, known as recursive marginal quantization is described. We show how RMQ can be adapted to ensure non-negativity of solutions. We further provide the basic details on how higher-order Itō-Taylor schemes may be implemented within the RMQ framework. Then, having treated the case of one-dimensional SDEs, we turn our attention to the coupled SDEs used in stochastic volatility models. We present Joint RMQ, an instance of a product quantizer. Finally we demonstrate the performance of the methods with some financial pricing problems. In particular we show the performance improvements of the higher order RMQ methods and provide numerical examples of contingent claim pricing for the quadratic volatility, Stein-Stein and Heston models. This paper is styled as a

J. Kienitz
Fachbereich Mathematik und Naturwissenschaften, Bergische Universität Wuppertal / The African Institute for Financial Markets and Risk Management (AIFMRM), University of Cape Town / Quaternion Risk Management GmbH, e-mail: jkienitz@uni-wuppertal.de

T.A. McWalter
The African Institute for Financial Markets and Risk Management (AIFMRM), University of Cape Town / Faculty of Science, Department of Statistics, University of Johannesburg

R. Rudd
The African Institute for Financial Markets and Risk Management (AIFMRM), University of Cape Town

E. Platen
Finance Discipline Group and School of Mathematical and Physical Sciences, University of Technology Sydney / The African Institute for Financial Markets and Risk Management (AIFMRM), University of Cape Town

M. Günther, W. Schilders (eds.), *Novel Mathematics Inspired by Industrial Challenges*, Mathematics in Industry 38, https://doi.org/10.1007/978-3-030-96173-2_11

"guided tour" throughout — introducing the reader to the basic concepts and then referring to the original sources that provide full detail.

1 Introduction

In this paper we outline how a technique originally developed in the context of information theory can be used to solve challenging problems in quantitative finance. In particular, we highlight the recent developments related to the application of quantization to the solutions of stochastic differential equations.

1.1 Finance and Stochastic Differential Equations

From as early as 1900, when Bachelier [2] developed his "Théorie de la spéculation", people have applied probabilistic concepts to study and analyze the movements of prices observed in financial markets. However, at that time, the approach did not attain the popularity and recognition that it has today. Only after the appearance of the seminal works of Black & Scholes [4] and Merton [41] did things change dramatically. They used the principle of *no-arbitrage* and the modern language of *stochastic differential equations* (SDEs), developed over the previous decades, to provide a coherent theory for pricing financial claims. The theory of integration with respect to stochastic processes and SDEs was initiated by Itō [23, 24, 25, 26] — for texts on the subject we refer to [22, 27, 28, 51]. It is now standard practice for option pricing problems to be solved in the no-arbitrage context — for texts on application to financial mathematics see, for instance, [30, 43].

In what follows we shall assume that W_t is a Brownian motion on some probability space, $(\Omega, \mathcal{F}, \mathbb{P})$, equipped with a filtration $(\mathcal{F}_t)_{t\in[0,T]}$, to which the Brownian motion is adapted. The latter can be seen as the mathematical definition of information available at time t and how this information is incrementally revealed. Then, in modern quantitative finance terms, Bachelier considered the dynamics given by the SDE

$$\mathrm{d}X_t = a(X_t,t)\,\mathrm{d}t + \sigma_{\mathrm{B}}\,\mathrm{d}W_t, \qquad X_0 = x_0 \in \mathbb{R}, \tag{1}$$

whereas Black & Scholes chose

$$\mathrm{d}X_t = a(X_t,t)\,\mathrm{d}t + \sigma_{\mathrm{BS}} X_t\,\mathrm{d}W_t, \qquad X_0 = x_0 \in \mathbb{R}_+, \tag{2}$$

to guarantee the positivity of X_t. Here, the constants $\sigma_{\mathrm{B}}, \sigma_{\mathrm{BS}} \in \mathbb{R}_+$ describe the volatility. The form of the drift function, $a : \mathbb{R} \times \mathbb{R}_+ \to \mathbb{R}$, must be chosen to fulfill several economic constraints, one of the most important being the absence of arbitrage. If we neglect interest rates for brevity, or correspondingly consider *forward prices* of X, then

$$a(X_t,t)=0.$$

Both dynamics specify model primitives that can be used to describe the evolution of market risk factors directly. In the cases of stock or exchange rate modelling, X may be directly interpreted as a market observable quantity, whereas, more generally, there exists a function $f:\mathbb{R}\times\mathbb{R}_+\to\mathbb{R}$ such that $f(X_t,t)$ is a market observable quantity, which is often the case for modelling of interest or inflation rates.

Under both models it is possible to derive closed-form prices for European call and put options, these being the right, but not the obligation to purchase (resp. sell) the market observable X for an exercise price K, also called the strike price, at an exercise time T. Thus, we are in a position to easily value these types of financial contracts if the underlying financial asset is described by either (1) or (2).

In practice, however, the dynamics required to realistically model the evolution of risk factors are much more complicated. Several phenomena such as *skews*, *smiles* or *jumps* in the quoted option values occur and lead to skewed or fat-tailed probability distributions for the underlying market observable. The prices corresponding to solutions of more sophisticated models, which incorporate these features, do not have closed-form solutions, and the modeller must rely on numerical methods.

As examples, we describe local volatility (LV) and stochastic volatility (SV) models. For LV models, the volatility is generalised to be a time and state dependent function, with dynamics described by

$$\mathrm{d}X_t=a(X_t,t)\,\mathrm{d}t+\sigma(X_t,t)\,\mathrm{d}W_t,\qquad X_0=x_0\in\mathbb{R}_+. \tag{3}$$

For SV models, the volatility is generalised further to be dependent on another stochastic factor, V_t, as described by the coupled SDEs

$$\begin{aligned}\mathrm{d}X_t&=a^x(X_t,t)\,\mathrm{d}t+\sigma^x(X_t,V_t,t)\,\mathrm{d}W_t, && X_0=x_0\in\mathbb{R}_+,\\ \mathrm{d}V_t&=a^v(V_t,t)\,\mathrm{d}t+\sigma^v(V_t,t)\,\mathrm{d}Z_t, && V_0=v_0\in\mathbb{R}_+,\end{aligned} \tag{4}$$

where Z_t is another Brownian motion, possibly correlated with W_t. The functions a^v and σ^v describe the drift and volatility of the new stochastic factor.

Since there is no universally agreed upon approach for modelling market observables, it is convention to quote market volatilities for European options with respect to a reference model — most often the Black-Scholes model (2), but sometimes the Bachelier model (1). This leads to the notion of an implied volatility surface.

The implied volatility surface is defined as the mapping of the variables strike and maturity, (K,T), to a unique number $\sigma_{\text{imp}}(K,T)$, used as the diffusion coefficient in either the Bachelier or a Black-Scholes-Merton model. Using the closed-form pricing equation for the relevant model with this diffusion coefficient leads to the recovery of the observed market price for the particular strike and maturity[1].

The marginal distributions of market observables can be determined by a continuum — assuming it exists — of European option prices. The link between the Black-Scholes implied volatility, $\sigma_{\text{imp}}(K,T)$, and the local volatility function, $\sigma(X_t,t)$ was

[1] This has lead to the well-known statement by Rebonato [52] that implied volatility is “the wrong number to put in the wrong formula to get the right price.”

established by seminal papers of Dupire [12] and Derman & Kani [8] — for an introduction to this approach see [17], for instance.

Knowledge of the marginal distribution of X at time T is, however, not enough. To fully determine the probabilistic description of a risk factor, we also require *transition probabilities*, i.e., the probability of reaching state x_T at time T having started in state x_t at time $t < T$. The processes that we are considering belong to the class of *Markov* processes, which, intuitively speaking, means that they have no memory — the manner in which the state (x_t, t) is reached is irrelevant to the law governing the further evolution to (x_T, T). For a rigorous treatment of Markov processes see for instance [58]. Knowing the transition probabilities is necessary when one considers the valuation of more exotic options traded in the financial markets. For instance, we shall consider *Bermudan options*, for which the decision to exercise is not restricted to time T, but to a discrete set of time points $0 < t_1 < t_2 < \ldots < t_n = T$. This may even be generalised to a continuum of time points $s \in [t, T]$, in which case the claim is called an *American* option. Other exotic options may include path dependencies, such as options on the minimum, maximum or average of the underlying, X_t, or barrier features, which allow the option to come into existence (knock-in) or expire worthless (knock-out) if the underlying reaches a certain barrier value.

There is a deep connection between SDEs and the probabilistic and analytic concepts described by the famous Kolmogorov equations [15, 33] — these were used to derive the connection between implied volatility and local volatility described earlier. Moreover, a result due to Feynman and Kac provides a link between SDEs and parabolic *partial differential equations* (PDEs) [15]. The classical methods for solving option pricing problems originate in PDEs or probabilistic simulation, the latter also known as *Monte Carlo simulation*. General references for PDE methods applied to finance include [10, 34], while [11, 18, 29] are dedicated to Monte Carlo methods. For numerous numerical techniques see [31], which also introduces pricing based on characteristic functions.

1.2 Quantization

A technique distinct from the aforementioned methods is *quantization*. Its origins lie in information theory and signal processing where it was conceived as a lossy compression technique. Quantization can be traced back to 1948 when Bennett [3] published results based on Gaussian processes — similar ideas have appeared in papers as long ago as 1898. For a detailed outline of quantization and its history we refer to [19, 20, 37, 46].

The idea is simple, yet powerful: replace a continuous signal or function with a discrete *quantized* version that "best" represents the original. The notion of "best" is quantified by a *distortion function* that measures the deviation between the original function and its quantized version. This allows the specification of an optimization problem, the result of which is a *quantizer*, consisting of N values, called *codewords*, with associated probability weights.

The theory has been applied to numerical problems appearing in probability — [45] provides a survey on the methods and ideas. Further reading includes [44, 49, 50]. More recently, quantization has been directly applied to the solution of SDEs. The technique, known as *recursive marginal quantization*, has been initiated by Pagès and Sagna [48] — see also [5, 6, 40, 47, 54].

Here a sequence of quantizers is generated representing the marginal distributions of the stochastic process at discrete times. Since the approach recursively applies an SDE time-discretisation scheme using an Itō-Taylor expansion, the method may be applied to very general settings. This enables the pricing of derivative instruments that use models for which there are no analytical solutions for the underlying SDE dynamics. Moreover, since a byproduct of computing the quantizers at each step is a transition density matrix, this makes the method particularly useful for pricing more exotic options, such as the aforementioned Bermudan and barrier options.

Since inception, the theory has been extended and made more efficient to tackle many challenging contingent claim pricing problems, with applications to both local and stochastic volatility models. The method is also efficient enough that it may be used for calibrating such models using market data [5].

1.3 Outline of the Paper

Our exposition is structured as follows. Section 2 introduces the original quantization method, namely *Vector Quantization* (VQ). The basic concepts required to formulate the problem and find optimal quantizers are presented. In particular after writing down the distortion function, we describe Lloyd's algorithm and an efficient Newton-Raphson-based method to solve the optimization problem. In Section 3 we describe *Recursive Marginal Quantization* (RMQ), which enables the quantization of the approximate marginal distributions of Itō-Taylor approximations of an SDE. After formulating the method, we show how the RMQ algorithm may be modified to ensure non-negativity of solutions as well as our higher-order extensions. Having explored RMQ for one-factor SDEs, we turn our attention to an efficient method for two-factor coupled SDEs. Since a general multi-dimensional formulation of RMQ must rely on versions of Lloyds' algorithm, which is computationally inefficient, we instead consider a product quantizer formulation. *Joint Recursive Marginal Quantization* (JRMQ) is presented in Section 4. In the penultimate section we provide numerical examples of the methods as applied to the geometric Brownian motion, quadratic volatility, Stein-Stein and Heston Models. In particular we demonstrate numerical efficiency and pricing examples for European, barrier and Bermudan options and demonstrate the efficacy of JRMQ for calibration of stochastic volatility models using non-vanilla options. Section 6 concludes.

2 Vector Quantization

Let X be a continuous random vector taking values in $\mathbb{R}^d$, with distribution F_X, and defined on a probability space $(\Omega, \mathcal{F}, \mathbb{P})$. The aim of this section is to answer the following question:

> How does one optimally approximate X, by a discrete random vector, $\widehat{X} : \Omega \to \Gamma$, where Γ is a finite set of elements in $\mathbb{R}^d$?

Vector quantization is a lossy compression technique that provides a way to encode a vector space using a discrete subspace.

A vector quantizer may be expressed as

$$\pi : \mathbb{R}^d \to \Gamma,$$

where π is known as the *quantization function* or *quantizer*, and the set

$$\Gamma = \{\gamma^1, \ldots, \gamma^N\}$$

is a subset of $\mathbb{R}^d$, with (at most) $N \geq 1$ elements, known as the *quantization grid*. Each element, γ^i, is known as an *elementary quantizer* or *codeword*. As a result of the wide application of vector quantization, terminology abounds: the set Γ is alternatively known as a *quantizer*, *codebook* or *code*.

In quantitative finance this approach is useful as it allows the efficient approximation of expectations of the kind that often arise in contingent claim pricing:

$$\mathbb{E}[H(X)] = \int_{\mathbb{R}^d} H(x)\, \mathrm{d}F_X(x) \approx \sum_{\gamma \in \Gamma} H(\gamma) \mathbb{P}\{\widehat{X} = \gamma\},$$

where H is a payoff function. In contrast to standard Monte Carlo or quasi-Monte Carlo procedures that have equally weighted samples, vector quantization differs fundamentally in that the codewords are selected optimally and are weighted with an appropriate probability that is computed using the region mapped to each codeword.

The *regions* $R^i(\Gamma) \subset \mathbb{R}^d$ are the subsets of the values of X that are mapped to each codeword γ^i:

$$R^i(\Gamma) := \{x \in \mathbb{R}^d : \pi(x) = \gamma^i\}.$$

Consider the pointwise error made when approximating an input vector X by $\pi(X)$, its projection onto Γ. It is clear that

$$|X - \pi(X)| \geq \inf_{\gamma \in \Gamma} |X - \gamma|.$$

Here, $|\cdot|$ is the Euclidean norm. Equality will only hold in the above expression when π is chosen to be the *nearest-neighbor projection operator*, $\pi_\Gamma : \mathbb{R}^d \to \Gamma$, defined by

$$\pi_\Gamma(X) := \{\gamma^i \in \Gamma : |X - \gamma^i| \leq |X - \gamma^j| \text{ for } j = 1, \ldots, N,\ j \neq i\}.$$

As a simple tie-breaking rule, when an input vector is an equal distance from two or more elementary quantizers, the elementary quantizer with the lowest index is selected. With the quantization function chosen in this way, the set of regions $\{R^i(\Gamma)\}_{i=1}^N$, forms a special Borel partition satisfying

$$\begin{aligned} R^i(\Gamma) &= \{x \in \mathbb{R}^d : \pi_\Gamma(x) = \gamma^i\}, \\ &= \{x \in \mathbb{R}^d : |x - \gamma^i| \leq |x - \gamma^j| \text{ for } j = 1, \ldots, N,\ j \neq i\}, \end{aligned}$$

known as a *Voronoi* or *nearest neighbour* partition.

2.1 Optimal Quantization Grids

Consider the approximation of X by $\widehat{X}$, a discrete random vector defined as the nearest-neighbour projection of X onto the quantization grid Γ,

$$\widehat{X} := \pi_\Gamma(X).$$

This is known as the *quantized* version of X. We now aim to find the quantization grid Γ, such that $\widehat{X}$ "best" approximates X. The optimality of a quantization grid is measured using the expected error made when reproducing the input vectors, known as the *distortion*. Although the theory of quantization can be established for a variety of norms, see [19], the Euclidean norm is most commonly used in probabilistic applications.

Definition 2.1 (Distortion) Let $X \in \mathscr{L}^2$. The $\mathbb{R}_+$-valued function D, defined on $(\mathbb{R}^d)^N$ by

$$D : \Gamma \to \|X - \pi_\Gamma(X)\|_2^2 = \mathbb{E}\left[\left|X - \widehat{X}\right|^2\right] = \int_{\mathbb{R}^d} |x - \pi_\Gamma(x)|^2 \, \mathrm{d}F_X(x) \tag{5}$$

is known as the *distortion* function.

Here, $\|\cdot\|_p$ indicates the $\mathscr{L}^p$-norm[2]. An optimal quantization grid is one that minimizes this distortion function. It can be shown that when considering the $\mathscr{L}^2$-norm and random vectors in $\mathbb{R}^d$, the distortion function will always attain a minimum with the resulting quantization grid having full size, i.e., pairwise-distinct components, and not inducing any Voronoi regions with zero measure. A proof of this result can be found in [45, Sec. 2.1].

To minimize the distortion, and find the optimal grid, the differentiability of the distortion function must be established.

Proposition 2.1 (Differentiability of Distortion) *Let $X \in \mathscr{L}^2$. If $\Gamma = \{\gamma^1, \ldots, \gamma^N\}$ has pairwise distinct components, the distortion function, D, is finite and differentiable at Γ and*

[2] If $X \in \mathscr{L}^p$ then $\int_{\mathbb{R}^d} |x|^p \, \mathrm{d}F_X(x) < +\infty$.

$$
\begin{aligned}
[\nabla D(\Gamma)]_i &= \frac{\partial D(\Gamma)}{\partial \gamma^i} \\
&= 2\mathbb{E}\left[(\gamma^i - X)\mathbb{I}_{\{X \in R^i(\Gamma)\}}\right] \\
&= 2\int_{R^i(\Gamma)} (\gamma^i - x)\,\mathrm{d}F_X(x)
\end{aligned} \tag{6}
$$

for $i = 1, \ldots, N$.

Proof See Pagès [45, Sec. 3.1] □

This result leads directly to the following corollary, which is central to all numerical applications of vector quantization.

Corollary 2.1 (Self-consistent Quantizers) *Let* $X \in \mathscr{L}^2$*. Any grid* Γ *that minimizes the distortion function* D *is a* self-consistent[3] *quantizer, i.e.,*

$$
\begin{aligned}
\widehat{X} &= \mathbb{E}\big[X \,\big|\, \widehat{X}\big] \\
&= \mathbb{E}\big[X \,\big|\, \pi_\Gamma(X) = \widehat{X}\big],
\end{aligned}
$$

or equivalently

$$
\begin{aligned}
\gamma^i &= \frac{\mathbb{E}\left[X\mathbb{I}_{\{X \in R^i(\Gamma)\}}\right]}{\mathbb{P}\{X \in R^i(\Gamma)\}} \\
&= \frac{\int_{R^i(\Gamma)} x\,\mathrm{d}F_X(x)}{\int_{R^i(\Gamma)} \mathrm{d}F_X(x)}
\end{aligned} \tag{7}
$$

for $i = 1, \ldots, N$.

Intuitively, knowledge of $\widehat{X}$ is knowledge of the *region* to which X belongs. The self-consistency condition states that, for an optimal quantization grid, averaging X over any region yields the codeword that generates that region.

Numerical methods for obtaining optimal quantization grids now proceed either by minimizing the distortion function directly, or by searching for self-consistent grids.

Searching for a self-consistent quantization grid is equivalent to constructing a centroidal Voronoi tessellation under the distribution of X, where the points that generate each of the Voronoi regions are simultaneously the *probability mass centroids*[4] of their regions. This can be seen directly from (7). This geometric view of the optimal quantization problem is thoroughly explored in [9].

---

[3] The self-consistency property is often known as *stationarity* in the literature. This term is not used here so as to avoid the potential confusion with stationarity as applied to stochastic processes.

[4] That is, γ_i is the conditional expectation of X over the region R^i.

2.2 Numerical Methods

Armed with the relevant definitions we highlight two of the most common numerical methods for obtaining optimal quantization grids. For further detail, specifically on how the methods may be optimized for quicker convergence, see [49].

2.2.1 Lloyd's Algorithm

A popular fixed-point algorithm that produces an optimal quantization grid is *Lloyd's algorithm*, originally due to [35]. It is based on recursively enforcing the necessary self-consistency condition of the quantization grid and proceeds as follows:

1. Select a set of N pairwise distinct points to form the initial quantization grid, Γ.
2. Construct the Voronoi partition, i.e., the collection of nearest-neighbour regions $\{R^i(\Gamma)\}_{i=1}^N$, associated with the current grid.
3. Compute the probability mass centroids of the Voronoi regions found in Step 2. Assign these centroids as the codewords for the new quantization grid.
4. If this new quantization grid meets a pre-specified convergence criterion, terminate. Otherwise, return to Step 2.

In this way, the necessary self-consistency condition for optimal quantization grids, (7), gives rise to an iterative fixed-point procedure given by

$$^{(l+1)}\gamma^i = \frac{\int_{R^i(^{(l)}\Gamma)} x \, \mathrm{d}F_X(x)}{\int_{R^i(^{(l)}\Gamma)} \mathrm{d}F_X(x)}, \tag{8}$$

for $1 \leq i \leq N$, with $0 \leq l < l_{\max}$ indicating the iteration index. It should be clear that computing the mass centroids in (8) may require high-dimensional integrals over Voronoi cells. In a probabilistic setting, Monte Carlo (or quasi-Monte Carlo) methods are used to compute these integrals. This variation is known as the *randomized* Lloyd's algorithm,

$$^{(l+1)}\gamma^i = \frac{\sum_{m=1}^M X_m \mathbb{I}_{\{X_m \in R^i(^{(l)}\Gamma)\}}}{\sum_{m=1}^M \mathbb{I}_{\{X_m \in R^i(^{(l)}\Gamma)\}}},$$

where X_m is a random or quasi-random sample generated from the distribution of X, and M is the number of samples generated per iteration of the algorithm. Convergence is established for Lloyd's algorithm in [13]. Lloyd's algorithm is illustrated in Figure 1 for a bivariate Gaussian distribution.

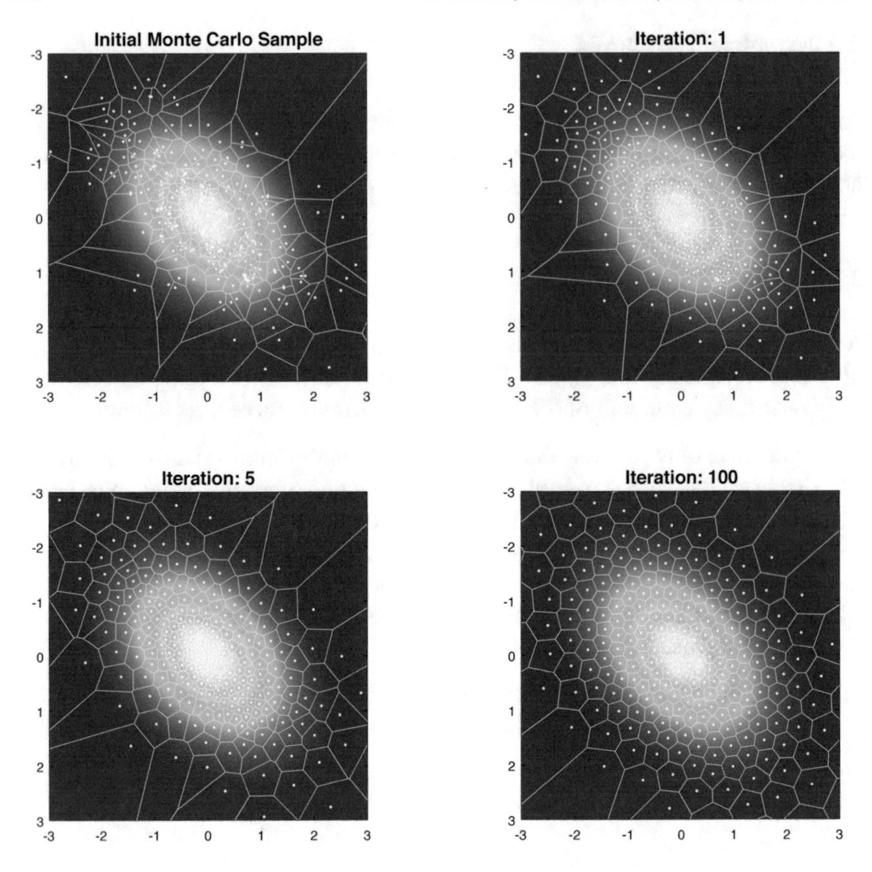

Fig. 1: The evolution of Lloyd's algorithm applied to a bivariate Gaussian distribution with correlation of 0.4. The dots represent the codewords and the lines represent the boundaries of the Voronoi regions.

2.2.2 The Newton-Raphson Algorithm

The special case when X is a one-dimensional random variable with a well-defined density function is relevant for many applications, including the recursive marginal quantization algorithm presented in the next section.

In this case, let $\boldsymbol{\Gamma}$ be a column vector representing the quantizer, then given the gradient and Hessian of the distortion, the optimal quantizer can be computed using the Newton-Raphson iteration,

$$^{(l+1)}\boldsymbol{\Gamma} = {}^{(l)}\boldsymbol{\Gamma} - \left[\nabla^2 D\left({}^{(l)}\boldsymbol{\Gamma}\right)\right]^{-1} \nabla D\left({}^{(l)}\boldsymbol{\Gamma}\right), \tag{9}$$

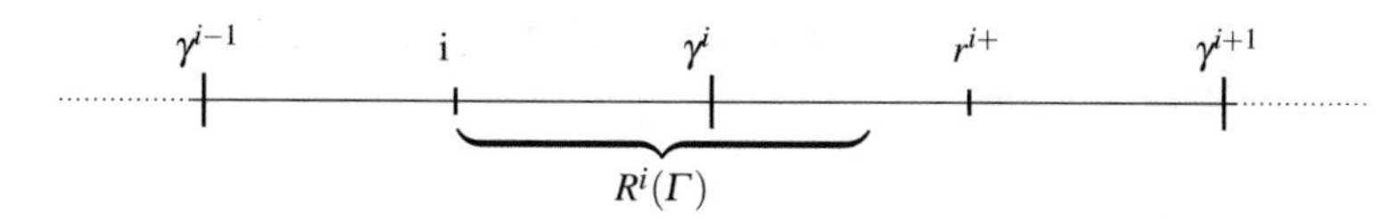

Fig. 2: Representation of the region $R^i(\Gamma)$ associated with codeword γ^i in the one-dimensional setting.

for $0 \leq l < l_{\max}$. Since the Voronoi regions are easy to specify in the one-dimensional case, the gradient and the Hessian of the distortion function can be explicitly computed for an arbitrary random variable.

The regions associated with a quantization grid may be defined directly as $R^i = \{x \in \mathbb{R} : r^{i-} < x \leq r^{i+}\}$ with

$$r^{i-} = \frac{\gamma^{i-1} + \gamma^i}{2} \qquad \text{and} \qquad r^{i+} = \frac{\gamma^i + \gamma^{i+1}}{2},$$

for $1 \leq i \leq N$, where, by definition, $r^{1-} = -\infty$ and $r^{N+} = \infty$. If the distribution under consideration is not defined over the whole real line, then r^{1-} and r^{N+} are adjusted to reflect the interval of support. Figure 2 shows a simple graphical representation of these regions.

Suppose f_X and F_X are the density and distribution functions of X, respectively. Define the p-th lower partial expectation as

$$M_X^p(x) := \mathbb{E}\left[X^p \mathbb{I}_{\{X<x\}}\right],$$

where $M_X^0(x) = F_X(x)$ represents the distribution function of X. Then, direct integration of the distortion function (5) gives

$$\begin{aligned}
D(\Gamma) &= \sum_{i=1}^{N} \int_{R^i(\Gamma)} \left|x - \gamma^i\right|^2 \mathrm{d}F_X(x) \\
&= \sum_{i=1}^{N} \int_{r^{i-}}^{r^{i+}} \left|x - \gamma^i\right|^2 f_X(x)\,\mathrm{d}x \\
&= \sum_{i=1}^{N} \Big[M_X^2(r^{i+}) - M_X^2(r^{i-}) - 2\gamma^i \left(M_X^1(r^{i+}) - M_X^1(r^{i-})\right) \\
&\qquad + (\gamma^i)^2 \left(F_X(r^{i+}) - F_X(r^{i-})\right) \Big].
\end{aligned}$$

Consequently, the elements of the vector $\nabla D(\boldsymbol{\Gamma})$ are given by

$$\frac{\partial D(\Gamma)}{\partial \gamma^i} = 2\gamma^i \left(F_X(r^{i+}) - F_X(r^{i-})\right) - 2\left(M_X^1(r^{i+}) - M_X^1(r^{i-})\right),$$

for $1 \leq i \leq N$.

Similarly, the tridiagonal Hessian matrix, $\nabla^2 D(\boldsymbol{\Gamma})$, may be computed. It has diagonal elements given by

$$\frac{\partial^2 D(\Gamma)}{\partial(\gamma^i)^2} = 2\left(F_X(r^{i+}) - F_X(r^{i-})\right) + \tfrac{1}{2}\left(f_X(r^{i+})(\gamma^i - \gamma^{i+1}) + f_X(r^{i-})(\gamma^{i-1} - \gamma^i)\right),$$

and super- and sub-diagonal elements given by

$$\frac{\partial^2 D(\Gamma)}{\partial\gamma^i\partial\gamma^{i+1}} = \tfrac{1}{2} f_X(r^{i+})(\gamma^i - \gamma^{i+1}) \quad \text{and} \quad \frac{\partial^2 D(\Gamma)}{\partial\gamma^i\partial\gamma^{i-1}} = \tfrac{1}{2} f_X(r^{i-})(\gamma^{i-1} - \gamma^i),$$

respectively. Note that the quantities required to compute a Newton-Raphson iteration (i.e., the gradient and Hessian) only require the density function, distribution function and first lower partial expectation to be known. The second lower partial expectation is required only if one wishes to compute the final distortion.

Having computed the quantizer $\boldsymbol{\Gamma}$, the associated row vector of probabilities, $\mathbf{p}$, is computed as

$$[\mathbf{p}]_i = \mathbb{P}\{\widehat{X} = \gamma^i\} = \mathbb{P}\{X \in R^i(\Gamma)\} = F_X(r^{i+}) - F_X(r^{i-}),$$

for $1 \leq i \leq N$. Defining $\mathbf{p}$ as a row vector is convenient since the expectation of H may be approximated by

$$\mathbb{E}[H(X)] \approx \sum_{i=1}^{N} H(\gamma^i)\mathbb{P}\{\widehat{X} = \gamma^i\} = \mathbf{p}H(\boldsymbol{\Gamma}),$$

where H is applied element-wise to $\boldsymbol{\Gamma}$.

As an example of the Newton-Raphson procedure, let X be a standard normal random variable with

$$\begin{aligned} f_X(x) &= \phi(x), \\ F_X(x) &= \Phi(x), \\ M_X^1(x) &= -\frac{1}{\sqrt{2\pi}} e^{-\frac{x^2}{2}} = -\phi(x), \end{aligned}$$

where $\phi(\cdot)$ and $\Phi(\cdot)$ are the standard normal density and distribution functions, respectively. Here, a good guess for the initial quantizer ${}^{(0)}\boldsymbol{\Gamma}$ is

$$\gamma^i = \frac{5.5i}{N+1} - 2.75,$$

for $1 \leq i \leq N$.

It has been shown in [7] that if Γ is an optimal quantizer for the random variable X of cardinality N, the mass of each of the elementary codewords is given by

$$\mathbb{P}\{X \in R^i(\Gamma)\} = \tfrac{1}{N} f_X^{2/3}(\gamma^i) \int_{\mathbb{R}} f_X^{2/3}(z)\,\mathrm{d}z + o\left(\tfrac{1}{N}\right). \tag{10}$$

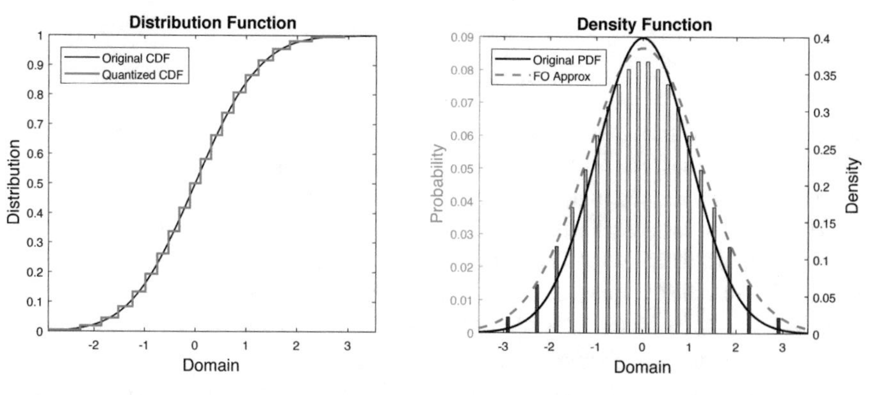

Fig. 3: Vector quantization of the standard Gaussian distribution using the Newton-Raphson method with $N = 20$ codewords.

As a first example, the optimal quantizer of a standard Gaussian random variable using 20 codewords is shown in Figure 3. The left panel shows the quantized cumulative distribution function, whereas the right panel shows the quantization grid itself, along with its accompanying probabilities. The first order (FO) term of the above formula, (10), is indicated by the dashed blue line in the right panel of Figure 3. For comparison, the standard Gaussian density is displayed with its own scaled axis on the right.

3 Recursive Marginal Quantization

Consider the continuous-time one-dimensional diffusion specified by the stochastic differential equation[5]

$$dX_t = a(X_t)\,dt + b(X_t)\,dW_t, \qquad X_0 = x_0 \in \mathbb{R}, \tag{11}$$

defined on the filtered probability space $(\Omega, \mathcal{F}, (\mathcal{F}_t)_{t\in[0,T]}, \mathbb{P})$, where W_t is a standard Brownian motion. Here a and b are assumed to be sufficiently smooth and bounded to ensure the existence of a strong solution. The question we aim to answer is:

> How does one optimally approximate $X_{t_k} : \Omega \to \mathbb{R}$, for some time discretisation point $t_k \in [0,T]$, when the distribution of X_{t_k} is unknown?

Usually this is achieved by performing a Monte Carlo experiment using a discrete-time approximation scheme for the SDE, the simplest being the Euler-Maruyama [38] update:

[5] This is slightly less general than the full local volatility model presented earlier, but suitable for parametric local volatility specifications.

$$\begin{aligned}\overline{X}_{k+1} &= \overline{X}_k + a(\overline{X}_k)\Delta t + b(\overline{X}_k)\sqrt{\Delta t} z_{k+1} \\ &=: \mathcal{U}(\overline{X}_k, z_{k+1}),\end{aligned}$$

for $0 \leq k < n$, where $\Delta t = T/n$ and independent $z_{k+1} \sim \mathcal{N}(0,1)$, with initial value $\overline{X}_0 = x_0$. The innovation of Pagès and Sagna [48] was to show that a recursive procedure based on quantizing these updates is possible.

The random variable $\overline{X}_1$ is an Euler update of the known quantity, $\overline{X}_0$, and thus has a Gaussian distribution. Therefore, it is possible to use vector quantization to obtain $\widehat{X}_1$, an optimal quantizer for the first step of the above scheme. This yields $\Gamma_1 = \{x_1^1, \ldots, x_1^N\}$ and its associated probabilities[6]. Note that we have chosen to use x_k^i to represent the i-th elementary quantizer of $\widehat{X}_k$.

One must, however, find a way to quantize the successive (marginal) distributions of $\overline{X}_{k+1}$. Given knowledge of the distribution of $\overline{X}_k$, the distortion of the quantizer Γ_{k+1} may be written as

$$\begin{aligned}\bar{D}(\Gamma_{k+1}) &= \mathbb{E}\left[\left|\overline{X}_{k+1} - \widehat{X}_{k+1}\right|^2\right] \\ &= \mathbb{E}\left[\mathbb{E}\left[\left|\overline{X}_{k+1} - \widehat{X}_{k+1}\right|^2 \middle| \overline{X}_k\right]\right] \\ &= \mathbb{E}\left[\mathbb{E}\left[\left|\mathcal{U}(\overline{X}_k, z_{k+1}) - \widehat{X}_{k+1}\right|^2 \middle| \overline{X}_k\right]\right] \\ &= \int_{\mathbb{R}} \mathbb{E}\left[\left|\mathcal{U}(x, z_{k+1}) - \widehat{X}_{k+1}\right|^2\right] \mathrm{d}\mathbb{P}\{\overline{X}_k \leq x\}. \end{aligned} \tag{12}$$

Unfortunately, the exact distribution of $\overline{X}_k$ is unknown for $k > 1$. The central idea of recursive marginal quantization is to approximate the continuous and unknown distribution of $\overline{X}_k$ with the previously *quantized* and discrete distribution of $\widehat{X}_k$. The main result of [48] shows that if one makes this approximation, this procedure converges.

In this way, the integral in (12) is represented as a sum over the codewords of quantizer Γ_k and their associated probabilities,

$$\begin{aligned}\bar{D}(\Gamma_{k+1}) &\approx \sum_{i=1}^{N} \mathbb{E}\left[\left|\mathcal{U}(x_k^i, z_{k+1}) - \widehat{X}_{k+1}\right|^2\right] \mathbb{P}\{\widehat{X}_k = x_k^i\} \\ &=: D(\Gamma_{k+1}).\end{aligned} \tag{13}$$

With this definition of a new distortion, $D(\Gamma_{k+1})$, the numerical methods outlined in Section 2.2.1 may now be used to compute the quantizer at time-step $k+1$. Note that the marginal distribution of the Euler scheme, $\overline{X}_{k+1}$, can be written explicitly as

[6] While it is possible to allow the cardinality of the quantizer to vary with time step, we shall assume that the cardinality is N for every time step.

$$
\begin{aligned}
F_{\overline{X}_{k+1}}(x) &= \int_{\mathbb{R}^d} \mathbb{P}\{\mathcal{U}(s, z_{k+1}) \le x\} \, \mathrm{d}\mathbb{P}\{\overline{X}_k \le s\} \\
&\approx \sum_{i=1}^{N} \mathbb{P}\{\mathcal{U}(x_k^i, z_{k+1}) \le x\} \mathbb{P}\{\widehat{X}_k = x_k^i\} \\
&=: F_{\widetilde{X}_{k+1}}(x),
\end{aligned} \tag{14}
$$

where $F_{\widetilde{X}_{k+1}}(x)$ is the approximate marginal distribution based on the quantizer at the previous step.

The above reasoning leads to the following recursive procedure:

$$
\begin{aligned}
\widehat{X}_1 &= \pi_{\Gamma_1}(\overline{X}_1) && \text{with } \overline{X}_1 = \mathcal{U}(x_0, z_1), \\
\widehat{X}_{k+1} &= \pi_{\Gamma_{k+1}}(\widetilde{X}_{k+1}) && \text{with } \widetilde{X}_{k+1} = \mathcal{U}(\widehat{X}_k, z_{k+1}),
\end{aligned}
$$

for $k = 1, \ldots, n-1$. As mentioned previously, $\widehat{X}_1$ is determined by quantizing the initial update, which is normally distributed. Subsequent quantizers are determined by quantizing the distribution associated with the random variable $\widetilde{X}_{k+1}$, which is a probability-weighted sum of normal distributions, given by (14).

Figure 4 provides an intuitive depiction of the process that occurs[7]. The top panel shows the quantizer at time step k. Conditional on each codeword, a Gaussian Euler update is propagated (second panel). In panel three, these updates are weighted by the probability of the associated originating codeword and summed to produce the implied marginal density at time step $k+1$, as shown in the final panel. The distribution associated with this marginal density is $F_{\widetilde{X}_{k+1}}$, which is quantized to produce the quantizer at time step $k+1$. This process is repeated until the quantizer at the final time is produced. The output of the algorithm is an inhomogeneous Markov chain, depicted as a tree in Figure 5.

3.1 Numerical Methods

Given the quantizer at time t_k, represented as a column vector $\boldsymbol{\Gamma}_k$, and the associated probabilities, $\mathbb{P}\{\widehat{X}_k = x_k^i\}$ for $1 \le i \le N$, the Newton-Raphson iteration that generates the quantizer $\boldsymbol{\Gamma}_{k+1}$, at time t_{k+1}, is given by

$$
{}^{(l+1)}\boldsymbol{\Gamma}_{k+1} = {}^{(l)}\boldsymbol{\Gamma}_{k+1} - \left[\nabla^2 D\left({}^{(l)}\boldsymbol{\Gamma}_{k+1}\right)\right]^{-1} \nabla D\left({}^{(l)}\boldsymbol{\Gamma}_{k+1}\right), \tag{15}
$$

where $0 \le l < l_{\max}$ is the iteration index.

The update may be written in affine form as

$$
\mathcal{U}(x_k^i, Z_{k+1}) = m_k^i Z_{k+1}^i + c_k^i, \tag{16}
$$

where

[7] Here, for easier visualization, we have shown the process using density functions, not distributions.

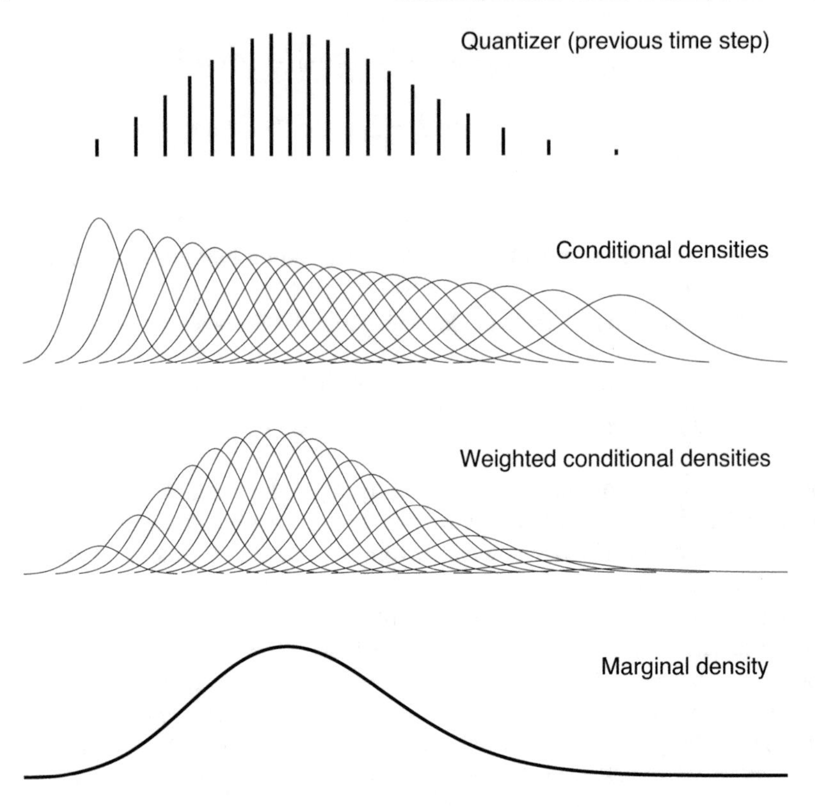

Fig. 4: Illustration of the RMQ algorithm.

$$m_k^i := b(x_k^i)\sqrt{\Delta t} \qquad \text{and} \qquad c_k^i := x_k^i + a(x_k^i)\Delta t,$$

with $Z_{k+1}^i \sim \mathcal{N}(0,1)$ identically distributed to z_{k+1}. The corresponding density and distribution functions are denoted by $f_{Z_{k+1}^i}$ and $F_{Z_{k+1}^i}$, respectively. Here, a new index i is introduced for the random variable Z_{k+1}^i anticipating that the distribution may depend on x_k^i. This is redundant in the case of the Euler update because Z_{k+1}^i is a standard Gaussian random variate irrespective of starting point, it is, however, required for RMQ based on higher order updates.

Given the affine form, it is now possible to derive explicit expressions for the elements of the gradient and Hessian of the distortion, as in the vector quantization section. This is performed in full detail in [40], with an efficient matrix presentation provided in [39]. In a manner similar to the vector quantization case, the final quantities are now *summations* over the density functions, cumulative distribution functions and first lower partial expectations of the random variables, Z_{k+1}^i.

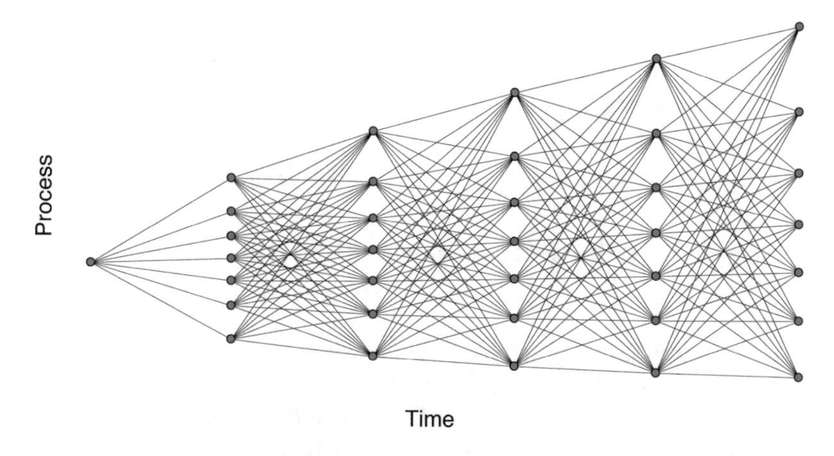

Fig. 5: RMQ produces an inhomogeneous Markov chain. The grid represented here has 5 time steps and 7 codewords per time step.

3.2 The Zero Boundary

Sometimes discrete-time approximations of an SDE may exhibit behaviour that is inconsistent with the true solution at the *zero boundary*. For example, an Euler-Maruyama approximation of geometric Brownian motion process can, under certain circumstances, generate negative values, even though the SDE specification guarantees non-negativity in each case. As a result, discrete-time Monte Carlo simulations are often modified to generate reflecting or absorbing behaviour at zero — see for example [36]. As shown in [40], the RMQ algorithm can be modified in a similar way to model either an *absorbing* or *reflecting* boundary at zero.

Intuitively, modelling an absorbing boundary requires truncating the implied marginal distribution, at the zero boundary, thus ensuring only positive codewords are generated at the next time-step. This results in a quantizer where the total probability does not sum to one. Instead, the remaining probability is accumulated by a *pseudo*-codeword at zero. Here, *pseudo* is used to indicate that the location of the codeword is not determined using the optimization algorithm. At each time-step, the probability accumulated at the pseudo-codeword at zero indicates the probability of the approximating process having attained zero by that time. The left panel of Figure 6 shows how the weighted updates are truncated at the zero boundary, shown in red.

Modelling a reflecting boundary instead requires that the implied marginal distribution is reflected at the zero boundary. This is equivalent to reflecting the conditional implied distribution of each update. The final Newton-Raphson algorithm is modified to instead use these reflected versions of the density, distribution and lower partial expectation functions [40].

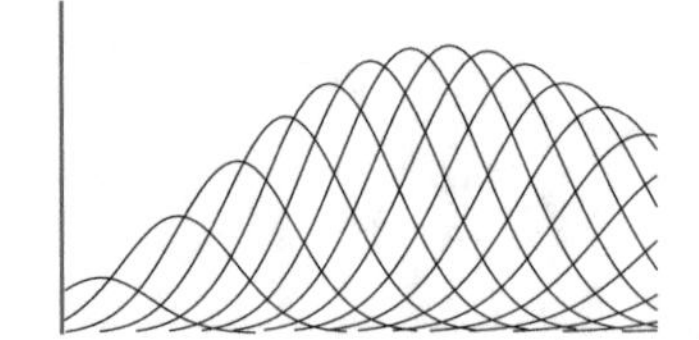
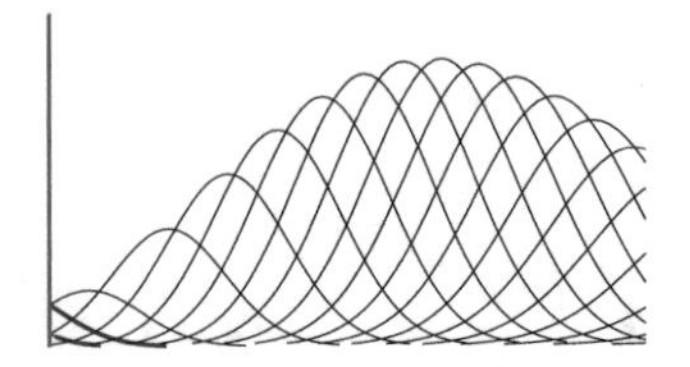

Fig. 6: Absorption (left) and reflection (right) at the zero boundary.

Note that modelling an absorbing boundary incurs no additional computational burden, whereas modelling a reflecting boundary does. This is because evaluating a reflected distribution function requires two evaluations of the original distribution function, and similarly for the density and lower partial expectation. The right panel of Figure 6 shows how the weighted updates are reflected (shown in blue) at the zero boundary.

3.3 Higher-order Updates

In [40] the RMQ technique is extended to higher-order discretisation schemes, specifically the Milstein [42] scheme and the simplified weak order 2.0 scheme [32].

The Milstein scheme is specified as

$$\overline{X}_{k+1} = \overline{X}_k + a(\overline{X}_k)\Delta t + b(\overline{X}_k)\sqrt{\Delta t} z_{k+1} + \tfrac{1}{2} b(\overline{X}_k) b'(\overline{X}_k)\Delta t (z_{k+1}^2 - 1),$$

while the simplified weak order 2.0 scheme is given by

$$\begin{aligned}\overline{X}_{k+1} = \overline{X}_k &+ a(\overline{X}_k)\Delta t + b(\overline{X}_k)\sqrt{\Delta t} z_{k+1} + \tfrac{1}{2} b(\overline{X}_k) b'(\overline{X}_k)\Delta t (z_{k+1}^2 - 1) \\ &+ \tfrac{1}{2}\left(a'(\overline{X}_k) b(\overline{X}_k) + a(\overline{X}_k) b'(\overline{X}_k) + \tfrac{1}{2} b''(\overline{X}_k) b^2(\overline{X}_k)\right)(\Delta t)^{3/2} z_{k+1} \\ &+ \tfrac{1}{2}\left(a(\overline{X}_k) a'(\overline{X}_k) + \tfrac{1}{2} a''(\overline{X}_k) b^2(\overline{X}_k)\right)(\Delta t)^2.\end{aligned}$$

Modifying RMQ to use higher-order discretisation schemes is straightforward, since both the above schemes are amenable to being written in the affine form (16). By completing the square on these formulae, it is possible to write the updates in terms of a random variable Z_{k+1}^i that, instead of being Gaussian, is noncentral chi-squared with one degree of freedom[8]. Consequently, the density, distribution and lower partial expectation functions associated with the noncentral chi-squared random variables are used in the Newton-Raphson algorithm. The rest of the algorithm proceeds as normal.

In [40] it is shown that these schemes can provide an advantage in accuracy and efficiency when compared to the Euler RMQ case — we reproduce some of

[8] Since the schemes are both noncentral chi-squared distributed, the random variable now depends explicitly on i through the non-centrality parameter.

these results in Section 5. In [53] a theoretical convergence result, similar to the convergence result for the Euler scheme [48], is shown for RMQ based on these higher order updates.

4 Recursive Marginal Quantization for Stochastic Volatility Models

We now consider stochastic volatility models of the form

$$\mathrm{d}X_t = a^x(X_t)\,\mathrm{d}t + b^x(X_t,V_t)\left(\rho\,\mathrm{d}Z_t + \sqrt{1-\rho^2}\,\mathrm{d}Z_t^{\perp}\right), \qquad X_0 = x_0 \in \mathbb{R}_+, \tag{17}$$

$$\mathrm{d}V_t = a^v(V_t)\,\mathrm{d}t + b^v(V_t)\,\mathrm{d}Z_t, \qquad V_0 = v_0 \in \mathbb{R}_+, \tag{18}$$

where Z_t and $Z_t^{\perp}$ are uncorrelated Brownian motions, and $\rho \in (-1,1)$ is a correlation constant. Here, when compared with (4), we have rewritten W_t in terms of Z_t and an orthogonal component so as to explicitly account for the correlation.

If we were to apply the standard RMQ algorithm, previously described in Section 3, to this coupled system, it would be necessary to use numerical methods such as randomized Lloyd's method or Competitive Learning Vector Quantization, the latter using stochastic gradient descent [45]. Unfortunately these methods are not nearly as efficient as the Newton-Raphson method described earlier.

An alternative approach is to use a so-called *product quantizer*. This can be described intuitively as follows: Consider the SDE for the V process above. It does not depend in any way on the X process. As a result we call the V process the *independent process*. We start by quantizing this process using the standard RMQ algorithm of Section 3 to find Γ_k^v at each time step k. We then perform the recursive marginal quantization of the X process at each successive time step $k+1$, denoted Γ_{k+1}^x, conditional on knowing the quantizers Γ_k^x and Γ_k^v. The grids formed by the Cartesian product $\Gamma_k^x \times \Gamma_k^v$ for each time step k, along with the associated joint probabilities and transition probabilities are then used to perform pricing.

Consider the Euler-Maruyama scheme for the above coupled SDEs

$$\begin{aligned}
\overline{X}_{k+1} &= \overline{X}_k + a^x(\overline{X}_k) + b^x(\overline{X}_k,\overline{V}_k)\sqrt{\Delta t}\left(\rho z_{k+1} + \sqrt{1-\rho^2} z_{k+1}^{\perp}\right), \qquad \overline{X}_0 = x_0,\\
&=: \mathcal{U}^x(\overline{X}_k,\overline{V}_k,z_{k+1},z_{k+1}^{\perp})\\
\overline{V}_{k+1} &= \overline{V}_k + a^v(\overline{V}_k) + b^v(\overline{V}_k)\sqrt{\Delta t}z_{k+1}, \qquad \overline{V}_0 = v_0,\\
&=: \mathcal{U}^v(\overline{V}_k,z_{k+1})
\end{aligned}$$

for $1 \leq k < n$, where $z_{k+1}, z_{k+1}^{\perp} \sim \mathcal{N}(0,1)$ are independent. Given this set-up, it can be shown that quantizing the Euler update $\mathcal{U}^x(\overline{X}_k,\overline{V}_k,z_{k+1},z_{k+1}^{\perp})$ is equivalent to quantizing an update specified by

$$\mathcal{U}(\overline{X}_k,\overline{V}_k,z) = \overline{X}_k + a^x(\overline{X}_k) + b^x(\overline{X}_k,\overline{V}_k)\sqrt{\Delta t}z, \tag{19}$$

where $z \sim \mathcal{N}(0,1)$ is **any** standard Gaussian distributed random variable [54].

Proposition 4.1 *Let Γ^x_{k+1} be the quantizer for $\overline{X}_{k+1}$, then the distortion function may be written as*

$$\mathbb{E}\Big[\big|\mathcal{U}^x(\overline{X}_k,\overline{V}_k,z_{k+1},z^{\perp}_{k+1}) - \widehat{X}_{k+1}\big|^2\Big] = \mathbb{E}\Big[\big|\mathcal{U}(\overline{X}_k,\overline{V}_k,z) - \widehat{X}_{k+1}\big|^2\Big]$$
$$=: \overline{D}(\Gamma^x_{k+1})$$

with $z \sim \mathcal{N}(0,1)$.

From the perspective of the distortion, this means that the correlation is irrelevant. More precisely, since

$$\big|\mathcal{U}^x(\overline{X}_k,\overline{V}_k,z_{k+1},z^{\perp}_{k+1}) - \widehat{X}_{k+1}\big|^2 \stackrel{d}{=} \big|\mathcal{U}(\overline{X}_k,\overline{V}_k,z) - \widehat{X}_{k+1}\big|^2,$$

and $\overline{D}(\Gamma^x_{k+1})$ is computed using an expectation, one need not consider the effect of the correlation when quantizing $\widehat{X}_{k+1}$.

Of course, when one determines the weights (probabilities) associated with the optimal quantizer once it is computed, then it is necessary to consider the effect of the correlation.

In an analogous fashion to standard RMQ, the distortion of the true Euler update is approximated by replacing the true joint distribution of the update at time k with the corresponding quantized distribution. Then, the approximate distortion, $D(\Gamma^y_{k+1})$, used in the computation of the time-$k+1$ quantizer is defined by

$$\begin{aligned}
\overline{D}(\Gamma^x_{k+1}) &= \int_{\mathbb{R}^2} \mathbb{E}\Big[\big|\mathcal{U}(x,v,z) - \widehat{X}_{k+1}\big|^2\Big]\, \mathrm{d}\mathbb{P}\{\overline{X}_k \le x, \overline{V}_k \le y\} \\
&\approx \sum_{i=1}^{N}\sum_{u=1}^{N} \mathbb{E}\big[(\mathcal{U}(x^i_k,v^u_k,z) - \widehat{X}_{k+1})^2\big]\,\mathbb{P}\{\widehat{X}_k = x^i_k, \widehat{V}_k = v^u_k\} \\
&=: D(\Gamma^y_{k+1}).
\end{aligned}$$

Once again, a Newton-Raphson iteration can be formulated using the gradient and Hessian of the distortion — see [54] for full details. Once the quantizer, $\boldsymbol{\Gamma}^x_{k+1}$ has been computed, the joint probability of $\overline{X}_{k+1}$ and $\overline{V}_{k+1}$ is approximated by

$$\begin{aligned}
&F_{\overline{X}_{k+1},\overline{V}_{k+1}}(x,v) \\
&= \int_{\mathbb{R}^2} \mathbb{P}\big\{\mathcal{U}(r,s,w_{k+1}) \le x, \mathcal{U}^v(s,z_{k+1}) \le v\big\}\, \mathrm{d}\mathbb{P}\{\overline{X}_k \le r, \overline{V}_k \le s\} \\
&\approx \sum_{i=1}^{N}\sum_{u=1}^{N} \mathbb{P}\big\{\mathcal{U}(x^i_k,v^u_k,w_{k+1}) \le x, \mathcal{U}^y(v^u_k,z_{k+1}) \le v\big\}\mathbb{P}\{\widehat{X}_k = x^i_k, \widehat{V}_k = v^u_k\} \\
&= F_{\widetilde{X}_{k+1},\widetilde{V}_{k+1}}(x,v),
\end{aligned}$$

where $w_{k+1} = \rho z_{k+1} + \sqrt{1-\rho^2} z_{k+1}^{\perp}$ is now used to correctly compute the bivariate normal distribution. The fact that the bivariate normal distribution must be used to compute the joint probability follows directly from the distributional properties of the coupled Euler system.

Unfortunately, the computation time associated with the bivariate normal distribution is onerous. Thus, in [54] an approximation based on the univariate normal distribution is provided — this has been independently derived in [6], where a less efficient conditioning argument is used to generate the quantizer for the X process. More recently [16] propose a multi-dimensional generalisation of product quantizers.

5 Numerical Results

In this section we provide a few numerical examples illustrating applications of recursive marginal quantization. We start by demonstrating the effectiveness of the higher-order RMQ schemes. This is achieved by providing numerical evidence of weak order convergence, showing the accuracy improvement in the RMQ implied marginal distribution and testing computational efficiency for the schemes. To ensure that we have closed-form solutions for comparison, we use geometric Brownian motion as the underlying process.

Next we provide results for the quadratic volatility model. This is an instance of a parametric local volatility model, which has characteristic pricing results for European options [1]. We compare the accuracy of RMQ with these results and price discrete up-and-out barrier options.

As examples of stochastic volatility models, we provide results for the Stein-Stein [57] and Heston [21] models. Again, European options may be priced using characteristic function methods, enabling comparison. We conclude by showing a preliminary calibration exercise for the Heston model, where the underlying options are Bermudan.

5.1 Numerical Convergence Results

To demonstrate the numerical convergence of recursive marginal quantization we start by presenting weak order convergence results. The Black-Scholes model, as defined by the SDE (2), was used with $a(X_t,t) = rX_t$, where the process-specific parameters were $X_0 = 100$, $r = 10\%$, $\sigma_{\text{BS}} = 20\%$, $T = 2$, and the RMQ-specific parameters were $N = 1\,000$ and n variable, depending on Δt. Figure 7 shows a base-two log-log plot of absolute error against time step size, Δt, for the first four moments of geometric Brownian motion. The gradient of each line indicates the weak order convergence rate. By inspection, the gradient of the lines representing the Euler and Milstein schemes, for all moments, is approximately one. This indicates weak order

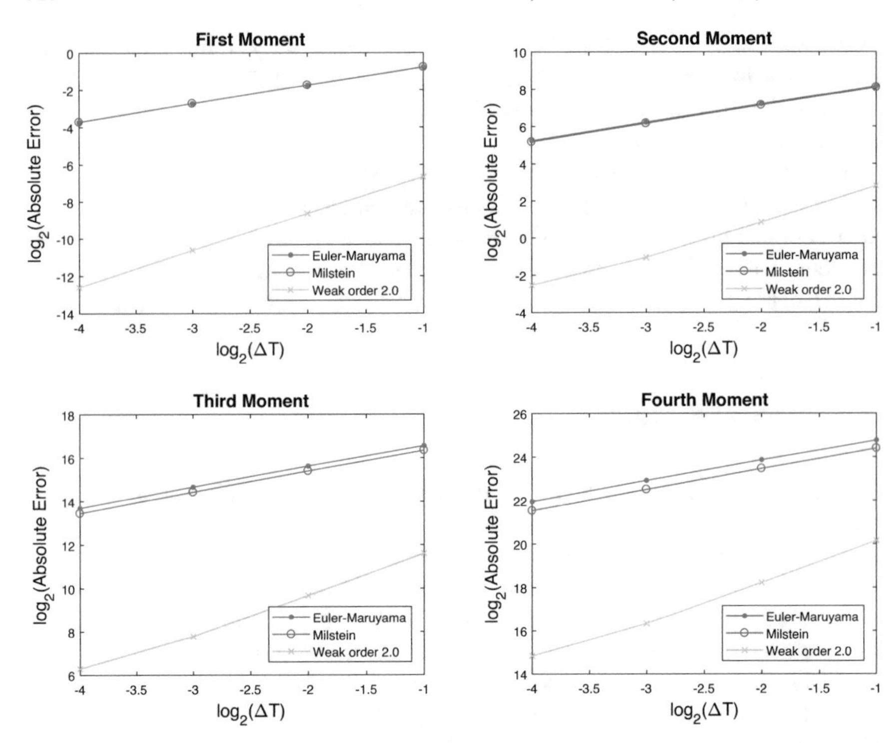

Fig. 7: Weak order convergence of the RMQ algorithms for the first four moments of geometric Brownian motion. In each case the approximate gradient of the lines is one for the Euler and Milstein schemes and two for the weak order 2.0 scheme as expected.

one convergence. For the weak order 2.0 scheme the gradients, for all moments, are approximately two as expected.

Further evidence of improved error performance is demonstrated in Figure 8, which shows the difference between the exact marginal distribution for GBM and the marginal distribution implied by the quantizer (14), at each step $1 \leq k < n = 12$. The quantizers are plotted as curves, which change colour from blue to green indicating the times from zero to T (time-step k). All three graphs are plotted at the same scale so as to show the improvement that the higher order methods are able to achieve.

While the improved error performance is demonstrated by the previous two figures, this comes at the cost of extra computation. Thus, it is important to demonstrate the efficiency of the methods. Figure 9 shows a base-two log-log plot of average absolute error against execution time. The error plotted here is calculated as the difference between a portfolio of options priced using RMQ and the same options priced using the Black-Scholes formula.

In the left graph the number of time steps is varied ($n \in [2, 4, 8, 16, 32]$, with $N = 250$), while in the right graph cardinality is varied ($N \in [50, 100, 200, 400, 800]$,

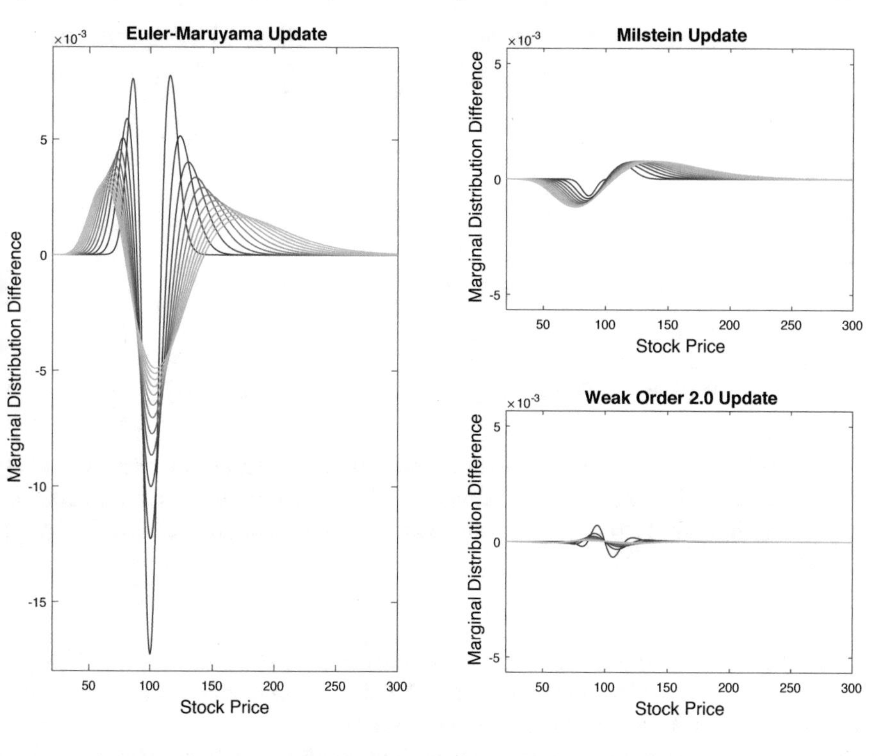

Fig. 8: The difference between the marginal distribution as approximated by RMQ and the exact marginal distribution for geometric Brownian motion. For comparison, the errors for all three methods are drawn on the same scale

with $n = 12$), with all other parameters kept constant. The graphs show that the weak order 2.0 method is the most efficient (lowest error as a function of execution time). The Milstein method, while not as accurate, is still better than the Euler scheme. This is a result of the fact that, while the two higher order schemes require roughly about 1.8 times the amount of computation time, they provide more than twice the accuracy.

5.2 An Example of a Local Volatility Model

We now turn our attention to a parametric local volatility model, the quadratic volatility (QV) model. The model is defined by

$$a(X_t) = 0 \qquad \text{and} \qquad b(X_t) = \sigma\left(qX_t + (1-q)x_0 + \frac{1}{2}s\frac{(x_t - x_0)^2}{x_0}\right),$$

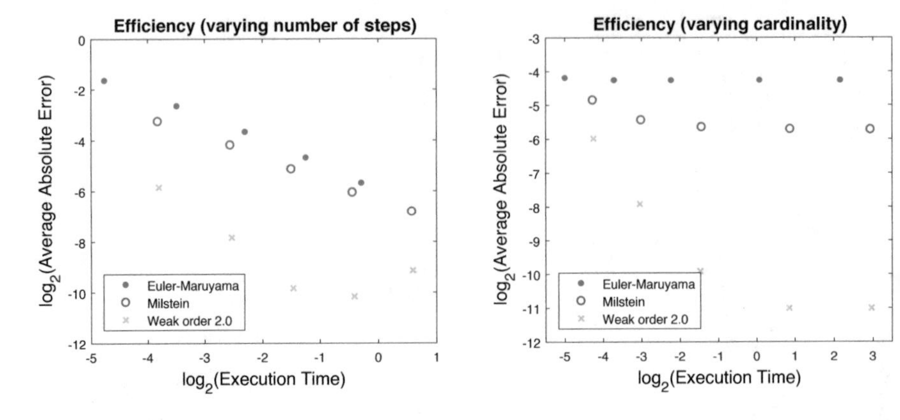

Fig. 9: Plots demonstrating computational efficiency. Both graphs present a log-log plot of average absolute error as a function of computation time. On the left, the number of time steps (n) is varied, while on the right the quantizer cardinality (N) is varied.

for at-the-money volatility $\sigma \in \mathbb{R}_+$, skew parameter $q \in \mathbb{R}$ and $s \in \mathbb{R}_+$, being a measure of the convexity of the skew. For full details on the model and characteristic function based pricing for European options see [1].

In Figure 10, the left panel shows the pricing error incurred by the RMQ schemes for European put options as a function of fixed-spot inverse moneyness[9]. Here the process-specific parameters used were $X_0 = 100$, $\sigma = 20\%$, $q = 0.5$, $s = 0.1$ and $T = 1$, with the RMQ-specific parameters $N = 200$ and $n = 12$. The right panel shows the difference between a high-resolution Monte Carlo simulation and RMQ prices for discrete up-and-out put options, with monthly monitoring, as a function of barrier level, specified as a multiple of strike price $K = 100$. Here the RMQ prices are computed using the so-called *transition kernel* formulation — see [40, 55] for details. The Monte-Carlo sample was generated using an Euler-Maruyama scheme with 1 200 intervals over $[0, T]$ and 1 000 000 sample paths. The black line shows the three-standard deviation boundary, indicating that the two higher-order schemes are statistically significantly better than the Euler scheme, at this sample size.

5.3 Stochastic Volatility Models

We now turn our attention to stochastic volatility models. The SDEs for the Stein-Stein model may be specified in the notation of (17) and (18) as

[9] Fixed-spot inverse moneyness is specified as the variable strike, K, over the fixed initial asset price, X_0.

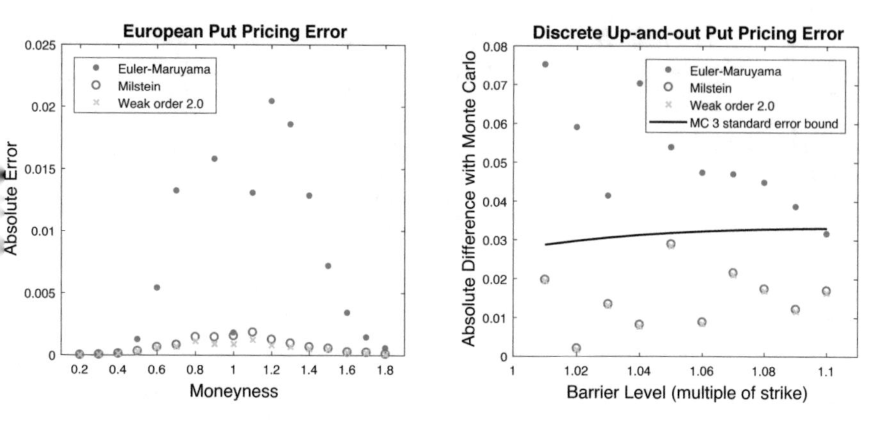

Fig. 10: Pricing error for European put prices (left) and discrete up-and-out put prices (right) under the quadratic volatility model.

$$a^x(X_t) = rX_t, \qquad b^x(X_t, V_t) = V_t X_t,$$
$$a^v(V_t) = \kappa(\theta - V_t), \qquad b^v(V_t) = \sigma,$$

for the parameters $\kappa = 4$, $\theta = 0.2$, $\sigma = 0.1$, $r = 0.0953$, $\rho = -0.5$, $x_0 = 0.2$ and $y_0 = 100$, with the maturity of the option set at one year. These parameters are from Table 1 in [56].

The left graph in Figure 11 displays the pricing error for four algorithms. The first is the JRMQ algorithm using the bivariate Gaussian distribution, the second is the JRMQ algorithm using the univariate joint probability approximation, the third is the stochastic volatility RMQ algorithm from [6] and the fourth is a two-dimensional standard Euler Monte Carlo simulation.

For the RMQ algorithms, $K = 12$ time steps was used, with $N^v = 30$ codewords at each step for the independent process and $N^x = 60$ codewords for the dependent process. The Monte Carlo simulation used 500000 sample paths with 120 time steps per path.

While the JRMQ algorithm, using the bivariate distribution, is the most accurate, this comes at computational cost. When one uses the univariate joint probability approximation instead, the JRMQ algorithm is an order of magnitude faster. The latter was also significantly faster than the algorithm of [6], which uses a less efficient conditioning argument, but produces results of almost exactly the same accuracy. Barring three values, both these algorithms price to within the three standard deviation bound of the significantly higher resolution Monte Carlo simulation.

The SDEs for the Heston model may be specified as

$$a^x(X_t) = rX_t, \qquad b^x(X_t, V_t) = \sqrt{V_t} X_t,$$
$$a^v(V_t) = \kappa(\theta - V_t), \qquad b^v(V_t) = \sigma\sqrt{V_t},$$

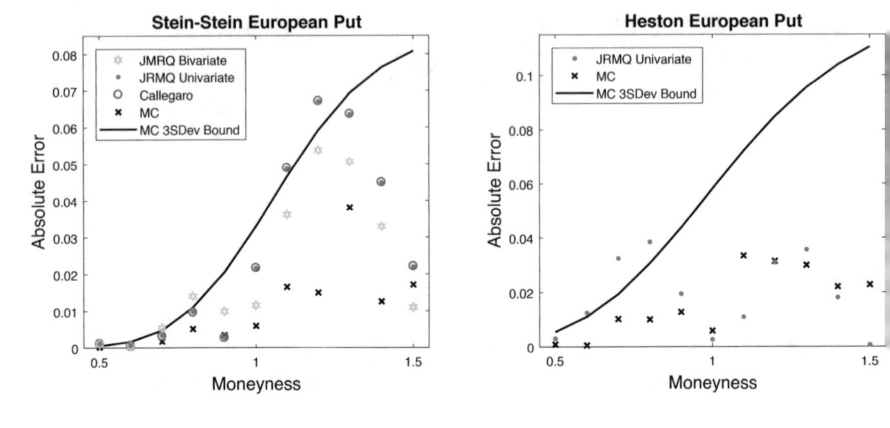

Fig. 11: Pricing error of European put prices under the Stein-Stein (left) and Heston (right) models.

and in the example considered the parameters chosen were $\kappa = 2$, $\theta = 0.09$, $\sigma = 0.4$, $r = 0.05$, $\rho = -0.3$, $x_0 = 0.09$ and $y_0 = 100$, with the maturity of the option set at one year. These parameters are based on the SV-I parameter set from Table 3 of [36], with σ adjusted from 1 to 0.4 to ensure that the Feller condition is satisfied for the square-root variance process.

The right graph in Figure 11 displays the pricing error for JRMQ compared with a two-dimensional fully truncated log-Euler scheme, suggested as the least-biased Monte Carlo scheme for stochastic volatility models [36]. For the JRMQ algorithm, $K = 12$ time steps were used with $N^x = N^v = 30$ codewords at each step for both processes. A reflecting zero-boundary was used to ensure positivity when computing the standard RMQ algorithm for the independent variance process. Compared to a Monte Carlo simulation, which used the same resolution as the previous simulation, the JRMQ algorithm performs very well despite the coarseness of the grid.

5.4 Calibration

An advantage of the RMQ algorithm, as with traditional tree methods, is the ability to price multiple options without needing to regenerate the underlying grid. Once the optimal quantization grid has been generated out to the longest option maturity, the computational cost of pricing options is negligible. An immediate application is the calibration of models directly to non-vanilla products (e.g. Bermudan/American options). RMQ has already been used to calibrate the quadratic normal volatility model using European options on the DAX index by [5].

The calibration problem can be formulated as

$$\Theta = \arg\min F(\Theta),$$

	Maturity	T	Options	r
T_1	18/06/2005	0.458	26	0.0274
T_2	21/01/2006	1.05	19	0.0311

Table 1: Summary of the calibration data for the Heston model calibrated to American put options on GOOG for 03/01/2005.

where F is the objective or error function and $\Theta = \{V_0, \sigma, \kappa, \theta, \rho\}$ is the parameter set for the Heston model. In [14] the *relative squared volatility error* is recommended as the objective function for calibrating the Heston model. It is defined as

$$F(\Theta) = \sum_{l=1}^{L} \left(\frac{\sigma_l^{\text{Model}}(\Theta) - \sigma_l^{\text{Market}}}{\sigma_l^{\text{Market}}} \right)^2,$$

where L is the number of calibration instruments used, $\sigma_l^{\text{Model}}(\Theta)$ is the Black-Scholes implied volatility that corresponds to pricing calibration instrument l with the model parameters Θ, and σ_l^{Market} is the implied volatility for that instrument in the market.

By minimising $F(\Theta)$, the Heston model is calibrated to American put options of two different maturities on the GOOG stock for 03/01/2005. The data used for the calibration is summarized in Table 1. The U.S. Department of Treasury rate was used as a proxy for the risk-free rate and has been interpolated to the necessary maturities. Table 1 also indicates the number of calibration instruments for each maturity. Only options with strikes within 50% of at-the-money were considered. The stock price on this day was $X_0 = 202.71$.

The calibrated parameters are summarized in Table 2, along with the final value of the objective function.

	V_0	σ	κ	θ	ρ	$F(\Theta)$
T_1	0.2	0.898	0.1	0.1	−0.63	0.0031
T_2	0.204	0.599	0.155	0.171	−0.582	0.0005

Table 2: Summary of the results for the Heston model calibrated to American put options on GOOG for 03/01/2005.

Figure 12 illustrates how the Heston model captures the shape of the market implied volatility curve. The calibrated parameters are fairly stable across the two maturities, indicating that the Heston model could be a candidate for calibration to the volatility surface.

Although all five parameters of the Heston model are calibrated using minimization in this example, it may be more appropriate to recover some of the parameters using filtering techniques. Kalman filtering is usually used to recover the hidden

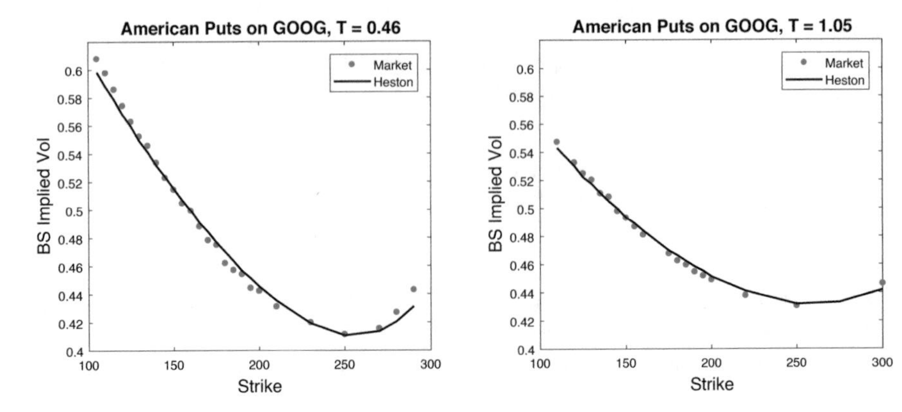

Fig. 12: Calibration of the Heston model directly to American put options on GOOG for 03/01/2005.

state, V_0, as well as parameters that have a stable interpretation through time, i.e., the rate of mean reversion, κ, and the mean reversion level, θ.

6 Conclusion

In this paper, we have highlighted the suitability of quantization for solving challenging problems in quantitative finance. We began by introducing the vector quantization of probability distributions and then proceeded to quantize the marginal distributions of one-dimensional stochastic differential equations using recursive marginal quantization [48].

We explained how the original RMQ algorithm can be extended to correctly model the zero boundary behaviour of SDEs with nonnegative solutions. By an analysis of the Milstein and weak order 2.0 Itō-Taylor schemes it was also shown that RMQ can by adapted to higher order updates, where instead of Gaussian distributions, noncentral chi-squared distributions are propagated.

While, in general, it is not possible to efficiently solve the optimization problems required for RMQ applied in two-dimensions, use of product quantizers is efficient. Using a margining argument, we introduced the JRMQ algorithm with both bivariate and efficient univariate approximations for the joint probabilities.

In the last section the improved weak order convergence and efficiency of the higher-order discretisation schemes was demonstrated for geometric Brownian motion. RMQ was also used to accurately price European options and discrete barrier options under the quadratic volatility model, an instance of a parametric local volatility model. The JRMQ algorithm was demonstrated by pricing European options under the popular Heston stochastic volatility model and compared to both Monte

Carlo simulation and a previous quantization algorithm. Lastly, a basic calibration exercise was performed for the Heston model using American option price data.

References

[1] Andersen L (2011) Option pricing with quadratic volatility: A revisit. Finance and Stochastics 15(2):191–219

[2] Bachelier L (1900) Théorie de la Spéculation. PhD Thesis, Ecole Polytechnique, In: Annales Scientifiques de l'École Normal Supérieur (3)

[3] Bennett W (1948) Spectra of quantized signals. Bell System Technical Journal 3(27):446–472

[4] Black F, Scholes M (1973) The pricing of options and corporate liabilities. Journal of Political Economy 81(3):637–659

[5] Callegaro G, Fiorin L, Grasselli M (2015) Quantized calibration in local volatility. Risk 28:62–67

[6] Callegaro G, Fiorin L, Grasselli M (2016) Pricing via recursive quantization in stochastic volatility models. Quantitative Finance 17(6):855–872

[7] Delattre S, Fort JC, Pagès G (2004) Local distortion and μ-mass of the cells of one dimensional asymptotically optimal quantizers. Communications in Statistics-Theory and Methods 33(5):1087–1117

[8] Derman E, Kani I (1994) Riding on a smile. Risk 7:32–39

[9] Du Q, Faber V, Gunzburger M (1999) Centroidal Voronoi tessellations: Applications and algorithms. SIAM Review 41(4):637–676

[10] Duffy D (2006) Finite Difference Methods in Financial Engineering. John Wiley & Sons

[11] Duffy D, Kienitz J (2009) Monte Carlo Frameworks — Building Customisable and High-performance C++ Applications. John Wiley & Sons

[12] Dupire B (1994) Pricing with a smile. Risk 7:18–20

[13] Emelianenko M, Ju L, Rand A (2008) Nondegeneracy and weak global convergence of the Lloyd algorithm in $\mathbb{R}^d$. SIAM Journal on Numerical Analysis 46(3):1423–1441

[14] Escobar M, Gschnaidtner C (2016) Parameters recovery via calibration in the Heston model: A comprehensive review. Wilmott Magazine 86:60–81

[15] Feller W (1949) On the Theory of Stochastic Processes, with Particular Reference to Applications. Proceedings of the (First) Berkeley Symposium on Mathematical Statistics and Probability

[16] Fiorin L, Pagès G, Sagna A (2017) Product Markovian quantization of an $\mathbb{R}^d$-valued Euler scheme of a diffusion process with applications to finance, `https://arxiv.org/abs/1511.01758v3`, preprint

[17] Gatheral J (2011) The Volatility Surface: A Practitioner's Guide. John Wiley & Sons

[18] Glasserman P (2004) Monte Carlo Methods in Financial Engineering. Springer

[19] Graf S, Luschgy H (2000) Foundations of Quantization for Probability Distributions. Springer
[20] Gray RM, Neuhoff DL (1998) Quantization. IEEE Transactions on Information Theory 44(6):2325–2383
[21] Heston SL (1993) A closed-form solution for options with stochastic volatility with applications to bond and currency options. Review of Financial Studies 6(2):327–343
[22] Ikeda N, Watanabe W (1981) Stochastic Differential Equations. North-Holland Amsterdam
[23] Ito K (1940) Stochastic integral. Proceedings of the Imperial Academy 20(8):519–524
[24] Ito K (1946) On a stochastic integral equation. Proceedings of the Japan Academy 22(2):32–35
[25] Ito K (1950) Stochastic differential equations in a differentiable manifold. Nagoya Mathematical Journal 1:35–47
[26] Ito K (1951) On a formula concerning stochastic differentials. Nagoya Mathematical Journal 3:55–65
[27] Ito K (1974) Foundations of Stochastic Differential Equations in Infinite Dimensional Spaces. Philadelphia: Society for Industrial and Applied Mathematics
[28] Ito K, McKean H (1974) Diffusion processes and their sample paths. Springer
[29] Jäckel P (2002) Monte Carlo Methods in Finance. John Wiley & Sons
[30] Karatzas I, Shreve SE (1998) Methods of Mathematical Finance. Springer
[31] Kienitz J, Wetterau D (2012) Financial modelling: Theory, Implementation and Practice with MATLAB Source. John Wiley & Sons
[32] Kloeden P, Platen E (1999) Numerical Solution of Stochastic Differential Equations. Springer
[33] Kolmogoroff A (1931) Ūber die analytischen methoden in der wahrscheinlichkeitsrechnung. Mathematische Annalen 104(1):415–458
[34] LeVeque RJ (2007) Finite Difference Methods for Ordinary and Partial Differential Equations. SIAM
[35] Lloyd SP (1982) Least squares quantization in PCM. IEEE Transactions on Information Theory 28(2):129–137
[36] Lord R, Koekkoek R, Van Dijk D (2010) A comparison of biased simulation schemes for stochastic volatility models. Quantitative Finance 10(2):177–194
[37] Luschgy H, Pagès G (2008) Functional quantization rate and mean regularity of processes with an application to Lévy processes. The Annals of Applied Probability 18(2):427–469
[38] Maruyama G (1955) Continuous Markov processes and stochastic equations. Rendiconti del Circolo Matematico di Palermo 4(1):48–90
[39] McWalter TA, Rudd R, Kienitz J, Platen E (2017) Appendix to recursive marginal quantization of higher-order schemes. Available at SSRN 3071201
[40] McWalter TA, Rudd R, Kienitz J, Platen E (2018) Recursive marginal quantization of higher-order schemes. Quantitative Finance 18(4):693–706

[41] Merton, R (1976) Option pricing when underlying stock returns are discontinuous. Journal of Financial Economics 3:125–144
[42] Milstein G (1975) Approximate integration of stochastic differential equations. Theory of Probability and Its Applications 19(3):557–562
[43] Musiela M, Rutkowski M (2004) Martingale Methods in Financial Modelling, 2nd edition. Springer
[44] Pagès G (1998) A space quantization method for numerical integration. Journal of Computational and Applied Mathematics 89(1):1–38
[45] Pagès G (2015) Introduction to optimal vector quantization and its applications for numerics. ESAIM: Proceedings and Surveys 48:29–79
[46] Pagès G, Pham H (2005) Optimal quantization methods for nonlinear filtering with discrete-time observations. Bernoulli 11(5):893–932
[47] Pagès G, Printems J (2009) Optimal quantization for finance: From random vectors to stochastic processes. Handbook of Numerical Analysis 15:595–648
[48] Pagès G, Sagna A (2015) Recursive marginal quantization of the Euler scheme of a diffusion process. Applied Mathematical Finance 22(5):463–498
[49] Pagès G, Pham H, Printems J (2003) Optimal quadratic quantization for numerics: The Gaussian case. Monte Carlo Methods and Applications 9(2):135–165
[50] Pagès G, Pham H, Printems J (2004) Optimal quantization methods and applications to numerical problems in finance. In: Handbook of Computational and Numerical Methods in Finance, Springer, pp 253–297
[51] Protter P (2004) Stochastic Integrals and Differential Equations. Springer
[52] Rebonato R (2005) Volatility and Correlation: The perfect Hedger and the Fox. John Wiley & Sons
[53] Rudd R (2018) Recursive Marginal Quantization: Extensions and Applications in Finance. PhD Thesis, University of Cape Town
[54] Rudd R, McWalter TA, Kienitz J, Platen E (2017) Fast quantization of stochastic volatility models. Available at SSRN 2956168
[55] Sagna A (2011) Pricing of barrier options by marginal functional quantization. Monte Carlo Methods and Applications 17(4):371–398
[56] Schöbel R, Zhu J (1999) Stochastic volatility with an Ornstein–Uhlenbeck process: an extension. European Finance Review 3(1):23–46
[57] Stein EM, Stein JC (1991) Stock price distributions with stochastic volatility: An analytic approach. Review of Financial Studies 4(4):727–752
[58] Stroock DW (2013) An Introduction to Markov Processes. Springer

Index

M. Günther, W. Schilders (eds.), *Novel Mathematics Inspired by Industrial Challenges*, Mathematics in Industry 38, https://doi.org/10.1007/978-3-030-96173-2

B

C

D

K

L

M

Printed in the United States
by Baker & Taylor Publisher Services